M.J. Grimble and V. Kučera (Eds)

Polynomial Methods for Control Systems Design

With 45 Figures

Springer

Professor Michael J. Grimble
Industrial Control Centre, Graham Hills Building,
University of Strathclyde, 50 George Street, Glasgow G1 1QE, UK

Professor Vladimir Kučera
Akademie ved Ceske Republiky, Ustav Teorie, Informace A Automatizace,
Pod vodarenskou vezi 4, 128 08 Praha 8, Prague, Czech Republic

ISBN 3-540-76077-6 Springer-Verlag Berlin Heidelberg New York

British Library Cataloguing in Publication Data
Polynomial methods for control systems design
1.Control theory 2.H∞ control
3.Polynomials
I.Grimble, Michael J. II.Kučera, Vladimir
629.8'312
ISBN 3540760776

Library of Congress Cataloging-in-Publication Data
Polynomial methods for control systems design / Michael J. Grimble and Vladimír Kučera, eds.
p. cm.
Includes bibliographical references (p.).
ISBN 3-540-76077-6 (pbk. : alk. paper)
1. Automatic control - - Mathematics. 2. System design.
3. Polynomials. I. Grimble, Michael J. II. Kučera, Vladimír, 1943- .
TJ213.P5644 1996 96-27441
629.8'312 - - dc20 CIP

Printed in Great Britain

Typesetting: Camera ready by editors
Printed and bound at the Athenæum Press Ltd, Gateshead
69/3830-543210 Printed on acid-free paper

Polynomial Methods for Control Systems Design

Springer
London
Berlin
Heidelberg
New York
Barcelona
Budapest
Hong Kong
Milan
Paris
Santa Clara
Singapore
Tokyo

Contents

Preface

This monograph was motivated by a very successful workshop held before the 3rd IEEE Conference on Decision and Control held at the Buena Vista Hotel, lake Buena Vista, Florida, USA. The workshop was held to provide an overview of polynomial system methods in LQG (or H_2) and H_∞ optimal control and estimation. The speakers at the workshop were chosen to reflect the important contributions polynomial techniques have made to systems theory and also to show the potential benefits which should arise in real applications.

An introduction to H_2 control theory for continuous-time systems is included in chapter 1. Three different approaches are considered covering state-space model descriptions, Wiener-Hopf transfer function methods and finally polynomial equation based transfer function solutions. The differences and similarities between the techniques are explored and the different assumptions employed in the solutions are discussed. The standard control system description is introduced in this chapter and the use of Hardy spaces for optimization. Both control and estimation problems are considered in the context of the standard system description. The tutorial chapter concludes with a number of fully worked examples.

Attention turns to the solution of H_∞ control problems using the standard system description in chapter 2. A frequency domain solution is considered and the use of this approach is justified relative to the more conventional state space solution methods. The importance of spectral factorization in the solution of H_∞ control problems is considered and the particular role of J-spectral factorization is discussed. Algorithmic procedures are described and illustrated by several examples. A MATLAB based polynomial toolbox is in development based upon the algorithms presented in this chapter.

In chapter 3 a more detailed polynomial matrix solution of the LQG regulation and tracking problems is discussed. Once again general system configurations are considered but in this case for discrete time systems. The stochastic approach to this problem is particularly relevant for applications like multivariable adaptive control. The solution is obtained using spectral factor and diophantine equations and a completing the squares type of solution procedure is used. The solution of multi- channel deconvolution problems is also considered and an example of a three degree of freedom control problem is presented. It is an interesting conclusion of the chapter that in the solution of the general LQG problem via polynomial equations there is an embedded solution of the

Kalman filtering problem. Thus, the two stage procedure involved is very reminiscent of the certainty equivalent principle used in solving state-space based LQG problems.

The solution of multivariable H_∞ control problems for continuous-time systems is considered using a game theory approach in chapter 4. To some extent this links the H_2 and H_∞ problems. In fact, a further link is found since J-spectral factorization is essential in the game theory solution method. A standard system description is again employed and an example is presented. The links between H_2 or LQG problems and H_∞ solutions is also emphasised when the form of the diophantine equation like equations is considered. The resulting numerical algorithms are relatively straightforward and have parallels with the solutions in chapter 2.

The polynomial approach applies equally well to linear estimation problems and these are considered in chapter 5. Robust filtering problems are introduced where the model uncertainties are parameterized using sets of random variables. The design of robust *cautious* Wiener filters is considered and topics such a deconvolution filtering (see also chapter 3) and the *dual* feedforward control problem, are explored. Applications problems in digital communication systems, such as decision feedback equalizers are discussed. Examples are provided including the very topical subject of narrowband detection in cellular digital mobile radio systems. Whilst demonstrating the benefits of the polynomial approach in filtering problems links are also established to the state-space based Kalman filtering techniques.

The inferential H_2 and H_∞ optimal control problems are considered in the final chapter which concentrates more on applications. In this case the polynomial analysis is close to the LQG polynomial methods even though H_∞ problems can be solved by a similar technique. The application was from an actual design study for a new generation of co-ordinate measuring machines. One of the advantages of the polynomial approach is that self-tuning controllers may easily be constructed since plants are often identified in ARMAX form and the controllers are readily available using this system description. The potential of the polynomial approach in this and other applications are clear from the example.

The IEEE Workshop was one of a number of attempts to introduce the advantages of the polynomial and frequency domain approaches to a wider audience. Engineers in industry often find frequency domain methods easier to utilise than state equation based techniques. However, state equation based algorithms are widely available because of a very large investment world-wide in this type of algorithmic tool. This was in part stimulated by the space race but also to some extent reflects the additional difficulty experienced in polynomial matrix algorithms. However, there is a growing recognition that the advantages offered by polynomial methods demand appropriate tools be made available. It is hoped that the original Workshop and the Monograph which has now followed will stimulate research in both algorithms and subsequent applications.

M.J. Grimble and V. Kučera
2nd April 1996

1

A Tutorial on H_2 Control Theory: The Continuous Time Case

Vladimír Kučera

1.1 Introduction

Modern control theory is a powerful tool for control system design. It facilitates the design task by providing analytical design procedures.

Optimal control provides analytical designs of a specially appealing type. The resulting control system is supposed to be the best possible system of a particular class.

Linear–quadratic methods set out to obtain linear optimal controls. The system that is controlled is assumed linear and a linear control law is achieved by working with quadratic performance indices (Anderson and Moore, 1990).

A number of justifications for linear–quadratic methods may be advanced. Many engineering systems are linear, linear optimal control problems have computable solutions, linear control laws are simple to implement, and will frequently suffice.

The best known and by now classical linear–quadratic method is the LQG control theory. It consists of controlling a linear stochastic system optimally under a quadratic integral functional. The system under control is driven by a Gaussian random process with known statistics.

The significance of the LQG theory derives from the appealing structure of the solution. The optimal control law is split into two distinct parts: the conditional mean estimate of the current state and the optimal feedback in the estimated state. This is the famous separation theorem of stochastic control, see Åström (1970).

Of more recent origin is the H_2 control theory. The control system is linear and its transfer function is considered as an element of the Hardy space H_2 equipped with a quadratic norm. The task is to minimize the norm while stabilizing the system.

The H_2 control theory is the subject matter of this contribution. Three different approaches, well known from the literature, are considered: a state space solution, a Wiener–Hopf analytic solution, and a transfer function solution based on Diophantine equations. A synthesis of these approaches is attempted, with the aim of providing useful insight.

It is shown that these three approaches are not entirely equivalent. Due to different mathematical tools applied, the problems are solved at different levels of generality under different assumptions. Therefore, each solution makes an attractive alternative in certain situations.

The tutorial value of the contribution is enhanced by further comparing the H_2 and LQG control theories. The nature of these problems is different: LQG is a stochastic problem while H_2 is a deterministic one. Yet they are very close to each other. In fact, the inifinite horizon LQG can be recast as an H_2 problem.

The notation used throughout the chapter is mostly standard. Special notation is introduced in the relevant sections. The list of references includes the original sources and a selection of the relevant textbooks. No attempt has been made to compile a list of related publications.

1.2 LQG control theory

1.2.1 Problem formulation

The aim of this section is to review briefly the LQG control theory. The problem considered is that of controlling a stochastic linear system optimally under the quadratic criteria. The system under control is driven by a Gaussian process with known statistics. The importance of Linear – Quadratic – Gaussian assumptions in obtaining the *separation property* of the optimal control is discussed. The exposition follows that given in Tse (1971).

The matrix notations used throughout the text are standard. For any matrix M, M' denotes the transpose of M. For a square matrix M, $\operatorname{Tr} M$ stands the trace of M and $\operatorname{Det} M$ is the determinant of M. The inequalities $M \geq 0$ and $M > 0$ mean that M is real symmetric non–negative and positive definite, respectively. The identity matrix is denoted by I.

We consider a linear system described by the stochastic differential equations

$$\begin{aligned} \mathrm{d}x &= Fx\mathrm{d}t + G_2u\mathrm{d}t + \mathrm{d}p \\ \mathrm{d}y &= H_2x\mathrm{d}t + \mathrm{d}q \end{aligned} \tag{1.1}$$

for $t \geq t_0$, where x is an n vector state, u is an m vector control and y is an l vector measurement. The stochastic processes $p(t)$ and $q(t)$ are independent Wiener processes (Gaussian processes with zero means and independent stationary increments) having the incremental covariances $Q_0 dt$ and $I dt$, respectively. The matrices F, G_2, H_2 are constant ones and $Q_0 \geq 0$.

The initial state $x(t_0)$ is assumed Gaussian with mean x_0 and covariance $S_0 \geq 0$. The stochastic processes $p(t)$ and $q(t)$ are independent of $x(t_0)$. The initial output is zero, $y(t_0) = 0$.

The performance criterion for controlling the system (1.1) is chosen as the expected loss

$$J = \mathrm{E}\{x'(t_1)S_1 x(t_1) + \int_{t_0}^{t_1} [x'(t)Q_1 x(t) + u'(t)u(t)]\, \mathrm{d}t\} \tag{1.2}$$

where $Q_1 \geq 0$, $S_1 \geq 0$ and E is the expectation taken over all underlying random quantities $x(t_0), p(t)$ and $q(t)$.

The admissible control can only depend on the past data,

$$u(t) = \phi(t, y(\tau)), \quad \tau \in [t_0, t] \tag{1.3}$$

and, for technical reasons, it is assumed that ϕ satisfies a Lipschitz condition.

The control problem is to find an admissible control u_{opt} of the form (1.3), which will minimize the loss (1.2) subject to the dynamic constraint (1.1).

The reader will have noticed that the statistics of the Wiener processes as well as the terms involved in the expected loss have a special form. In general, the Wiener processes $p(t)$ and $q(t)$ can be jointly distributed with the incremental covariance

$$\begin{bmatrix} Q_0 & T_0 \\ T_0' & R_0 \end{bmatrix} \mathrm{d}t$$

where $R_0 > 0$ and the expected loss can involve the non–negative definite quadratic form

$$[x' \ u'] \begin{bmatrix} Q_1 & T_1' \\ T_1 & R_1 \end{bmatrix} \begin{bmatrix} x \\ u \end{bmatrix}$$

with $R_1 > 0$. These cases, however, can be reduced to our formulation by a change of variables.

1.2.2 Finite horizon solution

Let us first consider the expected loss (1.2) and rewrite it as

$$\begin{aligned} J = {} & \mathrm{E}\{\mathrm{E}\{x'(t_1)S_1 x(t_1) \mid \mathcal{Y}_{t_1}\}\} \\ & + \mathrm{E}\{\int_{t_0}^{t_1} \mathrm{E}\{x'(t)Q_1 x(t) \mid \mathcal{Y}_t\}\, \mathrm{d}t\} \\ & + \mathrm{E}\{\int_{t_0}^{t_1} u'(t)u(t)\, \mathrm{d}t\} \end{aligned} \tag{1.4}$$

where $\mathcal{Y}_t$ is the σ algebra generated by the measurements $y(\tau), \tau \in [t_0, t]$. Define the conditional mean and covariance of the state by

$$\hat{x}(t) = \mathrm{E}\{x(t) \mid \mathcal{Y}_t\} \tag{1.5}$$

$$P_0(t) = \mathrm{E}\{[x(t) - \hat{x}(t)]\,[x(t) - \hat{x}(t)]' \mid \mathcal{Y}_t\}. \tag{1.6}$$

Consider the term

$$\begin{aligned} &\mathrm{E}\{\mathrm{E}\{x'(t)Q_1x(t) \mid \mathcal{Y}_t\}\} \\ &\quad = \mathrm{E}\{\mathrm{E}\{[x(t)-\hat{x}(t)+\hat{x}(t)]'Q_1[x(t)-\hat{x}(t)+\hat{x}(t)] \mid \mathcal{Y}_t\}\} \\ &\quad = \mathrm{E}\{\mathrm{E}\{[x(t)-\hat{x}(t)]'Q_1[x(t)-\hat{x}(t)] \mid \mathcal{Y}_t\}\} \\ &\qquad + 2\mathrm{E}\{\mathrm{E}\{[x(t)-\hat{x}(t)]'Q_1\hat{x}(t) \mid \mathcal{Y}_t\}\} \\ &\qquad + \mathrm{E}\{\mathrm{E}\{\hat{x}'(t)Q_1\hat{x}(t) \mid \mathcal{Y}_t\}\}\,. \end{aligned} \tag{1.7}$$

Using (1.5) we have

$$\begin{aligned} &\mathrm{E}\{[x(t)-\hat{x}(t)]'Q_1\hat{x}(t) \mid \mathcal{Y}_t\} \\ &\quad = \mathrm{E}\{x'(t)Q_1\hat{x}(t) \mid \mathcal{Y}_t\} - \mathrm{E}\{\hat{x}'(t)Q_1\hat{x}(t) \mid \mathcal{Y}_t\} \\ &\quad = \hat{x}'(t)Q_1\hat{x}(t) - \hat{x}'(t)Q_1\hat{x}(t) = 0\,. \end{aligned}$$

Taking the trace

$$[x(t)-\hat{x}(t)]'Q_1[x(t)-\hat{x}(t)] = \operatorname{Tr} Q_1[x(t)-\hat{x}(t)][x(t)-\hat{x}(t)]'$$

and using (1.6), the expression (1.7) becomes

$$\mathrm{E}\{\mathrm{E}\{x'(t)Q_1x(t) \mid \mathcal{Y}_t\}\} = \mathrm{E}\{\operatorname{Tr} Q_1P_0(t)\} + \mathrm{E}\{\hat{x}'(t)Q_1\hat{x}(t)\}\,. \tag{1.8}$$

Similarly, we have

$$\mathrm{E}\{\mathrm{E}\{x'(t_1)S_1x(t_1) \mid \mathcal{Y}_{t_1}\}\} = \mathrm{E}\{\operatorname{Tr} S_1P_0(t_1)\} + \mathrm{E}\{\hat{x}'(t_1)S_1\hat{x}(t_1)\}. \tag{1.9}$$

Substituting (1.8) and (1.9) into (1.4) one obtains

$$\begin{aligned} J = {} & \mathrm{E}\{\hat{x}'(_1)S_1\hat{x}(t_1) + \int_{t_0}^{t_1} [\hat{x}'(t)Q\hat{x}(t) + u'(t)u(t)]\,\mathrm{d}t\} \\ & + \mathrm{E}\{\operatorname{Tr} S_1P_0(t_1) + \int_{t_0}^{t_1} \operatorname{Tr} Q_1P_0(t)\,\mathrm{d}t\}. \end{aligned} \tag{1.10}$$

Next we shall derive equations for the conditional mean $\hat{x}$ and the conditional covariance P_0 of x given $\mathcal{Y}_t$. If the control u is zero, $\hat{x}$ is governed by the stochastic differential equation (Kalman and Bucy, 1961)

$$\begin{aligned} \mathrm{d}\hat{x} &= F\hat{x}\mathrm{d}t + L_0(t)(\mathrm{d}y - H_2\hat{x}\,\mathrm{d}t) \\ \hat{x}(t_0) &= x_0 \end{aligned}$$

where

$$L_0(t) = P_0(t)H_2' \tag{1.11}$$

and P_0 is given by the Riccati equation

$$\begin{aligned} \dot{P}_0(t) &= FP_0(t) + P_0(t)F' - P_0(t)H_2'H_2P_0(t) + Q_0 \\ P_0(t_0) &= S_0\,. \end{aligned} \tag{1.12}$$

If the control u is not known *a priori* but is admissible, the corresponding conditional distribution of the state x is Gaussian with conditional mean $\hat{x}$ given by

$$\begin{aligned} \mathrm{d}\hat{x} &= F\hat{x}\mathrm{d}t + G_2 u\mathrm{d}t + L_0(t)(\mathrm{d}y - H_2\hat{x}\,\mathrm{d}t) \\ \hat{x}(t_0) &= x_0 \end{aligned} \tag{1.13}$$

and conditional covariance P_0 given by (1.12).

From (1.12) we note that the covariance P_0 is independent of the measurements as well as control, therefore minimizing J is equivalent to minimizing

$$\bar{J} = \mathrm{E}\{\hat{x}'(t_1)S_1\hat{x}(t_1) + \int_{t_0}^{t_1} [\hat{x}'(t)Q_1\hat{x}(t) + u'(t)u(t)]\,\mathrm{d}t\}\,.$$

The process

$$w(t) = y(t) - H_2\hat{x}(t)$$

is called an *innovation process*, and it can be shown that $w(t)$ is a Wiener process with the incremental covariance $I dt$. Instead of the original stochastic control problem with inaccessible state, we have the following equivalent stochastic control problem with accessible state.

Minimize

$$\bar{J} = \mathrm{E}\{\hat{x}'(t_1)S_1\hat{x}(t_1) + \int_{t_0}^{t_1} [\hat{x}'(t)Q_1\hat{x}(t) + u'(t)u(t)]\,\mathrm{d}t\} \tag{1.14}$$

subject to the constraint

$$\mathrm{d}\hat{x} = F\hat{x}\mathrm{d}t + G_2 u\mathrm{d}t + L_0(t)\mathrm{d}w, \quad \hat{x}(t_0) = x_0 \tag{1.15}$$

where $w(t)$ is a Wiener process with known incremental covariance, $u(t)$ is admissible and $\hat{x}(t)$ is accessible. It is easy to see that the control which minimizes (1.14) subject to (1.15) is the optimum control for the original stochastic control problem, and is given (Wonham, 1968) by

$$u_{opt}(t) = -L_1(t)\hat{x}(t) \tag{1.16}$$

where

$$L_1(t) = G_2' P_1(t) \tag{1.17}$$

and P_1 is governed by the Riccati equation

$$\begin{aligned} -\dot{P}_1(t) &= P_1(t)F + F'P_1(t) - P_1(t)G_2G_2'P_1(t) + Q_1 \\ P_1(t_1) &= S_1. \end{aligned} \tag{1.18}$$

The optimal loss to go is given by

$$\bar{J}_t = \hat{x}'(t)P_1(t)\hat{x}(t) + \int_t^{t_1} \mathrm{Tr}\, L_0(\tau)L_0'(\tau)P_1(\tau)\,\mathrm{d}\tau.$$

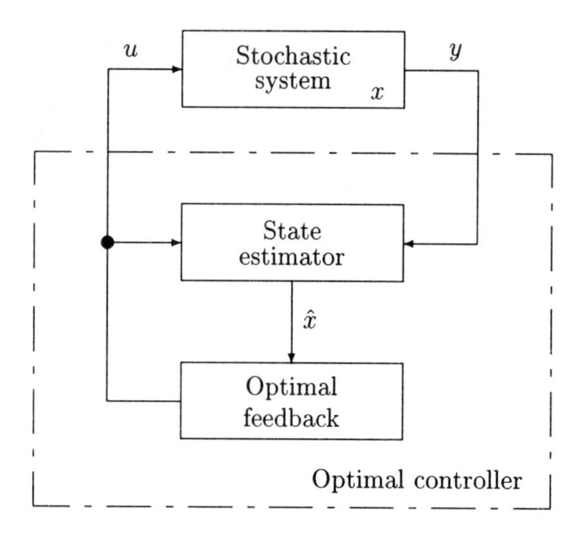

Figure 1.1: LQG control system.

Since J and $\bar{J}$ are related by

$$J = \bar{J} + \operatorname{Tr} S_1 P_0(t_1) + \int_{t_0}^{t_1} \operatorname{Tr} Q_1 P_0(t)\,\mathrm{d}t,$$

therefore the minimum expected loss in controlling (1.1) is

$$\begin{aligned} J_{opt} &= x_0' P_1(t_0) x_0 + \operatorname{Tr} S_1 P_0(t_1) \\ &\quad + \int_{t_0}^{t_1} \operatorname{Tr} Q_1 P_0(t)\,\mathrm{d}t + \int_{t_0}^{t_1} \operatorname{Tr} P_0(t) H_2' H_2 P_0(t) P_1(t)\,\mathrm{d}t \end{aligned} \tag{1.19}$$

The four terms of J_{opt} can be given distinct interpretations. The first term is the contribution due to the mean value of the initial state. The second term is the contribution due to the uncertainty of the initial state. The third term is due to the disturbances which are forcing the system and the fourth term is due to the uncertainty in the state estimate.

The structure of the overall LQG control system is described in figure 1.1.

The optimal control strategy can be split into two distinct procedures: (1) find the conditional mean estimate of the current state, and (2) find the optimal feedback as if the estimate is the true state of the system. This result is referred to as the *separation theorem*, which emphasizes the fact that the control problem is solved via two separate problems: state estimation and state regulation. The estimator does not depend on the loss function while the regulator does not depend on the noise statistics.

The separation theorem holds because of the fact that only the first conditional moment of the state is influenced by the control; the error covariance, which represents the performance of estimation, is not influenced by the control at all.

The optimal closed–loop system is described by the equations (1.1), (1.13) and (1.16). Introducing x and $x - \hat{x}$ as state variables, we find that these equations reduce to

$$\mathrm{d}\begin{bmatrix} x \\ x-\hat{x} \end{bmatrix} = \begin{bmatrix} F - G_2L_1(t) & G_2L_1(t) \\ 0 & F - L_0(t)H_2 \end{bmatrix} \begin{bmatrix} x \\ x-\hat{x} \end{bmatrix} \mathrm{d}t + \begin{bmatrix} \mathrm{d}p \\ \mathrm{d}p - L_0(t)\,\mathrm{d}q \end{bmatrix}$$

The dynamics of the closed–loop system are thus determined by the dynamics of the optimal state regulator $F - GL_1$ and the dynamics of the optimal state estimator $F - L_0H$.

Not only the LQG control problem can be split in two subproblems, state estimation and state regulation, but these two are *duals* of each other. That is to say, equations (1.11), (1.12) are the same as equations (1.17), (1.18) when one makes the following identifications:

$$t_0 \sim -t_1,\; F \sim F',\; G_2 \sim H_2',\; H_2 \sim G_2',\; Q_0 \sim Q_1,\; S_0 \sim S_1,\; P_0 \sim P_1 .$$

1.2.3 Infinite horizon solution

Although all the given data in the LQG problem formulation are constant, the optimal control strategy is time–varying. This is a consequence of the finite control horizon $t_1 - t_0$. Therefore one would expect the optimal control law to be time–invariant when extending the control horizon to infinity. This is indeed the case under additional assumptions.

In case $t_1 - t_0 \to \infty$, the expected loss (1.2) needs replacement. It is clear that there is no way that u and x will decay to zero as $t \to \infty$, if in (1.1), the random process $p(t)$ remains active. The expected loss (1.2) is therefore modified to "loss rate"

$$J^* = \lim_{\substack{t_0 \to -\infty \\ t_1 \to +\infty}} \frac{1}{t_1 - t_0} J . \tag{1.20}$$

For finite t_0 and t_1, the solutions of the two Riccati equations, $P_0(t)$ and $P_1(t)$, depend on the boundary conditions. To emphasize this, we denote $P_0(t)$ which satisfies (1.12) by $P_0(t; t_0, S_0)$ and $P_1(t)$ which satisfies (1.18) by $P_1(t; t_1, S_1)$. When $t_0 \to -\infty$ and $t_1 \to +\infty$ we must have $P_0(t; t_0, S_0)$ and $P_1(t; t_1, S_1)$ remain bounded and the overall system must be *stable* in order to reach the steady state. Then $x(t)$ and $u(t)$ will be stationary random processes and (1.20) can be rewritten as

$$J^* = \mathrm{E}\{x'(t)Q_1x(t) + u'(t)u(t)\} . \tag{1.21}$$

A bit of terminology is needed. A real square matrix is said to be *stable* if all its eigenvalues have negative real part. A pair of real matrices (A, B) is said to be *stabilizable* if there exists a real matrix M such that $A + BM$ is square and stable. A pair of real matrices (A, C) is *detectable* if there exists a real matrix N such that $A + NC$ is square and stable. Clearly (A, C) is detectable if and only if (A', C') is stabilizable.

Now we consider the following set of assumptions:

(LQG.1) (F, G_2) is stabilizable;

(LQG.2) (F, H_2) is detectable;

(LQG.3) (F, Q_0) is stabilizable;

(LQG.4) (F, Q_1) is detectable.

Then for all $S_0 \geq 0$,

$$\lim_{t_0 \to -\infty} P_0(t; t_0, S_0) = P_0$$

where P_0 is the (unique) symmetric non–negative definite solution of the algebraic Riccati equation

$$FP_0 + P_0F' - P_0H_2'H_2P_0 + Q_0 = 0$$

and $F - P_0H_2'H_2$ is stable. Similarly, for all $S_1 \geq 0$,

$$\lim_{t_1 \to +\infty} P_1(t; t_1, S_1) = P_1$$

where P_1 is the (unique) symmetric non–negative definite solution of the algebraic Riccati equation

$$P_1F + F'P_1 - P_1G_2G_2'P_1 + Q_1 = 0$$

and $F - G_2G_2'P_1$ is stable.

The separation theorem continues to hold for (1.20) and

$$L_0 = P_0H_2', \quad L_1 = G_2'P_1$$

define a time-invariant optimum controller

$$\begin{aligned} u_{opt} &= -L_1\hat{x} \\ d\hat{x} &= F\hat{x}dt + G_2udt + L_0(dy - H_2\hat{x}dt). \end{aligned}$$

The optimal expected loss rate follows from (1.19) and (1.20) as

$$J^*_{opt} = \operatorname{Tr} Q_1P_0 + \operatorname{Tr} P_0H_2'H_2P_0P_1.$$

1.3 H_2 control theory

1.3.1 Preliminaries

The subject matter of this section is the H_2 control problem. Roughly speaking, it consists of finding a stabilizing controller that minimizes the H_2 norm of the closed-loop transfer function.

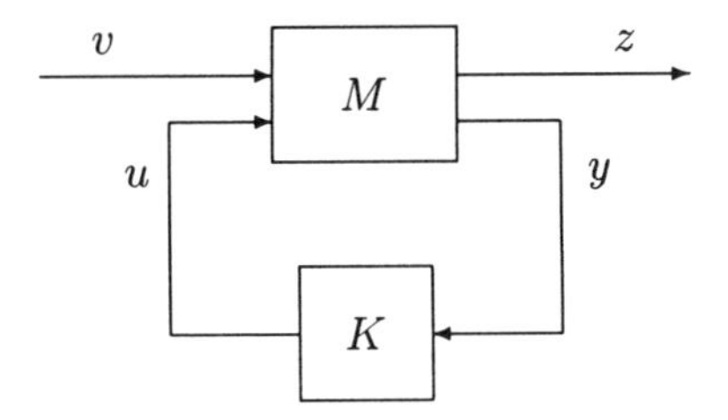

Figure 1.2: Standard control system.

We consider the feedback configuration given in figure 1.2,
where M is the *plant* and K is the *controller*. The plant accounts for all system components except for the controller. The signal v contains all external inputs, including disturbances, sensor noise, and commands. The output z is an error signal, y is the vector of measured variables, and u is the control input.

The conventional feedback and/or feedforward control configurations, known as the one-, two-, and three-degree-of-freedom configurations, are all special cases of this general control system. No matter what the number of independent multivariable sensors (degrees of freedom), all sensor outputs can be incorporated into y. This general approach eliminates the need for a number of theories with each being specific to only one control structure. Instead one compact solution is obtained which can be simplified or adjusted for every particular design. The equations which describe the feedback configuration of figure 1.2 are

$$\begin{bmatrix} z \\ y \end{bmatrix} = M(s) \begin{bmatrix} v \\ u \end{bmatrix}$$
$$u = K(s)y \tag{1.22}$$

where M and K are *real-rational* transfer matrices in the complex variable $s = \lambda + j\omega$. This means that only finite-dimensional linear time-invariant systems are considered. We write

$$M(s) = \begin{bmatrix} M_{11}(s) & M_{12}(s) \\ M_{21}(s) & M_{22}(s) \end{bmatrix} \tag{1.23}$$

where the partitioning is consistent with the dimensions of z, y and v, u. These will be denoted respectively by l_1, l_2 and m_1, m_2.

The *Hardy* space H_2 consists of square-integrable matrix functions $F(s)$ on the imaginary axis with analytic continuation into the right half-plane, whose norm is defined by

$$||F||_2 = \left(\frac{1}{2\pi} \int_{-\infty}^{\infty} \operatorname{Tr} F'(-j\omega)F(j\omega)\, d\omega \right)^{1/2}$$

Similarly $H_2^\perp$ will denote the Hardy space of square–integrable matrix functions on the imaginary axis with analytic continuation into the left half–plane. H_2 is isomorphic, via the Laplace transform and the Paley–Wiener theorem, with the Hilbert space L_{2+} of Lebesgue square–intebrable matrix functions on $[0,\infty)$. Similarly $H_2^\perp$ is isomorphic with L_{2-}, the Hilbert space of Lebesgue square-integrable matrix functions on $(-\infty,0]$. Thus H_2 as well as $H_2^\perp$ have the Hilbert space structure with the inner product defined by

$$\langle F, G \rangle = \frac{1}{2\pi}\int_{-\infty}^{\infty} \operatorname{Tr} G'(-j\omega)F(j\omega)\, d\omega.$$

A *real–rational* matrix $F(s)$ is in H_2 if and only if it is strictly proper (i.e., $F(\infty)=0$) and stable (i.e., $F(s)$ is analytic in $\operatorname{Re} s \geq 0$ where $\operatorname{Re} s$ denotes the real part of s). Similarly, $F(s)$ is in $H_2^\perp$ if and only if it is strictly proper and anti-stable (i.e., $F(s)$ is analytic in $\operatorname{Re} s \leq 0$).

It is convenient to define the *conjugate transpose* matrix of any real-rational $F(s)$ by

$$F^*(s) = F'(-s).$$

The H_2-norm of F then becomes

$$||F||_2 = \left(\frac{1}{2\pi j}\int \operatorname{Tr} F^*(s)F(s)\, ds\right)^{1/2} \tag{1.24}$$

where the integral is a contour integral up the imaginary axis and then around an infinite semicircle in the left half-plane.

A proper and stable real-rational matrix U is called *inner* if $U^*(s)U(s) = I$ or $U(s)U^*(s) = I$. The inner matrices have the norm-preserving property, $||UF||_2 = ||F||_2$ for every F in H_2.

The H_2 *control problem* to be studied in this section is then defined as follows. Given a plant M, find a controller K that internally stabilizes the control system of figure 1.2 and minimizes the H_2 norm of T, the transfer matrix from v to z.

The problem will be solved in the following subsections using three different techniques: state–space realizations, Wiener–Hopf optimization, and fractional representations leading to Diophantine equations. The three solutions are not entirely equivalent as they are obtained under different assumptions on M and K. A detailed discussion of the results will then follow in section 4 in order to provide a connected account of the H_2 control problem.

1.3.2 State space solution

The approach closest in nature to the LQG control problem is the state–space approach. We assume that a state–space model of the plant M is available and shall proceed to derive state–space formulas for the H_2–optimal controller K and the associated norm $||T||_2$. In this setting, the requirement of internal stability will mean that the states of M and K go to zero from all initial values when $v = 0$.

Our basic assumption is that both M and K are *proper* real–rational matrices. Associated with a proper rational M is a state-space realization (F, G, H, J), where

$$M(s) = H(sI - F)^{-1}\, G + J. \tag{1.25}$$

In the partitioned form, we have

$$G = \begin{bmatrix} G_1 \; G_2 \end{bmatrix}$$

$$H = \begin{bmatrix} H_1 \\ H_2 \end{bmatrix} \qquad J = \begin{bmatrix} J_{11} & J_{12} \\ J_{21} & J_{22} \end{bmatrix}$$

yielding the equations

$$\begin{aligned} \dot{x} &= Fx + G_1 v + G_2 u \\ z &= H_1 x + J_{11} v + J_{12} u. \\ y &= H_2 x + J_{21} v + J_{22} u. \end{aligned} \tag{1.26}$$

The controller realization will be denoted by

$$K(s) = \hat{H}(sI - \hat{F})^{-1}\, \hat{G} + \hat{J}.$$

It is further assumed that the realization of M is such that

(SS.1) (F, G_1) is stabilizable and (F, H_1) is detectable;

(SS.2) (F, G_2) is stabilizable and (F, H_2) is detectable;

(SS.3) $J_{11} = 0, \quad J_{22} = 0$;

(SS.4) $J_{12}' H_1 = 0, \quad J_{12}' J_{12} = I$;

(SS.5) $G_1 J_{21}' = 0, \quad J_{21} J_{21}' = I$.

Assumption (SS.2) is essential for achieving internal stability. Assumption (SS.1) together with (SS.2) guarantees that the two algebraic Riccati equations considered below possess the requisite solutions. Assumptions (SS.3), (SS.4) and (SS.5) are made in the sake of simplicity and their role will become clear from the connection between the H_2 and LQG control theories.

The state-space solution to be presented follows the paper by Doyle, Glover, Khargonekar, and Francis (1989). It is constructed via a separation argument from two special problems. The requirement of internal stability, however, is treated first.

Internal stability

The following lemma shows that internal stability of the control system in figure 1.2 can be achieved if and only if the plant is stabilizable from u and detectable from y.

Lemma 1.1 (Safonov *et al.*, 1987). *Let M have a state-space realization given by (1.26). Then there exists an internally stabilizing controller K for the closed-loop system in figure 1.2 if and only if (F, G_2) is stabilizable and (F, H_2) is detectable.*

Proof: If either stabilizability or detectability fails to hold, the closed-loop system in figure 1.2 will be internally unstable for all choices of K. If both stabilizability and detectability hold, a stabilizing K may be constructed with the aid of standard observer-based pole-placement methods. □

Thus, when the above conditions are satisfied, a controller K will internally stabilize M if it internally stabilizes M_{22}.

It is to be noted that the condition of lemma 1.1 is precisely assumption (SS.2).

Full information problem

This is a special H_2 control problem in which the controller is provided with full information, $y = \begin{bmatrix} x \\ v \end{bmatrix}$. Thus, the plant equations (1.26) read

$$\begin{aligned} \dot{x} &= Fx + G_1 v + G_2 u \\ z &= H_1 x + J_{12} u \\ y &= \begin{bmatrix} I \\ 0 \end{bmatrix} x + \begin{bmatrix} 0 \\ I \end{bmatrix} v \end{aligned} \tag{1.27}$$

and the following subset of assumptions (SS.1) - (SS.5) is relevant:

(FI.1) (F, H_1) is detectable;

(FI.2) (F, G_2) is stabilizable;

(FI.3) $J_{11} = 0, \quad J_{22} = 0$;

(FI.4) $J_{12}' H_1 = 0, \quad J_{12}' J_{12} = I$.

To find a stabilizing controller K which minimizes $||T||_2$ under these simplifications, we consider the algebraic Riccati equation

$$F' P_1 + P_1 F - P_1 G_2 G_2' P_1 + H_1' H_1 = 0. \tag{1.28}$$

Assumptions (FI.1) and (FI.2) are necessary and sufficient for (1.28) to have a unique soluton $P_1 \geq 0$ such that $F - G_2 L_1$ is stable, where

$$L_1 = G_2' P_1. \tag{1.29}$$

Theorem 1.2 (Doyle *et al.*, **1989).** *The unique optimal controller for the full information H_2 problem is*

$$K_{FI}(s) = [\,-L_1 \quad 0\,]. \tag{1.30}$$

Moreover,

$$\min ||T||_2^2 = ||N_1 G_1||_2^2 = \operatorname{Tr} G_1' P_1 G_1, \tag{1.31}$$

where

$$N_1(s) = (H_1 - J_{12}L_1)(sI - F + G_2L_1)^{-1}. \tag{1.32}$$

The proof requires a preliminary change of variables and a lemma. If we define a new control variable

$$\bar{u} = u + L_1 x,$$

the plant equations (1.27) become

$$\begin{aligned} \dot{x} &= (F - G_2L_1)x + G_1 v + G_2\bar{u} \\ z &= (H_1 - J_{12}L_1)x + J_{12}\bar{u} \end{aligned}$$

and the transfer function to z is

$$z = N_1(s)G_1 v + U(s)\bar{u} \tag{1.33}$$

where N_1 is given by (1.32) and

$$U(s) = N_1(s)G_2 + J_{12}. \tag{1.34}$$

The matrix U has two useful properties given in the following lemma.

Lemma 1.3. *U is inner and $N_1^* U$ belongs to H_2.*

Proof: Equation (1.28) in equivalent to

$$(F - G_2L_1)'P_1 + P_1(F - G_2L_1) + (H_1 - J_{12}L_1)'\,(H_1 - J_{12}L_1) = 0.$$

Adding and subtracting sP_1 one obtains

$$\begin{aligned} &(H_1 - J_{12}L_1)'(H_1 - J_{12}L_1) \\ &= P_1(sI - F + G_2L_1) + (-sI - F + G_2L_1)'P_1. \end{aligned} \tag{1.35}$$

Now, by explicit calculation,

$$\begin{aligned} &U^*(s)\,U(s) \\ &= J_{12}'J_{12} + J_{12}'(H_1 - J_{12}L_1)(sI - F + G_2L_1)^{-1}G_2 \\ &+ G_2'(-sI - F + G_2L_1)'^{-1}(H_1 - J_{12}L_1)'J_{12} \\ &+ G_2'(-sI - F + G_2L_1)'^{-1}(H_1 - J_{12}L_1)'(H_1 - J_{12}L_1) \\ &\quad \times (sI - F + G_2L_1)^{-1}G_2. \end{aligned}$$

By assumption (FI.4) and (1.29),

$$J_{12}'J_{12} = I, \quad J_{12}'(H_1 - J_{12}L_1) = -G_2'P_1$$

so that

$$\begin{aligned} & U^*(s)\, U(s) \\ &= I - G_2'P_1(sI - F + G_2L_1)^{-1}G_2 \\ &- G_2'(-sI - F + G_2L_1)'^{-1}P_1G_2 \\ &+ G_2'(-sI - F + G_2L_1)'^{-1}(H_1 - J_{12}L_1)' \\ &\quad \times (H_1 - J_{12}L_1)(sI - F + G_2L_1)^{-1}G_2. \end{aligned}$$

Substituting for $(H_1 - J_{12}L_1)'(H_1 - J_{12}L_1)$ from (1.35) gives $U^*(s)U(s) = I$. Since U is proper and stable, it is inner.

To prove the other claim, we observe that

$$N_1^*(s)U(s) = N_1^*(s)N_1(s)G_2 + N_1^*(s)J_{12}$$

where

$$\begin{aligned} & N_1^*(s)N_1(s) \\ &= (-sI - F + G_2L_1)'^{-1}(H_1 - J_{12}L_1)'(H_1 - J_{12}L_1)(sI - F + G_2L_1) \\ &= (-sI - F + G_2L_1)'^{-1}P_1 + P_1(sI - F + G_2L_1)^{-1} \end{aligned}$$

on using (1.35) and

$$N_1^*(s)J_{12} = -(-sI - F + G_2L_1)'^{-1}P_1G_2$$

by assumption (FI.4) and (1.29). Hence

$$N_1^*(s)U(s) = P_1(sI - F + G_2L_1)^{-1}G_2$$

is indeed in H_2. □

Proof of theorem 1.2 Let K be any stabilizing controller for M and look at how $\bar{u}$ is generated form v:

$$\begin{aligned} \dot{x} &= Fx + G_1v + G_2u \\ \bar{u} &= L_1x + u \\ y &= \begin{bmatrix} x \\ w \end{bmatrix}. \end{aligned}$$

These equations can be represented by a diagram shown in figure 1.3.

Denoting $\bar{T}$ the transfer function from v to $\bar{u}$, we obtain from (1.33) the following expression for T, the transfer function from v to z in figure 1.2:

$$T(z) = N_1(s)G_1 + U(s)\bar{T}(s).$$

Now $\bar{T}$ is in H_2 since K stabilizes M if and only if K stabilizes $\bar{M}$ (the two closed-loop systems have identical state-transition matrices). By lemma 1.3, N_1 is orthogonal to $U\bar{T}$ for every $\bar{T}$ in H_2. Hence

$$||T||_2^2 = ||N_1G_1||_2^2 + ||U\bar{T}||_2^2.$$

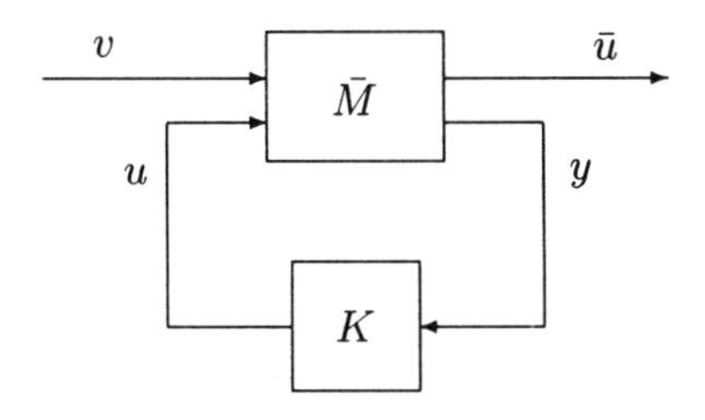

Figure 1.3: The system generating $\bar{u}$.

By lemma 1.3 again, U is inner and therefore norm-preserving:

$$||U\bar{T}||_2 = ||\bar{T}||_2$$

for any $\bar{T}$ in H_2. As a result,

$$||T||_2^2 = ||N_1G_1||_2^2 + ||\bar{T}||_2^2. \tag{1.36}$$

Now N_1G_1 does not depend on K and $\bar{T}$ can be made identically zero by setting $u = -L_1x$ so that $\bar{u} = 0$. This uniquely minimizes $||T||_2$ and results in

$$\min ||T||_2 = ||N_1G_1||_2$$

and $K_{FI}(s) = [\,-L_1 \quad 0\,]$. □

The H_2 optimal controller in this case is a *state* feedback: K_{FI} uses just x, with v providing only redundant information. Moreover, K_{FI} is a *constant* gain obtained from P_1, a well defined soluton of the algebraic Riccati equation (1.28).

Output estimation problem

This is another special H_2 control problem, that of estimating the output z given the measurement y. Thus, in the plant equations (1.26), $J_{12} = I$ and $G_2 = 0$. Having in mind the original objective, namely solving the output feedback problem, we shall not eliminate the effect of u through feedback and keep a non-zero G_2. This additional loop will have to be stabilized, of course. The plant equations are therefore

$$\begin{aligned} \dot{x} &= Fx + G_1v + G_2u \\ z &= H_1x + u \\ y &= H_2x + J_{21}v \end{aligned} \tag{1.37}$$

and assumptions (SS.1) to (SS.5) are modified as follows:

(OE.1) (F, G_1) is stabilizable;

(OE.2) (F, H_2) is detectable;

(OE.3) $J_{11} = 0, \quad J_{22} = 0;$

(OE.4) $G_1 J_{21}' = 0, \quad J_{21} J_{21}' = I;$

(OE.5) $F - G_2 H_1$ is stable.

To find a stabilizing controller K which minimizes $||T||_2$ under these assumptions, we consider the algebraic Riccati equation

$$FP_0 + P_0 F' - P_0 H_2' H_2 P_0 + G_1 G_1' = 0. \tag{1.38}$$

Assumptions (OE.1) and (OE.2) are necessary and sufficient for (1.38) to have a unique solution $P_0 \geq 0$ such that $F - L_0 H_2$ is stable, where

$$L_0 = P_0 H_2'. \tag{1.39}$$

Theorem 1.4 (Doyle *et al.*, **1989).** *The unique optimal controller for the ouput estimation H_2 problem is*

$$K_{OE}(s) = -H_1(sI - F + L_0 H_2 + G_2 H_1)^{-1} L_0. \tag{1.40}$$

Moreover,

$$\min ||T||_2^2 = ||H_1 N_2||_2^2 = \operatorname{Tr} H_1 P_0 H_1', \tag{1.41}$$

where

$$N_2(s) = (sI - F + L_0 H_2)^{-1} \; (G_1 - L_0 J_{21}). \tag{1.42}$$

The proof requires a dual formulation of the problem and a lemma. The *dual* of the estimation problem is defined by the equations

$$\begin{aligned} \dot{x} &= Fx + G_1 v + G_2 u \\ z &= H_1 x + J_{12} u \\ y &= H_2 x + v \end{aligned} \tag{1.43}$$

where F, G_1, G_2, H_1, H_2 and J_{12} correspond respectively to the F', H_1', H_2', G_1', G_2' and J_{21}' of (1.37). Assumptions (OE.1) – (OE.4) for (1.37) turn out to be (FI.1) – (FI.4) for (1.43). The dual of the additional assumption (OE.5) will be therefore denoted as

(FI.5) $F - G_1 H_2$ is stable.

The important observation is that the dual of the output estimation problem is equivalent to the full information control problem. Denote the plant for the full information problem as M_{FI} and the plant for the dual of the output estimation problem as $\hat{M}_{OE}$. Suppose we have controllers K_{FI} and $\hat{K}_{OE}$ and let T_{FI} and $\hat{T}_{OE}$ denote the closed-loop transfer functions in figure 1.4.

We shall assume that any controller realizations are stabilizable and detectable. The equivalence is shown in the following lemma.

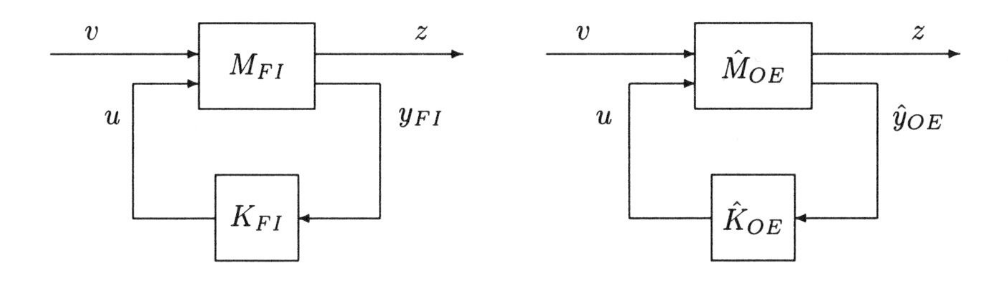

Figure 1.4: Two equivalent problems.

Lemma 1.5. *The controller K_{FI} internally stabilizes M_{FI} if and only if $\hat{K}_{OE}$ internally stabilizes $\hat{M}_{OE}$, where $\hat{K}_{OE}$ is the transfer function from $\hat{y}_{OE}$ to u of the system*

$$\begin{aligned}
\dot{\hat{x}} &= (F - G_1 H_2)\hat{x} + G_1 \hat{y}_{OE} + G_2 u \\
y_{FI} &= \begin{bmatrix} \hat{x} \\ \hat{y}_{OE} - H_2 \hat{x} \end{bmatrix} \\
u &= K_{FI}\, y_{FI}.
\end{aligned} \tag{1.44}$$

Furthermore, in this case $T_{FI} = \hat{T}_{OE}$.

Proof: Apply $\hat{K}_{OE}$ to $\hat{M}_{OE}$. The overall equations, (1.43) and (1.44) combined, in terms of $e = x - \hat{x}$ and $\hat{x}$ are

$$\begin{aligned}
\dot{e} &= (F - G_1 H_1)e \\
\dot{\hat{x}} &= F\hat{x} + G_1 v + G_2 u + G_1 H_2 e \\
z &= H_1 \hat{x} + J_{12} u + H_1 e \\
y_{FI} &= \begin{bmatrix} \hat{x} \\ v + H_2 e \end{bmatrix} \\
u &= K_{FI}\, y_{FI}.
\end{aligned} \tag{1.45}$$

Now apply K_{FI} to M_{FI}. The corresponding equations

$$\begin{aligned}
\dot{x} &= Fx + G_1 v + G_2 u \\
z &= H_1 x + J_{12} u \\
y_{FI} &= \begin{bmatrix} x \\ v \end{bmatrix} \\
u &= K_{FI}\, y_{FI}
\end{aligned} \tag{1.46}$$

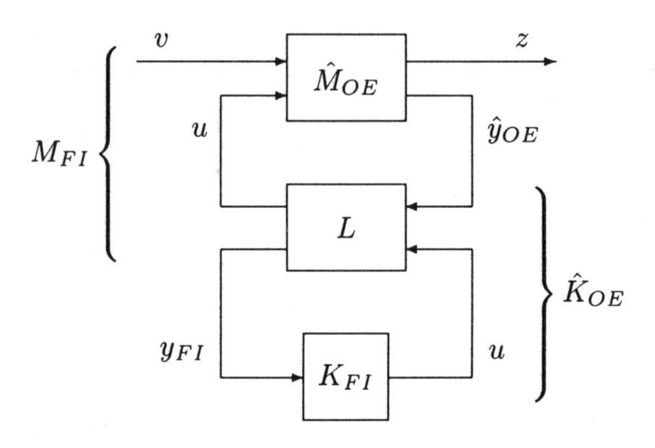

Figure 1.5: The system generating M_{FI} and $\hat{K}_{OE}$.

have the same structure as (1.45) has, the only difference being the terms involving e. So as long as the matrix $F - G_1 H_2$ governing the dynamics of e is stable, the controller K_{FI} internally stabilizes M_{FI} if and only if $\hat{K}_{OE}$ internally stabilizes $\hat{M}_{OE}$. Assumption (FI.5) addresses this requirement.

The resulting system is represented in figure 1.5, where L has the state-space representation (1.44).

We observe that K_{FI} and L generate $\hat{K}_{OE}$ whereas L and $\hat{M}_{OE}$ combine to give M_{FI}. This proves the transfer-function equivalence of the two configurations shown in figure 1.4. Hence $T_{FI} = \hat{T}_{OE}$. □

Proof of theorem 1.4: We shall now apply lemma 1.5 to produce the solution of the dual estimation problem from theorem 1.2. Since $K_{FI} = (-L_1 \; 0)$, one has from (1.44)

$$\hat{K}_{OE}(s) = -L_1(sI - F + G_1 H_2 + G_2 L_1)^{-1} G_1. \tag{1.47}$$

Since $T_{FI} = \hat{T}_{OE}$, one obtains from (1.31)

$$\min ||\hat{T}_{OE}||_2^2 = ||N_1 G_1||_2^2 = \operatorname{Tr} G_1' P_1 G_1.$$

Now, the solution of the primal output estimation problem is easily recovered by duality. The algebraic Riccati equation (1.38) is dual to (1.28) and (1.39) replaces (1.29). The controller (1.40) follows from (1.47) and the expressions (1.41), (1.42) are the duals of (1.31), (1.32). □

The H_2 controller in this case is a *dynamic output* feedback whose structure reflects the duality of full information control and output estimation problems.

H_2 optimal controller

We are now in a position to solve the H_2 control problem defined in subsection 3.2.1. The problem involves the plant (1.26) satisfying assumptions (SS.1) – (SS.5) and an *output* feedback controller. The solution will be obtained by combining the two special H_2 problems solved in subsections 3.2.2 and 3.2.3.

Theorem 1.6 (Doyle *et al.*, **1989).** *The unique optimal controller for the H_2 control problem is*

$$K_{opt}(s) = -L_1(sI - F + G_2L_1 + L_OH_2)^{-1}L_O, \tag{1.48}$$

where L_1 is given by (1.28), (1.29) and L_2 by (1.38), (1.39). Moreover,

$$\begin{aligned}\min ||T||_2^2 &= ||N_1G_1||_2^2 + ||L_1N_2||_2^2 \\ &= ||N_1L_O||_2^2 + ||H_1N_2||_2^2,\end{aligned} \tag{1.49}$$

where N_1 is given by (1.32) and N_2 by (1.42).

Proof: We define a new control variable $\bar{u} = u + L_1x$ as in the full information problem and follow the steps in the proof of theorem 1.2 through (1.36). Hence

$$\min ||T||_2^2 = ||N_1G_1||_2^2 + \min ||\bar{T}||_2^2.$$

At this point we observe that it is not possible to make $\bar{T}$ identically zero by setting $\bar{u} = 0$, i.e., $u = -L_1x$, since the state x is not available for feedback. We have to find the control variable $\bar{u}$ that minimizes $||\bar{T}||_2$ given the measurement y only. However, this is exactly the output estimation problem where $z = H_1x + u$ is identified with $\bar{u}$. The solution can thus be obtained by setting $H_1 = L_1$ in theorem 1.4. The norm $||\bar{T}||_2$ is uniquely minimized by the controller (1.48) and $\min ||\bar{T}||_2 = ||L_1N_2||_2$. □

This theorem reveals the separation property of the H_2 optimal controller. The gain L_1 is the optimal state feedback in the full information problem and L_O is the optimal output injection in the output estimation problem. The controller equations can be written in standard observer form as

$$\begin{aligned}\dot{\hat{x}} &= F\hat{x} + G_2u + L_O(y - H_2\hat{x}) \\ u &= -L_1\hat{x}.\end{aligned}$$

1.3.3 Wiener-Hopf solution

The Wiener-Hopf approach is an analytical design technique for systems which are described by transfer functions. It is an inherently open loop approach. We assume that a transfer matrix of the plant M (hence M_{11}, M_{12}, M_{21} and

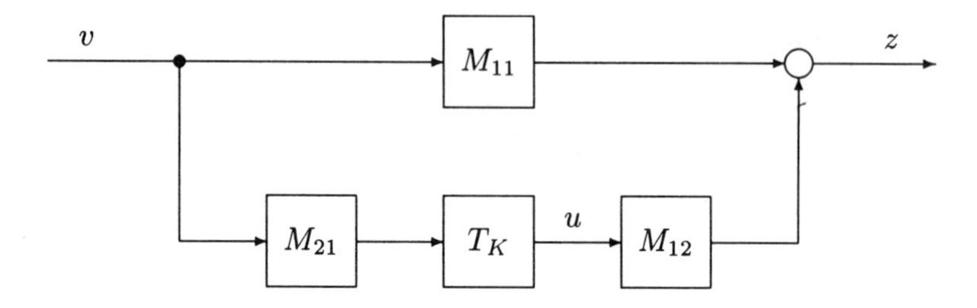

Figure 1.6: Cascade compensator.

M_{22}) is available and shall proceed to find the transfer matrix of the H_2-optimal equivalent cascade compensator T_K, shown in figure 1.6.

The transfer matrix of the corresponding feedback controller K is then calculated from T_K. In this setting, the requirement of internal stability will mean that the characteristic polynomial of the closed loop system shown in figure 1.2 is *strictly Hurwitz* (i.e., free of zeros in $\operatorname{Re} s \geq 0$).

Our basic assumption is that M and K are *real-rational* matrices, not necessarily proper. It follows from (1.22) and (1.23) that

$$z = [M_{11} + M_{12}(I - KM_{22})^{-1}KM_{21}]v$$

so that

$$T = M_{11} + M_{12}T_K M_{21} \tag{1.50}$$

where

$$T_K = (I - KM_{22})^{-1}K \tag{1.51}$$

is the equivalent cascade compensator. The corresponding feedback controller is then

$$K = T_K(I + M_{22}T_K)^{-1}. \tag{1.52}$$

There are many cascade compensators that yield both finite norm $||T||_2$ and internally stable closed loop system of figure 1.2. Included among these is an optimal one which minimizes $||T||_2$. To delineate this set, it is first necessary to solve the intermediate problem of finding all admissible cascade compensators. The solution to be presented follows the paper by Park and Bongiorno (1989).

Admissible cascade compensators

A rational matrix T_K is said to be *admissible* for the given plant M if there exists a controller K which realizes it as the designated closed-loop system transfer matrix given by (1.51) and the closed loop system, shown in figure 1.2, is internally stable.

The following assumption states the stabilizability condition for the given plant M. For any rational matrix, say G, the *characteristic denominator* of G is the monic least common multiple of the denominators of all minors of G, each minor assumed expressed as the ratio of two coprime polynomials. For any polynomial ψ, we denote by ψ^+ the monic polynomial that absorbs all the zeros of $\psi(s)$ in $\mathrm{Re}\, s \geq 0$.

(WH.1) The plant M is free of hidden (i.e., uncontrollable and/or unobservable) poles in $\mathrm{Re}\, s \geq 0$ and

$$\psi_M^+ = \psi_{M_{22}}^+$$

where ψ_M and $\psi_{M_{22}}$ are the characteristic denominators of M and M_{22}, respectively.

Under this assumption, one can characterize every admissible matrix T_K for a prescribed plant M in terms of an arbitrary real–rational matrix parameter $W(s)$ analytic in $\mathrm{Re}\, s \geq 0$. The parameterization is based on the following factorizations of M_{22},

$$A_L^{-1}B_L = M_{22} = B_R A_R^{-1} \tag{1.53}$$

where A_L, B_L is any left coprime and A_R, B_R is any right coprime pair of *polynomial* matrices. Associated with these matrices are polynomial matrices X_L, Y_L and X_R, Y_R such that

$$\begin{bmatrix} X_L & Y_L \\ -B_L & A_L \end{bmatrix} \begin{bmatrix} A_R & -Y_R \\ B_R & X_R \end{bmatrix} = I, \tag{1.54}$$

which means that both factors in (1.54) are unimodular. One refers to (1.54) as to the *Bézout identity*.

Theorem 1.7 (Park and Bongiorno, 1989). *A matrix T_K is admissible for the plant M if and only if*

$$T_K = -A_R(Y_L + WA_L) \tag{1.55}$$

where $W(s)$ is any real–rational matrix analytic in $\mathrm{Re}\, s \geq 0$ such that $X_L - WB_L$ is non-singular. The associated controller transfer matrix that realizes T_K is then given by

$$K = -(X_L - WB_L)^{-1}(Y_L + WA_L). \tag{1.56}$$

The proof requires a lemma.

Lemma 1.8 (Youla *et al.*, **1976).** *Let χ_M and χ_K denote the characteristic polynomials of the plant M and controller K, respectively. Then*

$$\chi = \chi_M \chi_K \, Det(I - KM_{22}) \tag{1.57}$$

is the characteristic polynomial for the control system shown in figure 1.2.

Proof: It is convenient to arrange the inputs and output of M and K in figure 1.2 into the vectors

$$\bar{y} = \begin{bmatrix} z \\ y \\ u \end{bmatrix}, \quad \bar{u} = \begin{bmatrix} v \\ u \\ y \end{bmatrix}$$

so that

$$\bar{y} = \bar{M}\bar{u} \tag{1.58}$$

where

$$\bar{M} = \begin{bmatrix} M_{11} & M_{12} & 0 \\ M_{21} & M_{22} & 0 \\ 0 & 0 & K \end{bmatrix}.$$

The interconnections in figure 1.2 are then expressed as

$$\bar{u} = P\bar{y} + Qv \tag{1.59}$$

where

$$P = \begin{bmatrix} 0 & 0 & 0 \\ 0 & 0 & I \\ 0 & I & 0 \end{bmatrix}, \quad Q = \begin{bmatrix} I \\ 0 \\ 0 \end{bmatrix}.$$

Denote x_M and x_K the states of M and K, respectively. The input-state equations for M and K are given by

$$\begin{aligned} F_M(s)x_M &= G_M(s)\bar{u} \\ F_K(s)x_K &= G_K(s)\bar{u} \end{aligned} \tag{1.60}$$

where F_M, G_M and F_K, G_K are some real polynomial matrices. The characteristic polynomials of M and K are, to within a scaling factor,

$$\chi_M = \text{Det}\, F_M, \; \chi_K = \text{Det}\, F_K.$$

Eliminating $\bar{u}$ in (1.60) with the help of (1.58) and (1.59), one obtains the overall input-state equation

$$\begin{bmatrix} F_M & 0 & -G_M P \\ 0 & F_K & -G_K P \\ 0 & 0 & I - \bar{M}P \end{bmatrix} \begin{bmatrix} x_M \\ x_K \\ \bar{y} \end{bmatrix} = \begin{bmatrix} G_M \\ G_K \\ \bar{M} \end{bmatrix} Qv. \tag{1.61}$$

The characteristic polynomial, χ, of the control system is then given by the determinant of the coefficient matrix on the left-hand side of (1.61). Consequently,

$$\chi = \chi_M \chi_K \,\text{Det}\,(I - \bar{M}P).$$

Since

$$I - \bar{M}P = \begin{bmatrix} I & 0 & -M_{12} \\ 0 & I & -M_{22} \\ 0 & -K & I \end{bmatrix}$$

one obtains (1.57). □

Proof of theorem 1.7 To prove necessity, suppose that T_K is an admissible cascade compensator for M. By lemma 1.8, the characteristic polynomial of the control system shown in figure 1.2 is given by (1.57) and it is strictly Hurwitz. In terms of the coprime polynomial factorizations

$$M_{22} = B_R A_R^{-1}, \; K = -P_L^{-1} Q_L,$$

we have

$$\begin{aligned}\chi &= \chi_M \chi_K \operatorname{Det}(I + P_L^{-1} Q_L B_R A_R^{-1}) \\ &= \frac{\chi_M}{\operatorname{Det} A_R} \frac{\chi_K}{\operatorname{Det} P_L} \operatorname{Det}(P_L A_R + Q_L B_R).\end{aligned}$$

To within a multiplicative constant, the polynomials $\operatorname{Det} A_R$ and $\operatorname{Det} P_L$ are the characteristic denominators for M_{22} and K, respectively. Thus, $\operatorname{Det} P_L$ divides χ_K and $\chi_M / \operatorname{Det} A_R$ is nonzero in $\operatorname{Re} s \geq 0$ because of (WH.1). Consequently, the polynomial $\operatorname{Det}(P_L A_R + Q_L B_R)$ must be strict Hurwitz when χ is.

It now follows that

$$\begin{aligned}T_K &= (I - K M_{22})^{-1} K \\ &= -(I + P_L^{-1} Q_L B_R A_R^{-1})^{-1} P_L^{-1} Q_L \\ &= -A_R (P_L A_R + Q_L B_R)^{-1} Q_L \\ &= -A_R N \qquad (1.62)\end{aligned}$$

where $N(s)$ is analytic in $\operatorname{Re} s \geq 0$. Defining $W(s)$ by means of the equation

$$N = Y_L + W A_L$$

one gets (1.55) and it remains to be shown that $W(s)$ is analytic in $\operatorname{Re} s \geq 0$.

The analyticity of

$$Y_L + W A_L = (P_L A_R + Q_L B_R)^{-1} Q_L$$

in $\operatorname{Re} s \geq 0$ is obvious. Also, recalling (1.54) and (1.62), it follows that

$$\begin{aligned}X_L - W B_L &= X_L - (N - Y_L) A_L^{-1} B_L \\ &= (X_L + Y_L B_R A_R^{-1}) - N M_{22} \\ &= A_R^{-1} + A_R^{-1} T_K M_{22} \qquad (1.63) \\ &= A_R^{-1} + A_R^{-1} (I - K M_{22})^{-1} K M_{22} \\ &= A_R^{-1} (I - K M_{22})^{-1} \qquad (1.64) \\ &= (P_L A_R + Q_L B_R)^{-1} P_L.\end{aligned}$$

Therefore, $X_L - W B_L$ is also analytic in $\operatorname{Re} s \geq 0$ and the identity

$$(Y_L + W A_L) X_R - (X_L - W B_L) Y_R = Y_L X_R - X_L Y_R + W$$

immediately establishes the requisite analyticity of W. Also note that, from (1.64), $X_L - WB_L$ is non-singular.

To prove sufficiency, let

$$T_K = -A_R(Y_L + WA_L)$$

where $W(s)$ is a real–rational matrix analytic in $\operatorname{Re} s \geq 0$. Suppose also that $X_L - WB_L$ is non-singular. The task is to prove that T_K is admissible for the given M. From (1.52),

$$\begin{aligned} K &= T_K(I + M_{22}T_K)^{-1} \\ &= -A_R(Y_L + WA_L)[I - B_R(Y_L + WA_L)]^{-1} \\ &= -A_R[I - (Y_L + WA_L)B_R]^{-1}(Y_L + WA_L) \\ &= -A_R(X_LA_R - WA_LB_R)^{-1}(Y_L + WA_L) \\ &= -(X_L - WB_L)^{-1}(Y_L + WA_L). \end{aligned} \tag{1.65}$$

Let $W = U_L^{-1}V_L$ be any left coprime polynomial factorization of W. Then

$$K = -(U_LX_L - V_LB_L)^{-1}(U_LY_L + V_LA_L).$$

The pair $U_LX_L - V_LB_L$, $U_LY_L + V_LA_L$ is left coprime, as it is obtained from the left coprime pair U_L, V_L by a unimodular transformation:

$$[U_LX_L - V_LB_L \quad U_LY_L + V_LA_L] = [U_L \quad V_L] \begin{bmatrix} X_L & Y_L \\ -B_L & A_L \end{bmatrix}.$$

Hence

$$\chi_K = \operatorname{Det}(U_LX_L - V_LB_L)$$

for any minimal realization of K. Moreover, since

$$\begin{aligned} I - KM_{22} &= I + (U_LX_L - V_LB_L)^{-1}(U_LY_L + V_LA_L)B_RA_R^{-1} \\ &= (U_LX_L - V_LB_L)^{-1}[U_L(X_LA_R + Y_LB_R) \\ &\qquad - V_L(B_LA_R - A_LB_R)]A_R^{-1} \\ &= (U_LX_L - V_LB_L)^{-1}U_LA_R^{-1}, \end{aligned}$$

one finds that

$$\begin{aligned} \chi &= \chi_M\chi_K \operatorname{Det}(I - KM_{22}) \\ &= \frac{\chi_M}{\operatorname{Det} A_R} \, \frac{\chi_K}{\operatorname{Det}(U_LX_L - V_LB_L)} \operatorname{Det} U_L. \end{aligned} \tag{1.66}$$

The first factor in (1.66) is strictly Hurwitz because (WH.1) holds for M, the second because K is realized minimally, and the third because $W(s)$ is analytic in $\operatorname{Re} s \geq 0$. Hence, the control system shown in figure 1.2 is internally stable and the proof of (1.55) is complete.

The claim (1.56) results from (1.65). □

The significance of theorem 1.7 lies in the fact that the minimization of the norm $||T||_2$ can be carried out over the set of real-rational matrices $W(s)$ analytic in $\operatorname{Re} s \geq 0$ instead of over the set of all controller transfer matrices $K(s)$.

The optimal controller

In addition to assumption (WH.1) we impose the following restrictions on the plant:

(WH.2) $M_{11}(s)$ is strictly proper;

(WH.3) $M_{12}(s)$ has full column rank and $M_{21}(s)$ has full row rank.

The square matrices $\Lambda(s)$ and $\Omega(s)$ are used to designate the *spectral factors*

$$A_R^* M_{12}^* M_{12} A_R = \Lambda^* \Lambda \tag{1.67}$$

and

$$A_L M_{21} M_{21}^* A_L^* = \Omega \Omega^*. \tag{1.68}$$

The spectral factors are real-rational matrices which are analytic, together with their inverses, in $\operatorname{Re} s \geq 0$. The following assumption guarantees that the matrices Λ^{-1} and Ω^{-1} are actually analytic in $\operatorname{Re} s \geq 0$.

(WH.4) $\Lambda^{-1}(s)$ and $\Omega^{-1}(s)$ are analytic on $\operatorname{Re} s = 0$.

We define the matrix

$$\Gamma = \Lambda^{*-1} A_R^* M_{12}^* M_{11} M_{21}^* A_L^* \Omega^{*-1} \tag{1.69}$$

which is real-rational and strictly proper. Indeed, from (1.67) and (1.68),

$$\begin{aligned} (M_{12} A_R \Lambda^{-1})^* \ (M_{12} A_R \Lambda^{-1}) &= I \\ (\Omega^{-1} A_L M_{21})(\Omega^{-1} A_L M_{21})^* &= I \end{aligned} \tag{1.70}$$

so that both $M_{12} A_R \Lambda^{-1}$ and $\Omega^{-1} A_L M_{21}$ are bounded at $s = \infty$. Consequently,

$$\Gamma = (M_{12} A_R \Lambda^{-1})^* \ M_{11} \ (\Omega^{-1} A_L M_{21})^*$$

is strictly proper because so is M_{11} by assumption (WH.2).

Two additional assumptions, which concerne the matrix Γ, will be needed in the sequel:

(WH.5) The difference $\operatorname{Tr} M_{11}^* M_{11} - \operatorname{Tr} \Gamma^* \Gamma$ is analytic on $\operatorname{Re} s = 0$;

(WH.6) The matrix $\Gamma - \Lambda A_R^{-1} Y_R \Omega$ is analytic on $\operatorname{Re} s = 0$.

The optimal controller will be obtained by minimizing $||T||_2$ with respect to the free parameter matrix W of (1.55). A standard variational method will be employed. However, the formula obtained by this method is merely a candidate for the optimal solution, unless it yields the minimum finite norm. The above assumptions are sufficient to prove the optimality.

The following notation is standard. In the partial fraction expansion of any real-rational matrix $G(s)$, the contribution from all poles in $\text{Re}\, s \leq 0$, $\text{Re}\, s > 0$, and at $s = \infty$ are denoted by $\{G\}_+, \{G\}_-$ and $\{G\}_\infty$, respectively. Thus

$$G = \{G\}_+ + \{G\}_- + \{G\}_\infty$$

and $\{G\}_+$ is strictly proper and analytic in $\text{Re}\, s > 0$, $\{G\}_-$ is strictly proper and analytic in $\text{Re}\, s \leq 0$, and $\{G\}_\infty$ is polynomial.

Theorem 1.9 (Park and Bongiorno, 1989). *Suppose that assumptions (WH.1) – (WH.6) are satisfied. Then,*

(a) the set of all admissible cascade compensators T_K that yield finite norm $||T||_2$ is generated by the formula

$$T_K = -A_R \Lambda^{-1}(\{\Gamma\}_+ + \{\Lambda A_R^{-1} Y_R \Omega\}_- + V)\Omega^{-1} A_L \tag{1.71}$$

where $V(s)$ is an arbitrary strictly proper real-rational matrix which is analytic in $\text{Re}\, s \geq 0$ and such that $I + T_K M_{22}$ is non-singular;

(b) the admissible T_K that minimizes $||T||_2$ is given by

$$T_{K_{opt}} = -A_R \Lambda^{-1}(\{\Gamma\}_+ + \{\Lambda A_R^{-1} Y_R \Omega\}_-)\Omega^{-1} A_L \tag{1.72}$$

and corresponds to the choice $V = 0$;

(c) the norm associated with V is

$$||T||_2^2 = \min\ ||T||_2^2 + ||V||_2^2. \tag{1.73}$$

Proof: The first step of the proof is to express $||T||_2$ in terms of the free parameter W. From (1.50),

$$T = M_{11} + M_{12} T_K M_{21}$$

and from theorem 1.7, the admissible matrices T_K are given by

$$T_K = -A_R(Y_L + W A_L)$$

where W is real-rational, analytic in $\text{Re}\, s \geq 0$ and such that $X_L - W B_L$ is non-singular. Hence

$$T = (M_{11} - M_{12} A_R Y_L M_{21}) - (M_{12} A_R) W (A_L M_{21}).$$

Using (1.24),

$$||T||_2^2 = \frac{1}{2\pi j} \int \text{Tr}\, T^*(s) T(s)\, ds$$

$$= \frac{1}{2\pi j} \int \text{Tr}\,(\Xi - 2W^*\Theta + W^*\Phi W \Psi)\, ds \tag{1.74}$$

where

$$
\begin{aligned}
\Xi &= (M_{11} - M_{12}A_RY_LM_{21})^*(M_{11} - M_{12}A_RY_LM_{21}) \\
\Theta &= (M_{12}A_R)^*(M_{11} - M_{12}A_RY_LM_{21})(A_LM_{21})^* \\
\Phi &= (M_{12}A_R)^*(M_{12}A_R) \\
\Psi &= (A_LM_{21})(A_LM_{21})^*.
\end{aligned}
\tag{1.75}
$$

The second step of the proof is a variational argument. It is assumed that a real-rational $W = W_0$ analytic in $\mathrm{Re}\, s \geq 0$ exists for which $||T||_2$ attains a minimum value. Then W_0 is replaced by $W = W_0 + \epsilon W_1$ where W_1 is real-rational and analytic in $\mathrm{Re}\, s \geq 0$ but otherwise arbitrary and ϵ is the variational parameter. Taking the partial derivative of $||T||_2$ with respect to ϵ, one obtains

$$\frac{\partial}{\partial\epsilon}||T||_2^2 = \frac{1}{2\pi j}\int \mathrm{Tr}\,(-2W_1^*\Theta + 2W_1^*\Phi W_0\Psi + 2\epsilon W_1^*\Phi W_1\Psi)\,\mathrm{d}s. \tag{1.76}$$

The necessary condition for W_0 to yield a minimum value of the norm is

$$\left.\frac{\partial}{\partial\epsilon}||T||_2^2\right|_{\epsilon=0} = 0.$$

Applying this condition to (1.76), one obtains

$$\int \mathrm{Tr}\, W_1^*(-\Theta + \Phi W_0\Psi)\mathrm{d}s = 0. \tag{1.77}$$

By Cauchy's theorem, the matrix

$$\Delta = -\Theta + \Phi W_0\Psi$$

must be analytic in $\mathrm{Re}\, s \leq 0$ in order to satisfy (1.77). Using (1.67) and (1.68), Φ and Ψ can be spectrally factored as

$$\Phi = \Lambda^*\Lambda, \;\; \Psi = \Omega\Omega^*.$$

Let

$$
\begin{aligned}
\bar{\Delta} &= \Lambda^{*\,-1}\Delta\Omega^{*\,-1} \\
&= -\Lambda^{*\,-1}\Theta\Omega^{*\,-1} + \Lambda W_0\Omega
\end{aligned}
\tag{1.78}
$$

and note that $\bar{\Delta}$ is analytic in $\mathrm{Re}\, s \leq 0$. The first term on the right-hand side of (1.78) has the partial fraction expansion

$$\Lambda^{*\,-1}\Theta\Omega^{*\,-1} = \{\Lambda^{*\,-1}\Theta\Omega^{*\,-1}\}_+ + \{\Lambda^{*\,-1}\Theta\Omega^{*\,-1}\}_- + \{\Lambda^{*\,-1}\Theta\Omega^{*\,-1}\}_\infty$$

and (1.78) can be written as

$$
\begin{aligned}
&\bar{\Delta} + \{\Lambda^{*\,-1}\Theta\Omega^{*\,-1}\}_- + \{\Lambda^{*\,-1}\Theta\Omega^{*\,-1}\}_\infty \\
&\quad = -\{\Lambda^{*\,-1}\Theta\Omega^{*\,-1}\}_+ + \Lambda W_0\Omega.
\end{aligned}
\tag{1.79}
$$

The right-hand side of (1.79) is analytic in $\mathrm{Re}\, s > 0$ and the left-hand side is analytic in $\mathrm{Re}\, s \leq 0$. Therefore, the right-hand side of (1.79) must be analytic for every finite s, which implies that it is equal to some polynomial matrix P. Setting the right-hand side of (1.79) equal to P and solving for W_0, one obtains

$$W_0 = \Lambda^{-1}(P + \{\Lambda^{*\,-1}\Theta\Omega^{*\,-1}\}_+)\Omega^{-1}.$$

Using (1.75) and (1.69),

$$\Lambda^{*\,-1}\Theta\Omega^{*\,-1} = \Gamma - \Lambda A_R^{-1} Y_R \Omega$$

and one obtains the following candidate formula for the optimal parameter matrix:

$$W_0 = \Lambda^{-1}(P + \{\Gamma - \Lambda A_R^{-1} Y_R \Omega\}_+)\Omega^{-1}. \tag{1.80}$$

The corresponding matrix T_{K0} is obtained from (1.55) to be

$$\begin{aligned} T_{K0} = -A_R\Lambda^{-1}(\{\Gamma\}_+ &+ \{\Lambda A_R^{-1} Y_R \Omega\}_- \\ &+ \{\Lambda A_R^{-1} Y_R \Omega\}_\infty + P)\Omega^{-1} A_L. \end{aligned} \tag{1.81}$$

The third step of the proof is to check whether the candidate W_0 is analytic in $\mathrm{Re}\, s \geq 0$ and yields a finite norm. It is obvious from (1.80) that W_0 is analytic in $\mathrm{Re}\, s > 0$. It will be analytic also on the imaginary axis $\mathrm{Re}\, s = 0$ if and only if $\Gamma - \Lambda A_R^{-1} Y_R \Omega$ is. Assumption (WH.6) addresses this requirement.

Next the requirement of finite norm is investigated. Using (1.50), the integrand of (1.74) can be written as

$$\begin{aligned} \mathrm{Tr}\, T^*T = \mathrm{Tr}\, &M_{11}^* M_{11} \\ &+ \mathrm{Tr}\,(M_{11}^* M_{12} T_K M_{21} + M_{21}^* T_K^* M_{12}^* M_{11}) \\ &+ \mathrm{Tr}\, M_{21}^* T_K^* M_{12}^* M_{12} T_K M_{21}. \end{aligned} \tag{1.82}$$

Substituting T_{K0} for T_K and recalling (1.67), (1.68) and (1.69), one has

$$\begin{aligned} \mathrm{Tr}\, T^*T = {} & \mathrm{Tr}\, M_{11}^* M_{11} \\ & - \mathrm{Tr}\,[\Gamma^*(\{\Gamma\}_+ + \{\Lambda A_R^{-1} Y_R \Omega\}_- + \{\Lambda A_R^{-1} Y_R \Omega\}_\infty + P) \\ & \qquad + (\{\Gamma\}_+ + \{\Lambda A_R^{-1} Y_R \Omega\}_- + \{\Lambda A_R^{-1} Y_R \Omega\}_\infty + P)^*\Gamma] \\ & + \mathrm{Tr}\,(\{\Gamma\}_+ + \{\Lambda A_R^{-1} Y_R \Omega\}_- + \{\Lambda A_R^{-1} Y_R \Omega\}_\infty + P)^* \\ & \qquad \times(\{\Gamma\}_+ + \{\Lambda A_R^{-1} Y_R \Omega\}_- + \{\Lambda A_R^{-1} Y_R \Omega\}_\infty + P) \\ = {} & \mathrm{Tr}\, M_{11}^* M_{11} - \mathrm{Tr}\,\Gamma^*\Gamma \\ & + \mathrm{Tr}\,(\{\Gamma\}_+ + \{\Lambda A_R^{-1} Y_R \Omega\}_- + \{\Lambda A_R^{-1} Y_R \Omega\}_\infty + P - \Gamma)^* \\ & \qquad \times(\{\Gamma\}_+ + \{\Lambda A_R^{-1} Y_R \Omega\}_- + \{\Lambda A_R^{-1} Y_R \Omega\}_\infty + P - \Gamma) \\ = {} & \mathrm{Tr}\, M_{11}^* M_{11} - \mathrm{Tr}\,\Gamma^*\Gamma \\ & + \mathrm{Tr}\,(P + \{\Lambda A_R^{-1} Y_R \Omega\}_\infty - \{\Gamma - \Lambda A_R^{-1} Y_R \Omega\}_-)^* \\ & \qquad \times(P + \{\Lambda A_R^{-1} Y_R \Omega\}_\infty - \{\Gamma - \Lambda A_R^{-1} Y_R \Omega\}_-). \end{aligned} \tag{1.83}$$

This integrand yields finite norm if and only if it is analytic on the imaginary axis and all its terms are products of strictly proper matrices. The last term in (1.83) is obviously analytic on $\operatorname{Re} s = 0$, hence the entire integrand is so if and only if the term $\operatorname{Tr} M_{11}^* M_{11} - \operatorname{Tr} \Gamma^* \Gamma$ is analytic on $\operatorname{Re} s = 0$. Assumption (WH.5) takes care of this requirement. Now M_{11} and Γ are strictly proper while the last term in (1.83) has a polynomial part $P + \{\Lambda A_R^{-1} Y_R \Omega\}_\infty$. Therefore the integrand (1.83) yields a finite norm if and only if this polynomial term vanishes. Using this result in (1.81) gives (1.72).

The fourth step of the proof is to verify the optimality of (1.72). Consider the set of all matrices T_K generated by the formula (1.71) with V an arbitrary real-rational matrix. Such a T_K is admissible if and only if $V(s)$ is analytic in $\operatorname{Re} s \geq 0$. For if one sets the right-hand side of (1.55) equal to that of (1.71), V and W are seen to be related as

$$V = \Lambda W \Omega + \{\Lambda A_R^{-1} Y_R \Omega\}_\infty - \{\Gamma - \Lambda A_R^{-1} Y_R \Omega\}_+$$

and it is apparent, in view of assumption (WH.6), that W is analytic in $\operatorname{Re} s \geq 0$ if and only if V is so. Substituting T_K given by (1.71) into (1.82) and following the same steps as those leading to (1.83), one obtains

$$\begin{aligned} \operatorname{Tr} T^* T = {} & \operatorname{Tr} M_{11}^* M_{11} - \operatorname{Tr} \Gamma^* \Gamma \\ & + \operatorname{Tr} (V - \{\Gamma - \Lambda A_R^{-1} Y_R \Omega\}_-)^* (V - \{\Gamma - \Lambda A_R^{-1} Y_R \Omega\}_-). \end{aligned} \tag{1.84}$$

Since the first two terms make a finite contribution to the norm of T and the last term is analytic on the imaginary axis, the norm of T is finite if and only if V is strictly proper.

Now it only remains to show which matrices T_K in (1.71) are such that $X_L - W B_L$ is non-singular. From (1.63),

$$A_R (X_L - W B_L) = I + T_K M_{22}$$

so $I + T_K M_{22}$ is to be non-singular. This proves the claim (a).

It easily follows from (1.84) that

$$\begin{aligned} ||T||_2^2 = {} & \min ||T||_2^2 + \frac{1}{2\pi j} \int \operatorname{Tr} V^* V ds \\ & - \frac{1}{2\pi j} \int 2 \operatorname{Tr} V^* \{\Gamma - \Lambda A_R^{-1} Y_R \Omega\}_- ds. \end{aligned} \tag{1.85}$$

Since both V^* and $\{\Gamma - \Lambda A_R^{-1} Y_R \Omega\}_-$ are analytic in $\operatorname{Re} s \leq 0$ and strictly proper, by Cauchy's theorem, the second integral on the right-hand side of (1.85) vanishes. This verifies the optimality of T_{Kopt} given by (1.72). For if $V \neq 0$, $||T||_2 > \min ||T||_2$ and the claims (b) and (c) follow. □

Theorem 1.9 contains three results. Firstly, it describes all admissible cascade compensators that yield a finite $||T||_2$ in terms of a parameter matrix V. Second, it shows that the optimal cascade compensator, which yields a minimum $||T||_2$,

to the choice $V = 0$. Third, it provides a meaningful interpretation
meter: $||V||_2$ defines the norm excess of the admissible cascade
(1.71).
optimal cascade compensator (1.72) has been found, the corre-
imal feedback controller K is calculated from (1.52). Equivalently, one can substitute the parameter

$$W_{opt} = \Lambda^{-1}(\{\Gamma - \Lambda A_R^{-1} Y_R \Omega\}_+ - \{\Lambda A_R^{-1} Y_R \Omega\}_\infty)\Omega^{-1} \tag{1.86}$$

for W in (1.56) and obtain the optimal K directly.

1.3.4 Diophantine equations solution

Another alternative to solve the H_2 control problem is based on proper-stable fractional representations of rational transfer functions. This approach makes it possible to avoid impulsive modes in the inernally stable control system. The variational argument is replaced by the completion of squares, thus avoiding some technicalities. The partial fraction expansions are dispensed with; the corresponding quantities are obtained by solving linear equations in proper-stable rational functions, called Diophantine equations.

The solution to be presented provides the optimal feedback controller K in figure 1.2 directly, without the intermediate step of finding an equivalent cascade compensator first. Our basic assumption is that u, v and y, z, the inputs and outputs appearing in figure 1.2, are all *scalar* quantities ($l_1 = l_2 = m_1 = m_2 = 1$). Accordingly, $M_{11}, M_{12}, M_{21}, M_{22}$ and K are assumed to be *real-rational functions*, not necessarily proper. This dimensionality assumption limits the utility of the results but it makes it possible to study the system properties in full detail.

In this setting, the requirement of internal stability for the control system shown in figure 1.2 will mean that whenever bounded inputs are injected at u, v and y then all the outputs u, y and z are bounded in amplitude.

The solution will proceed in three steps. Firstly we shall parameterize the set of all controllers K that internally stabilize M. Within this set we shall identify, once again in parametric form, all controllers that make the overall transfer function T belong to H_2. Finally we shall pick the controller that minimizes $||T||_2$. The presentation follows the papers by Kučera (1986; 1992).

Internal stability

A system is bounded-input bounded-output stable if and only if its (real-rational) transfer function is proper (i.e., analytic at $s = \infty$) and stable (i.e., analytic in $\text{Re}\, s \geq 0$). To study this type of stability, it is therefore convenient to express the rational functions of M and K as ratios of *proper and stable rational* functions.

The set of all proper and stable rational functions has the algebraic structure of a ring; in particular a principal ideal domain (Vidyasagar, 1985). This means that the addition, subtraction and multiplication of two proper and stable rational functions results in a proper and stable rational function. The proper and

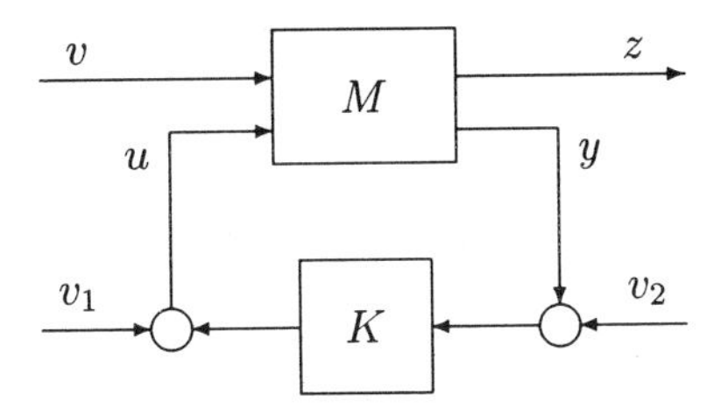

Figure 1.7: Input and output signals.

stable rational functions whose inverse is also proper and stable are called *units* (or bi-proper and bi-stable rational functions). Two proper and stable rational functions A and B are said to be *coprime* if there exist proper and stable rational functions P and Q such that $AP + BQ$ is a unit. With C and D two proper and stable rational functions, we say that D *divides* C if $C = DR$ for some proper and stable rational function R. Thus, any rational function can be expressed as a quotient of two coprime, proper and stable rational functions and these are unique to within multiplication by a unit.

In view of the above discussion, we write for the plant

$$M_{11} = \frac{E}{A}, \ M_{12} = \frac{D}{A}$$

$$M_{21} = \frac{C}{A}, \ M_{22} = \frac{B}{A} \tag{1.87}$$

where A, B, C, D and E is a quintuple of coprime, proper and stable rational functions. Thus, A is the least common denominator of the four entries of M. To avoid trivia, we make the following assumption:

(DE.1) $C \neq 0, \ D \neq 0$.

The controller will be written in a compatible form as

$$K = -\frac{Q}{P} \tag{1.88}$$

where P, Q is a pair of coprime, proper and stable rational functions.

Consider the control system depicted in figure 1.2 and inject signals v_1 and v_2 around the loop as shown in figure 1.7.

We definition, the closed loop system is *internally stable* if the nine transfer functions from the inputs v_1, v_2, v to the internal outputs u, y, z are all proper and stable. These are given by

$$T_{v_1,u} = \frac{AP}{AP+BQ} \qquad T_{v_2,u} = -\frac{AQ}{AP+BQ}$$

$$T_{v_1,y} = \frac{BP}{AP+BQ} \qquad T_{v_2,y} = -\frac{BQ}{AP+BQ}$$

$$T_{v,y} = \frac{CP}{AP+BQ} \qquad T_{v,u} = -\frac{CQ}{AP+BQ}$$
$$T_{v_1,z} = \frac{DP}{AP+BQ} \qquad T_{v_2,z} = -\frac{DQ}{AP+BQ}$$

and

$$T_{v,z} = \frac{E}{A} - \frac{CD}{AP+BQ}\frac{Q}{A} \tag{1.89}$$
$$= \frac{EP}{AP+BQ} + \frac{BE-CD}{A}\frac{Q}{AP+BQ}. \tag{1.90}$$

Lemma 1.10. *The control system shown in figure 1.2 is internally stable if and only if the following two conditions hold:*

(a) $AP+BQ$ is bi-proper and bi-stable;

(b) A divides $BE-CD$.

Proof: The sufficiency of (a) and (b) is obvious. To see that the conditions are necessary we observe that $AP+BQ$ must divide A, B, C and D in order for the first eight transfer functions to be proper and stable. Since $AP+BQ$ cancels in the second term of (1.89), it must cancel, being a factor of A, also in the first term. Therefore, $AP+BQ$ divides all A, B, C, D and E. This quintuple is coprime, however, which proves (a). For (1.90) to be proper and stable, it remains that A is absorbed in its second term. As A and Q are coprime we obtain (b). □

While the first condition (a) can be met by choosing a controller provided A and B are coprime, the second condition (b) is an inherent restriction for the existence of an internally stabilizing controller. The following theorem determines the set of all controllers K that (internally) stabilize the plant M in the configuration of figure 1.2. This set will be denoted by $\mathcal{S}_M$.

Theorem 1.11 (Kučera, 1992). *Let the plant M in figure 1.2 be given. Then,*

(a) there exists a controller K that stabilizes M if and only if

$$A \text{ and } B \text{ are coprime} \tag{1.91}$$
$$A \text{ divides } BE-CD; \tag{1.92}$$

(b) the set $\mathcal{S}_M$ of all controllers that stabilize M is generated by the formula

$$K = -\frac{Q}{P} \tag{1.93}$$

where P, Q with $P \neq 0$ is the proper and stable rational solution class of the equation

$$AP+BQ = 1. \tag{1.94}$$

Proof: The claim (a) is a restatement of lemma 1.10, as A and B are coprime if and only if $AP + BQ$ is a unit for some proper and stable rational P and Q.

To prove (b), we observe that the controllers defined in (1.93), (1.94) are indeed stabilizing by lemma 1.10. To see that the entire set $\mathcal{S}_M$ is generated in this way, consider a stabilizing controller

$$K_0 = -\frac{Q_0}{P_0}$$

where P_0 and Q_0 are proper and stable rational functions that do not satisfy (1.94). Then, by lemma 1.10,

$$AP_0 + BQ_0 = U$$

for some bi-proper and bi-stable rational function U. Define

$$P = \frac{P_0}{U}, \quad Q = \frac{Q_0}{U}.$$

Then P and Q are proper and stable and satisfy the equation (1.94). Furthermore,

$$K_0 = -\frac{Q_0}{P_0} = -\frac{Q}{P}$$

has the form (1.93), which shows that K_0 is in $\mathcal{S}_M$. □

Theorem 1.11 (b) is instrumental for the solution of the H_2 control problem. It provides a *parameterization* of $\mathcal{S}_M$. If $\bar{P}, \bar{Q}$ is a particular solution of equation (1.94) then (Kučera, 1991) the entire solution class of this equation is generated by

$$P = \bar{P} - BW, \quad Q = \bar{Q} + AW \tag{1.95}$$

where W is a free parameter, which is an arbitrary proper and stable rational function. Then, by (1.93), any controller in $\mathcal{S}_M$ has the form

$$K = -\frac{\bar{Q} + AW}{\bar{P} - BW} \tag{1.96}$$

where the parameter W varies over the ring of proper and stable rational functions while satisfying $\bar{P} - BW \neq 0$. This condition is not very restrictive, as $\bar{P} - BW$ can vanish for at most one choice of W.

Finite norm

We shall now focus on achieving a finite H_2 norm of the transfer function, $T(s)$, of the internally stable closed-loop control system shown in figure 1.2. The norm is given by (1.24) as

$$||T||_2^2 = \frac{1}{2\pi j} \int T^*(s)T(s)\mathrm{d}s \tag{1.97}$$

where $T^*(s) = T(-s)$ and it exists whenever T is strictly proper (it is already stable as the control system is internally stable). We shall proceed through a series of auxiliary results in order to give (1.97) a form suitable for our purposes.

The notation $S_C(s)$ and $S_D(s)$ is used to designate the *spectral factors* defined by

$$C^*C = S_C^*S_C, \quad D^*D = S_D^*S_D. \tag{1.98}$$

The spectral factors are proper and stable real-rational functions whose inverses are analytic in $\mathrm{Re}\, s > 0$. The spectral factors are unique up to the sign and enjoy the following property.

Lemma 1.12 (Kučera, 1986). *The rational functions C/S_C and D/S_D are proper and stable.*

Proof: The zeros of C^*C and $S_C^*S_C$ are symmetrically distributed with respect to the imaginary axis. Thus, if S_C has a zero lying on $\mathrm{Re}\, s = 0$ or one at $s = \infty$, this zero is shared by C and the ratio C/S_C is proper and stable. The argument can be repeated for D and S_D. □

The following technical result is crucial.

Lemma 1.13 (Kučera, 1986). *Let (1.91) and (1.92) hold and define a proper and stable rational function F by*

$$F = -\frac{BE - CD}{A}. \tag{1.99}$$

Then the pair of equations

$$\begin{aligned} S_C^*S_D^*X - Z^*B &= C^*D^*F \\ S_C^*S_D^*Y + Z^*A &= C^*D^*E \end{aligned} \tag{1.100}$$

has a unique, proper and stable rational solution triple X, Y and Z such that Z/S_CS_D is strictly proper.

Proof: First we shall prove the solvability of (1.100). Write the pair of equations in the matrix form

$$S_C^*S_D^*[X \quad Y] + Z^*[-B \quad A] = C^*D^*[F \quad E]$$

and postmultiply it by a bi-proper and bi-stable matrix

$$\begin{bmatrix} A & -Q \\ B & P \end{bmatrix}$$

where $AP + BQ = 1$. This results in the decoupled equations

$$AX + BY = S_C S_D \tag{1.101}$$

$$Z^* = S_C^* S_D^*(QX - PY) + C^* D^*(PE - QF) \tag{1.102}$$

where (1.92) has been used to obtain the first equation. By (1.91), there exists a proper and stable rational pair X, Y that satisfies (1.101) and any two such pairs X, Y and $\bar{X}, \bar{Y}$ are related by

$$X = \bar{X} - BW, \quad Y = \bar{Y} + AW \tag{1.103}$$

for some proper and stable rational W. Substituting (1.103) in (1.102) one obtains

$$Z^* = \bar{Z}^* - S_C^* S_D^* W \tag{1.104}$$

where

$$\begin{aligned} \bar{Z}^* &= S_C^* S_D^*(Q\bar{X} - P\bar{Y}) + C^* D^*(PE - QF) \\ &= S_C^* S_D^*[Q\bar{X} - P\bar{Y} + \frac{S_C S_D}{CD}(PE - QF)] \end{aligned} \tag{1.105}$$

on using (1.98). The definition of the spectral factors and lemma 1.12 imply that S_C^*/C^* and S_D^*/D^* are both proper and stable, so that the last term in (1.105) can be decomposed as

$$\frac{S_C S_D}{CD}(PE - QF) = \Phi + \Psi^* \tag{1.106}$$

where Φ is proper and stable while Ψ^* is strictly proper and anti–stable. Then the choice $W = Q\bar{X} - P\bar{Y} + \Phi$ yields a proper and stable Z, namely

$$Z = S_C S_D \Psi. \tag{1.107}$$

Next we shall show that there is a unique solution triple X, Y and Z satisfying $Z/S_C S_D$ strictly proper. To this end we deduce from (1.103) and (1.104) that any two solution triples X, Y, Z and $\bar{X}, \bar{Y}, \bar{Z}$ for (1.100) are related by

$$\begin{aligned} X &= \bar{X} - BW \\ Y &= \bar{Y} + AW \\ Z &= \bar{Z} - S_C S_D W^* \end{aligned} \tag{1.108}$$

where W as well as W^* must be proper and stable. Thus, W is a constant. Now suppose that the two solution triples satisfy the requirement that $Z/S_C S_D$ and $\bar{Z}/S_C S_D$ are both strictly proper. Then

$$\frac{\bar{Z}}{S_C S_D} - \frac{Z}{S_C S_D} = W. \tag{1.109}$$

Evaluating (1.109) at $s = \infty$ yields $W = 0$. Hence, using (1.108), $X = \bar{X}, Y = \bar{Y}, Z = \bar{Z}$ and the solution triple having the requisite property is unique.

We note that it is exactly this triple that has been constructed in (1.107). □

Let us now suppose the norm (1.97) is finite. Using (1.90),

$$T = \frac{E}{A} - \frac{CD}{A}Q.$$

Using (1.98),

$$||T||_2 = ||U||_2,$$

where

$$U = \frac{C^*D^*E}{S_C^*S_D^*A} - \frac{S_CS_D}{A}Q. \tag{1.110}$$

Let X, Y and Z be the solution triple of (1.100) such that Z/S_CS_D is strictly proper. Using the second equation in (1.100), the first term of (1.110) can be decomposed as follows

$$\frac{C^*D^*E}{S_C^*S_D^*A} = \frac{Z^*}{S_C^*S_D^*} + \frac{Y}{A}$$

thus yielding

$$U = \frac{Z^*}{S_C^*S_D^*} - V, \tag{1.111}$$

where

$$V = \frac{S_CS_D}{A}Q - \frac{Y}{A}. \tag{1.112}$$

We shall show that both Z/S_CS_D and V are strictly proper and stable rational functions. Indeed, Z/S_CS_D is strictly proper by definition and, using (1.101),

$$\begin{aligned} V &= \frac{S_CS_D}{A}Q - \frac{Y}{A}(AP + BQ) \\ &= \frac{S_CS_D - BY}{A}Q - YP \\ &= XQ - YP \end{aligned}$$

is proper and stable. Now finite $||T||_2$ implies that U is strictly proper and analytic on $\mathrm{Re}\, s = 0$. Then (1.111) implies that V is strictly proper and that Z/S_CS_D is analytic on $\mathrm{Re}\, s = 0$, hence stable. Both Z/S_CS_D and V being strictly proper and stable, the following integral vanishes:

$$\frac{1}{2\pi j}\int \frac{Z}{S_CS_D}V\mathrm{d}s = 0.$$

In view of this fact, the cross-terms contribute nothing to the norm and

$$||T||_2^2 = ||\frac{Z^*}{S_C^*S_D^*} - V||_2^2 = ||\frac{Z}{S_CS_D}||_2^2 + ||V||_2^2. \tag{1.113}$$

Having completed the square, we can prove the main result.

Theorem 1.14 (Kučera, 1992). *Let $\mathcal{S}_M$ be non-empty and let X, Y, Z be the proper and stable rational solution triple of equations (1.100) such that $Z/S_C S_D$ is strictly proper. Then,*

(a) there exists a controller K in $\mathcal{S}_M$ that renders $||T||_2$ finite if and only if

$$X \text{ and } Y \text{ are strictly proper whenever } S_C S_D \text{ is so} \tag{1.114}$$

$$\frac{EX-FY}{S_C S_D} \text{ is strictly proper;} \tag{1.115}$$

(b) the set $\mathcal{S}_{MT}$ of all stabilizing controllers that render $||T||_2$ finite is generated by the formula

$$K = -\frac{Q}{P} \tag{1.116}$$

where P, Q along with some V is the proper and stable rational solution class of the equations

$$\begin{aligned} S_C S_D P + BV &= X \\ S_C S_D Q - AV &= Y \end{aligned} \tag{1.117}$$

constrained to $P \neq 0$ and V strictly proper;

(c) the associated norm is given by

$$||T||_2^2 = ||\frac{Z}{S_C S_D}||_2^2 + ||V||_2^2.$$

Proof: To prove (a), suppose a controller exists in $\mathcal{S}_{MT}$. It gives rise to some proper and stable P, Q defined by (1.88) and to some strictly proper and stable V defined by (1.113). It follows from (1.112) that

$$Q = \frac{Y + AV}{S_C S_D}$$

and on substituting in (1.94),

$$P = \frac{X - BV}{S_C S_D}.$$

This implies (1.114). Furthermore, using (1.90),

$$\begin{aligned} T &= EP - FQ \\ &= \frac{EX - FY}{S_C S_D} + \frac{CD}{S_C S_D} V. \end{aligned} \tag{1.118}$$

By lemma 1.12 the last term is strictly proper. Hence T strictly proper implies (1.115).

On the other hand, suppose that (1.114) and (1.115) hold. Consider any proper and stable rational $\bar{P}, \bar{Q}$ such that $A\bar{P} + B\bar{Q} = 1$ and $\bar{P} \neq 0$. Denote $\bar{V} = X\bar{Q} - Y\bar{P}$. Then $\bar{V}$ is proper and stable. Let α be the value of $S_C S_D$ at $s = \infty$. If $\alpha = 0$ then (1.114) implies that $\bar{V}$ is strictly proper and we set $P = \bar{P}, Q = \bar{Q}$ and $V = \bar{V}$. If $\alpha \neq 0$ then we denote β the value of $\bar{V}$ at $s = \infty$ and put

$$
\begin{aligned}
P &= \bar{P} + \frac{\beta}{\alpha} B \\
Q &= \bar{Q} - \frac{\beta}{\alpha} A \\
V &= \bar{V} - \frac{\beta}{\alpha} S_C S_D.
\end{aligned}
$$

Then $P \neq 0$ since $\bar{P}$ and B are coprime, and V is strictly proper by construction. Now define a controller K via (1.88). This K belongs to $\mathcal{S}_M$ since

$$AP + BQ = A\bar{P} + B\bar{Q} = 1.$$

Moreover, the associated T is given by (1.118) where the last term is strictly proper. If (1.115) holds, T is strictly proper and K belongs to $\mathcal{S}_{MT}$.

To prove (b), we recall that any controller K in $\mathcal{S}_{MT}$ gives rise to a triple of proper and stable rational functions P, Q and V that satisfy

$$P = \frac{X - BV}{S_C S_D}, \quad Q = \frac{Y + AV}{S_C S_D}.$$

Therefore P, Q and V satisfy the pair of equations (1.117). The solution class of this equation constrained to $P \neq 0$ and V strictly proper generates all elements of $\mathcal{S}_{MT}$, for these constraints are necessary in order that K is well defined by (1.116) and T is strictly proper.

For the claim (c), see (1.113). □

Theorem 1.14 is the key result of our development. It delivers a *parameterization* of $\mathcal{S}_{MT}$. Any two solution triples P, Q, V and $\bar{P}, \bar{Q}, \bar{V}$ of the pair of equations (1.117) are related by

$$
\begin{aligned}
P &= \bar{P} - BW \\
Q &= \bar{Q} + AW \\
V &= \bar{V} + S_C S_D W
\end{aligned}
$$

where W is a proper and stable rational parameter. Since both V and $\bar{V}$ are constrained to be strictly proper, then $S_C S_D W$ must be strictly proper. As any P, Q satisfying (1.117) also satisfy (1.94), $\mathcal{S}_{MT}$ is a subset of $\mathcal{S}_M$. All such pairs P, Q are in turn parameterized by V; this parameter is not free, however.

Optimal controller

The set $\mathcal{S}_{MT}$ having been parameterized, it is a simple matter to pick the parameter that minimizes the H_2 norm of T.

Theorem 1.15 (Kučera, 1992). *Let $\mathcal{S}_{MT}$ be non-empty and let X, Y, Z be the proper and stable rational solution triple of equations (1.100) such that $Z/S_C S_D$ is strictly proper. Then,*

(a) there exists a controller K in $\mathcal{S}_{MT}$ that minimizes $||T||_2$ if and only if

$$X \neq 0 \tag{1.119}$$

$$X/S_C S_D \text{ and } Y/S_C S_D \text{ are proper and stable;} \tag{1.120}$$

(b) this optimal controller is unique and given by

$$K_{opt} = -\frac{Y}{X}; \tag{1.121}$$

(c) the associated minimal norm is

$$\min ||T||_2 = ||\frac{Z}{S_C S_D}||_2.$$

Proof: The norm associated with any controller K in $\mathcal{S}_{MT}$ is given by (1.113). Observe that $Z/S_C S_D$ does not depend on the controller. The best one can do to minimize $||T||_2$ is to set $V = 0$. The corresponding P, Q are given by (1.117) as

$$P = \frac{X}{S_C S_D}, \quad Q = \frac{Y}{S_C S_D}.$$

Clearly $P \neq 0$ if and only if (1.119) holds and P, Q are proper and stable if and only if (1.120) is verified. The claims (a), (b) and (c) then follow from theorem 1.14 (b) (c). □

Conditions (1.119) and (1.120) secure that the requirement of minimum norm is compatible with that of closed-loop stability. Theorem 1.15 (c) identifies the role of the parameter V: it defines the norm excess for a particular controller with respect to the optimal one.

1.4 Comparison and examples

The LQG is a stochastic control problem where assumptions about the system environment are essential. On the other hand, the H_2 control problem is a deterministic one. Though different in nature, the two problems are closely related.

The main goal of this section is to compare the three approaches to the solution of the H_2 control problem. They are not entirely equivalent because slightly different assumptions are imposed in each case.

1.4.1 The LQG as an H_2 problem

The LQG control problem involves a linear plant, quadratic cost and Gaussian environment. It is the specifics of the exogenous random signals (stationarity, gaussianess) that leads to a linear time-invariant feedback control strategy.

The H_2 control problem also involves a linear plant and a quadratic norm to be minimized. The nature of the exogenous signals is irrelevant, however. That is why the control system structure (linearity, feedback) must be postulated at the outset.

Although formally different, the two problems are close to each other. In fact, the infinite-horizon LQG problem can be recast as an H_2 problem. To see this, we let

$$\begin{bmatrix} Q_0 & 0 \\ 0 & I \end{bmatrix} = \begin{bmatrix} G_1 \\ J_{21} \end{bmatrix} \begin{bmatrix} G_1' & J_{21}' \end{bmatrix}$$

and

$$\begin{bmatrix} Q_1 & 0 \\ 0 & I \end{bmatrix} = \begin{bmatrix} H_1 & J_{12} \end{bmatrix} \begin{bmatrix} H_1' \\ J_{12}' \end{bmatrix}.$$

Then we define a new random source v by

$$\begin{bmatrix} p \\ q \end{bmatrix} = \begin{bmatrix} G_1 \\ J_{21} \end{bmatrix} v$$

and the error process z by

$$z = \begin{bmatrix} H_1 & J_{12} \end{bmatrix} \begin{bmatrix} x \\ u \end{bmatrix}.$$

The system equations (1.1) then read

$$\begin{aligned} dx &= Fx dt + G_1 dv + G_2 u dt \\ z &= H_1 x + J_{12} u \\ dy &= H_2 x dt + J_{21} dv \,. \end{aligned}$$

We observe that these stochastic differential equations correspond to the plant model (1.26) in which v and y are replaced by their derivatives, $\dot{v}$ and $\dot{y}$, respectively. Also $J_{11} = 0$, $J_{22} = 0$ as required by assumption (SS.3).

The expected loss rate (1.21) becomes

$$kJ^* = \mathrm{E}\{z'(t)\ z(t)\}.$$

Applying Parseval's theorem,

$$J^* = \frac{1}{2\pi} \int_{-\infty}^{\infty} \mathrm{Tr}\, \Theta(\omega)\, d\omega$$

where Θ denotes the spectral density matrix of the random process $z(t)$ in steady state.

Let $T(s)$ be the transfer function from v to z in the closed-loop system involving the plant (1.26). Then

$$\Theta(\omega) = T(j\omega)T'(-j\omega)$$

since $v(t)$ is a Wiener process with incremental covariance Idt. It follows that J^* is related to the H_2 norm of T as

$$J^* = ||T||_2^2.$$

To summarize, the infinite-horizon LQG control problem can be recast as an H_2 control problem. The estimator-based LQG controller is then one particular state-space realization of the H_2 controller transfer function K_{opt}. The other realizations of K_{opt} yield the same J^*; however, they do not generate the conditional mean estimate of the state.

1.4.2 Internal stability

The reader will have noticed that each solution of the H_2 control problem involves a different definition of internal stability. The state-space solution is based on the Lyapunov asymptotic stability, requiring the state of the control system to go to zero from any initial values. The Wiener-Hopf approach makes use of the polynomial factorizations of transfer functions, so it is natural to require the characteristic polynomial of the control system to be strictly Hurwitz. These two formulations are equivalent.

It is highly instructive, however, to compare the form of the stabilizability conditions used in these two approaches. Assumption (SS.2) requires the plant to be stabilizable by the control input u and detectable from the measured output y. This means that all unstable modes of the plant can be excited by u and observed at y. And this is exactly what assumption (WH.1) states. Thus lemma 1.1 indirectly shows the necessity of (WH.1) in achieving internal stability.

There is a deeper reason for the two different formulations, however. The Wiener-Hopf approach allows a more general plant model than the state-space approach: the plant transfer-function matrix need not be proper. This means that its state-space realization may lead to equations such as

$$\begin{aligned} E\dot{x} &= Fx + G_1 v + G_2 u \\ z &= H_1 x + J_{11} v + J_{12} u \\ y &= H_2 x + J_{21} v + J_{22} u \end{aligned}$$

where x lives in a generalized state space. These plants are efficiently handled using transfer functions.

This comparison immediately reveals a deficiency of the Wiener-Hopf approach. The optimal state trajectories, though approaching zero from any initial values, may contain impulses. The impulsive modes are invisible in the characteristic polynomial and must be eliminated, if so desired, by other ways and means.

That is why the third solution, based on Diophantine equations, involves the bounded-input bounded-output stability. True, this type of stability is equivalent to Lyapunov asymptotic stability if and only if the control system is free of unstable exponential modes that are uncontrollable and/or unobservable. If such modes are present, however, the system cannot be stabilized by output feedback anyway. So bounded-input bounded-output stability is a convenient tool to guard against impulsive modes, yet no loss of generality is incurred for output feedback control systems.

In this respect, it is instructive to see how assumption (WH.1) corresponds to the conditions of lemma 1.10. By definition, ψ^+_{M22} is the monic polynomial that absorbs all the zeros in $\mathrm{Re}\, s \geq 0$ of $A/\Delta_{A,B}$ where $\Delta_{A,B}$ denotes any greatest common divisor of A and B. Similarly ψ^+_M is the monic polynomial that absorbs all the zeros in $\mathrm{Re}\, s \geq 0$ of $A^2/\Delta_{A,BE-CD}$ where $\Delta_{A,BE-CD}$ denotes any greatest common divisor of A and $BE - CD$. Now $A/\Delta_{A,B}$ is a factor of A while $A^2/\Delta_{A,BE-CD}$ contains A as factor; the two are equal if and only if $\Delta_{A,B} = 1$ and $\Delta_{A,BE-CD} = A$. Therefore $\psi^+_M = \psi^+_{M22}$ if and only if A and B are coprime and A divides $BE - CD$. In this case, ψ^+_M and ψ^+_{M22} absorb the zeros of A in $\mathrm{Re}\, s \geq 0$.

1.4.3 Solvability assumptions

Different techniques used to solve the H_2 control problem naturally require different assumptions. This is due to a different level of generality of the formulation as well as to the nature of the mathematics employed.

All the three solutions presented assume a linear time-invariant plant with a rational transfer-function matrix. The specific assumptions are as follows: the state-space solution requires a proper rational plant transfer function while the Diophantine equation approach has been presented for the case of scalar inputs and outputs. Both assumptions can be relaxed but the generalization of the state-space solution is a highly non-trivial task.

There is a number of further, rather technical assumptions in each approach. These are primarily related to the solvability of requisite equations. In the state-space approach, assumptions (SS.4) and (SS.5) correspond to no cross–correlation ($T_0 = 0$) and no cross–weighting ($T_1 = 0$) in the LQG control problem formulation, and in fact are not essential. Under these assumptions, however, (SS.1) and (SS.2) are necessary and sufficient to guarantee the existence of unique, stabilizing, symmetric, non-negative definite solutions to the two algebraic Riccati equations (1.28) and (1.38).

Analogous assumptions in the Wiener-Hopf approach are made in order to secure the existence of spectral factors (1.67) and (1.68). In particular, assumptions (WH.3) and (WH.4) guarantee that the spectral factors are non-singular and their inverses are analytic on the finite part of the imaginary axis.

It is interesting to note that these assumptions are not *necessary* for the H_2 optimal controller to exist. They are the best testable sufficient conditions available, however. Any attempt to obtain conditions that are necessary as well results in tests whose complexity is equivalent to calculating the optimal controller.

Let us illustrate this fine point by the following example.

Example 1.1 Consider the plant (1.26) where

$$F = \begin{bmatrix} 0 & 1 \\ 0 & 1 \end{bmatrix} \quad G_1 = \begin{bmatrix} 3 & 0 \\ -1 & 0 \end{bmatrix} \quad G_2 = \begin{bmatrix} 0 \\ 1 \end{bmatrix}$$

$$H_1 = \begin{bmatrix} 0 & 0 \end{bmatrix} \quad J_{11} = \begin{bmatrix} 0 & 0 \end{bmatrix} \quad J_{12} = 1$$

$$H_2 = \begin{bmatrix} 1 & 3 \end{bmatrix} \quad J_{21} = \begin{bmatrix} 0 & 1 \end{bmatrix} \quad J_{22} = 0.$$

Assumptions (SS.1) and (WH.3) fail to hold. Nevertheless the H_2 control problem has a solution.

Applying the state-space method, the algebraic Riccati equation (1.38) has a unique non-negative definite stabilizing solution

$$P_0 = \frac{1}{2}\begin{bmatrix} 11 & -5 \\ -5 & 3 \end{bmatrix}$$

yielding the estimator gain

$$L_0 = \begin{bmatrix} -2 \\ 2 \end{bmatrix}.$$

The algebraic Riccati equation (1.28), however, has *two* non-negative definite solutions,

$$P_1 = \begin{bmatrix} 0 & 0 \\ 0 & 2 \end{bmatrix}, \quad \bar{P}_1 = \begin{bmatrix} 0 & 0 \\ 0 & 0 \end{bmatrix}$$

of which none is stabilizing. The regulator loop is not stable and the control problem seems to have no solution.

Let us try the Wiener-Hopf approach. We calculate the plant transfer functions

$$M_{11}(s) = 0 \qquad M_{12}(s) = 1$$

$$M_{21}(s) = \begin{bmatrix} \dfrac{4}{s^2 - s} & 1 \end{bmatrix} \qquad M_{22}(s) = \frac{3s+1}{s^2 - s}$$

the spectral factors (1.67) and (1.68)

$$\Lambda(s) = s^2 + s, \quad \Omega(s) = s^2 + 3s + 4$$

and the matrix (1.69)

$$\Gamma(s) = 0.$$

The Bézout identity (1.54) is satisfied by

$$A_L(s) = A_R(s) = s^2 - s$$

$$B_L(s) = B_R(s) = 3s + 1$$

and

$$X_L(s) = X_R(s) = \tfrac{9}{4}.$$

$$Y_L(s) = Y_R(s) = -\tfrac{3}{4}s + 1.$$

Hence

$$\{\Gamma\}_+ = 0, \quad \{\Lambda A_R^{-1} Y_R \Omega\}_- = \frac{4}{s-1}$$

and (1.72) gives

$$T_{K_{opt}}(s) = -\frac{4}{s+1}\,\frac{s^2 - s}{s^2 + 3s + 4}.$$

This results in the optimal controller

$$K_{opt}(s) = -\frac{4}{s+5}.$$

Why has the state-space method failed? The regulator feedback is not stable. However, if one chooses the solution P_1, the regulator gain

$$L_1 = [0 \quad 2]$$

yields the controller

$$\dot{\hat{x}} = \begin{bmatrix} 2 & 7 \\ -2 & -7 \end{bmatrix} \hat{x} + \begin{bmatrix} -2 \\ 2 \end{bmatrix} y$$

$$u = \begin{bmatrix} 0 & -2 \end{bmatrix} \hat{x}$$

which has an uncontrollable mode associated with the eigenvalue 1. The controller transfer function is

$$K(s) = -\frac{4}{s+5}$$

and when minimally realized it is optimal. Thus the unstabilizable mode is not needed to achieve optimality and can be removed from the plant at the outset. This situation, however, cannot be detected before the design is complete. That is why necessary and sufficient solvability conditions are not efficient. □

The remaining set of assumptions is related to the existence of a strictly proper and stable closed-loop system transfer function, hence the existence of finite norm. The strongest assumption is made in the state-space approach. Assumption (SS.3) requires the plant transfer functions M_{11} and M_{22} to be strictly proper. The Wiener-Hopf approach copies the assumption regarding M_{11}, see (WH.2), but relaxes the requirement on M_{22}. Since the plant is not necessarily proper in this approach, two more technical assumptions (WH.5) and (WH.6) are needed to secure the existence of a stabilizing controller that yields finite H_2 norm.

The reader will have noticed that only one assumption, namely (DE.1), is made in order to obtain the H_2 optimal controller via Diophantine equations. This assumption is analogous to the rank assumptions on M_{12} and M_{21} and guards against the trivial case of the zero transfer function to be spectrally factored. No further assumptions are imposed. The fact that the solution is restricted to scalar inputs and outputs makes the problem simple enough to relax in return all the unnecessary assumptions. Theorems 1.11, 1.14 and 1.15 provide necessary and sufficient conditions for the existence of stabilizing, finite-norm and optimal controllers, respectively. These conditions cannot be checked using the given data only. However, they provide some valuable insight as discussed in the next subsection.

1.4.4 Non-proper plants

The utility of the Diophantine equation approach consists in a systematic search for the set $\mathcal{S}_M$ of all stabilizing controllers for M, then for the subset $\mathcal{S}_{MT}$ of the stabilizing controllers that yield finite H_2 norm of T, and finally for the optimal controller. When the plant M involves any rational, not necessarily proper, transfer functions one observes the following phenomena: in some cases no stabilizing controller sends T to H_2 (see theorem 1.14 (a)) and even if some do, none of them need be optimal (theorem 1.15 (a)). A similar situation arises when M is proper but M_{12} and/or M_{21} are not bi-proper. In the LQG interpretation, this corresponds to the matrices R_0 and/or R_1 being singular.

Here are some examples to illustrate these points.

Example 1.2 Consider the plant

$$M(s) = \begin{bmatrix} 1 & s+1 \\ 2 & s+1 \end{bmatrix}$$

and define

$$A(s) = \frac{1}{s+1}, \qquad E(s) = \frac{1}{s+1}, \qquad D(s) = 1$$

$$C(s) = \frac{2}{s+1}, \qquad B(s) = 1.$$

The set $\mathcal{S}_M$ is generated by the controllers

$$K(s) = -\frac{Q(s)}{P(s)}$$

where

$$P(s) = 1 - W(s)$$

$$Q(s) = \frac{s}{s+1} + \frac{1}{s+1} W(s)$$

for any proper and stable rational $W(s) \neq 1$. Calculate $F(s) = 1$ and

$$S_C(s) = \frac{2}{s+1}, \qquad S_D(s) = 1$$

and solve equations (1.100) to obtain

$$X(s) = 1, \qquad Y(s) = \frac{1}{s+1}, \qquad Z(s) = 0.$$

Now $S_C S_D$ is strictly proper while X is not. Theorem 1.14 (a) implies that the set $\mathcal{S}_{MT}$ is empty. Indeed, V in (1.117) can never be strictly proper. Thus internal stability and finite norm are two antagonistic requirements here. □

Example 1.3 Given the plant

$$M(s) = \begin{bmatrix} 1 & \frac{s}{s+1} \\ 1 & \frac{s}{s+1} \end{bmatrix}.$$

Taking

$$A(s) = 1, \quad E(s) = 1, \quad D(s) = \frac{s}{s+1}$$
$$C(s) = 1, \quad B(s) = \frac{s}{s+1}$$

the set $\mathcal{S}_M$ is generated by

$$K(s) = -\frac{Q(s)}{P(s)}$$

where

$$P(s) = \frac{1}{s+1} - \frac{s}{s+1}W(s)$$
$$Q(s) = 1 + W(s)$$

for any proper and stable rational W.

We calculate $F(s) = 0$ and

$$S_C(s) = 1, \qquad S_D(s) = \frac{s}{s+1}.$$

Equations (1.100) yield

$$X(s) = 0, \qquad Y(s) = 1, \qquad Z(s) = 0$$

and the solution class of equations (1.117) reads

$$P(s) = \frac{1}{s+1} - \frac{s}{s+1}W(s)$$
$$Q(s) = 1 + W(s)$$
$$V(s) = -\frac{1}{s+1} + \frac{s}{s+1}W(s)$$

where W is any strictly proper and stable rational function. The set $\mathcal{S}_{MT}$ is obtained from $\mathcal{S}_M$ by restricting W to be strictly proper.

Now $X=0$ and $Y/S_C S_D$ is not stable. Theorem 1.15 (a) implies that there is no optimal controller. Indeed,

$$T(s) = P(s) = -V(s)$$

and $||T||_2 = 0$ for $V(s) = 0$. This value is not admissible, however, as it corresponds to an unstable parameter, $W(s) = 1/s$, and yields $P(s) = 0$. The norm can attain any positive value, however small, but it can never be reduced to zero. □

For non-proper plant transfer functions, the Diophantine equation solution provides an impulse-free optimal control system. This is not guaranteed by the Wiener-Hopf solution unless additional assumptions are imposed. This point is illustrated by the following simple example.

Example 1.4 Consider the plant

$$\dot{x}_1 = x_2 - v \quad z = x_1$$

$$0 = x_1 - u \quad y = x_2$$

with the transfer function

$$M(s) = \begin{bmatrix} 0 & 1 \\ 1 & s \end{bmatrix}.$$

We set $A_L = A_R = 1$, $B_L = B_R = s$ and obtain $X_L = X_R = 1$, $Y_L = Y_R = 0$. Then

$$\Lambda = 1, \quad \Omega = 1, \quad \Gamma = 0$$

and all the assumptions (WH.1) – (WH.6) are satisfied. Theorem 1.9 then yields

$$T_{K_{opt}} = 0, \quad K_{opt} = 0, \quad ||T||_2 = 0.$$

The Wiener-Hopf method thus produces an optimal system which is stable, with the characteristic polynomial $\chi(s) = 1$. Yet $y = x_2$ will contain an impulse at $t = 0$ whenever $x_1(0-) \neq 0$.

On the other hand, setting

$$A(s) = \frac{1}{s+1}$$

one obtains

$$S_C(s) = S_D(s) = \frac{1}{s+1}$$

and the Diophantine equations (1.100) yield the solution triple

$$X(s) = \frac{1}{s+1}, \qquad Y(s) = 0, \qquad Z(s) = 0.$$

Now $X/S_C S_D$ is not proper and, by theorem 1.15 (a), there is no optimal controller. The optimality, in this case, involves bounded–input bounded-output stability and this puts the controller K=0 outside the set $\mathcal{S}_{MT}$.
□

The case of state-space systems, whose transfer functions are proper rational, is recovered when A is bi-proper and E is strictly proper. If, in addition, neither C nor D has a zero on $\mathrm{Re}\, s = 0$ and at $s = \infty$, then $S_C S_D$ is bi-proper and bi-stable. Hence X is bi-proper and Y is strictly proper. It follows that the optimal controller exists whenever the closed-loop system can be stabilized. These restrictions correspond to assumptions (SS.1) – (SS.5) of the state-space solution.

1.4.5 Design examples

The three approaches to the H_2 control theory will be further related through several illustrative design examples. The first example serves to compare the state-space and the Wiener-Hopf methods, the second will illustrate the parallels between the Wiener-Hopf and the Diophantine equation approaches, and the third example will complete the circle by relating the Diophantine equations with the state space techniques.

Example 1.5 Consider the plant

$$F = \begin{bmatrix} 0 & 1 \\ 0 & 0 \end{bmatrix} \quad G_1 = \begin{bmatrix} \sqrt{2} & 0 & 0 \\ 0 & 1 & 0 \end{bmatrix} \quad G_2 = \begin{bmatrix} 0 \\ 1 \end{bmatrix}$$

$$H_1 = \begin{bmatrix} 2 & 0 \\ 0 & 0 \end{bmatrix} \qquad\qquad J_{12} = \begin{bmatrix} 0 \\ 1 \end{bmatrix}$$

$$H_2 = \begin{bmatrix} 1 & 0 \end{bmatrix} \quad J_{21} = \begin{bmatrix} 0 & 0 & 1 \end{bmatrix}$$

that satisfies (SS.1) – (SS.5) and design the H_2 optimal controller using the state space approach.

To solve the algebraic Riccati equation (1.38), write

$$P_0 = \begin{bmatrix} x_{11} & x_{12} \\ x_{12} & x_{22} \end{bmatrix}.$$

Then (1.38) is equivalent to the set of quadratic equations

$$x_{11}^2 - 2x_{12} - 2 = 0$$
$$x_{11}x_{12} - x_{22} = 0$$
$$x_{12}^2 - 1 = 0.$$

Since P_0 is to be non-negative definite, we obtain

$$P_0 = \begin{bmatrix} 2 & 1 \\ 1 & 2 \end{bmatrix}.$$

This gives the estimator gain

$$L_0 = \begin{bmatrix} 2 \\ 1 \end{bmatrix}.$$

Now we write

$$P_1 = \begin{bmatrix} y_{11} & y_{12} \\ y_{12} & y_{22} \end{bmatrix}$$

and convert the algebraic Riccati equation (1.28) to the set of equations

$$\begin{aligned} y_{12}^2 &= 0 \\ y_{12}y_{22} - y_{11} &= 0 \\ y_{22}^2 - 2y_{12} &= 0. \end{aligned}$$

This results in

$$P_1 = \begin{bmatrix} 4 & 2 \\ 2 & 2 \end{bmatrix},$$

the unique symmetric non-negative definite solution of (1.28). The feedback gain (1.29) becomes

$$L_1 = [2 \quad 2].$$

The optimal controller (1.48) is given in state–space form as $(\hat{F}, \hat{G}, \hat{H}, 0)$, where

$$\begin{aligned} \hat{F} = F - G_2L_1 - L_0H_2 &= \begin{bmatrix} -2 & 1 \\ -3 & -2 \end{bmatrix} \\ \hat{G} &= L_0 \\ \hat{H} &= -L_1 \end{aligned}$$

and has the transfer function

$$K_{opt}(s) = -\frac{6s+2}{s^2+4s+7}.$$

In order to apply the Wiener-Hopf approach, we first obtain the plant transfer-function matrix (1.23), where

$$M_{11}(s) = \begin{bmatrix} \frac{\sqrt{6}}{s} & \frac{2}{s^2} & 0 \\ 0 & 0 & 0 \end{bmatrix} \qquad M_{12}(s) = \begin{bmatrix} \frac{2}{s^2} \\ 1 \end{bmatrix}$$

$$M_{21}(s) = \begin{bmatrix} \frac{\sqrt{2}}{s} & \frac{1}{s^2} & 1 \end{bmatrix} \qquad M_{22}(s) = \frac{1}{s^2}.$$

The polynomial factorization (1.53) is

$$\begin{aligned} A_L(s) = A_R(s) &= s^2 \\ B_L(s) = B_R(s) &= 1 \end{aligned}$$

and the associated polynomial matrices satisfying the Bézout identity (1.54) can be taken as

$$X_l(s) = X_R(s) = 0$$

$$Y_L(s) = Y_R(s) = 1.$$

The spectral factors (1.67) and (1.68) are calculated to be

$$\Lambda(s) = s^2 + 2s + 2$$

$$\Omega(s) = s^2 + 2s + 1$$

and (1.69) yields

$$\Gamma(s) = \frac{4 - 8s^2}{(s^2 - 2s + 1)s^2(s^2 - 2s + 2)}.$$

All the assumptions (WH.1) – (WH.6) are satisfied. We proceed by computing

$$\{\Gamma\}_+ = \frac{6s + 2}{s^2}, \qquad \{\Lambda A_R^{-1} Y_R \Omega\}_- = 0$$

so that the optimal cascade compensator is given by (1.72) as

$$T_{K_{opt}} = -\frac{s^2(6s + 2)}{(s^2 + 2s + 1)(s^2 + 2s + 2)}.$$

The corresponding feedback controller (1.53) reads

$$K_{opt}(s) = -\frac{6s + 2}{s^2 + 4s + 7}$$

as already found using the state–space formula.

This optimal controller can also be obtained via (1.56),

$$K_{opt} = -\frac{1 + s^2 W_{opt}}{-W_{opt}}$$

where the optimal parameter is given by (1.86) as

$$W_{opt}(s) = -\frac{s^2 + 4s + 7}{(s^2 + 2s + 1)(s^2 + 2s + 2)}.$$

We note that the state estimator dynamics is given by $\operatorname{Det}\Omega(s)$,

$$\chi_{F - L_0 H_2} = s^2 + 2s + 1$$

and the full information feedback (state regulation) dynamics by $\operatorname{Det}\Lambda(s)$,

$$\chi_{F - G_2 L_1} = s^2 + 2s + 2.$$

□

Example 1.6. Suppose the plant is described by the equations

$$\dot{x} = -2x + v - 3u$$
$$z = x + u$$
$$y = x + u.$$

Assumptions (SS.3) – (SS.5) are violated and the state space approach is not applicable. The Wiener-Hopf approach is an alternative.

The plant possesses the transfer functions

$$M_{11}(s) = \frac{1}{s+2}, \qquad M_{12}(s) = \frac{s-1}{s+2}$$

$$M_{21}(s) = \frac{1}{s+2}, \qquad M_{22}(s) = \frac{s-1}{s+2}.$$

Taking

$$A_L(s) = A_R(s) = s + 2$$
$$B_L(s) = B_R(s) = s - 1$$

the Bézout identity (1.54) is satisfied with

$$X_L(s) = X_R(s) = \tfrac{1}{3}$$
$$Y_L(s) = Y_R(s) = \tfrac{2}{3}.$$

We calculate the spectral factors

$$\Lambda(s) = s + 1, \qquad \Omega(s) = 1$$

and

$$\Gamma(s) = \frac{1}{s-1}\,\frac{s+1}{s+2}, \qquad \Lambda(s)A_R^{-1}(s)Y_R(s)\Omega(s) = -\frac{1}{3}\,\frac{s+1}{s+2}.$$

Assumptions (WH.1) – (WH.6) are satisfied. The requisite partial fraction expansions are found to be

$$\{\Gamma\}_+ = \frac{1}{3}\,\frac{1}{s+2}, \qquad \{\Lambda A_R^{-1} Y_R \Omega\}_- = 0$$

and (1.72) yields the optimal cascade compensator

$$T_{K_{opt}}(s) = -\frac{1}{3}\,\frac{s+2}{s+1}.$$

The optimal feedback controller results from (1.52) as

$$K_{opt}(s) = -\frac{1}{2}.$$

Equivalently,

$$K_{opt}(s) = -\frac{-\frac{1}{3} + (s+2)W_{opt}}{\frac{1}{3} - (s-1)W_{opt}}$$

where

$$W_{opt} = \frac{1}{3}\frac{1}{s+1}.$$

The method based on Diophantine equations is applicable also. We choose the proper and stable fractional representation (1.87) for the plant as

$$A(s) = 1, \qquad B(s) = D(s) = \frac{s-1}{s+2}, \qquad C(s) = E(s) = \frac{1}{s+2}.$$

Then we calculate the spectral factors (1.98)

$$S_C(s) = \frac{1}{s+2}, \qquad S_D(s) = \frac{s+1}{s+2}$$

and the factor (1.99)

$$F(s) = 0.$$

The pair of Diophantine equations (1.100) reads

$$-\frac{1}{s-2}\frac{s-1}{s-2}X - Z^*\frac{s-1}{s+2} = 0$$

$$-\frac{1}{s-2}\frac{s-1}{s-2}Y + Z^* = -\frac{1}{s-2}\frac{s+1}{s-2}\frac{1}{s+2}$$

and has the unique solution triple

$$X(s) = \frac{2}{3}\frac{1}{s+2}$$

$$Y(s) = \frac{1}{3}\frac{1}{s+2}$$

$$Z(s) = -\frac{2}{3}\frac{1}{s^2+4s+4}$$

such that $Z/S_C S_D$ is strictly proper.

Assumption (DE.1) is satisfied and so are all the conditions of theorems 1.11, 1.14 and 1.15. The optimal controller is then given by (1.121), namely

$$K_{opt} = -\tfrac{1}{2}.$$

The set of stabilizing controllers $\mathcal{S}_M$ is given by (1.96) as

$$K(s) = -\frac{W(s)}{1 - \frac{s-1}{s+2}W(s)}$$

where W is a free proper and stable rational parameter. The set $\mathcal{S}_{MT}$ of all stabilizing controllers that render the H_2 norm of the system transfer function finite is generated using (1.117) and it is seen to coincide with $\mathcal{S}_M$. The optimal controller corresponds to the choice

$$W(s) = \frac{1}{3}\frac{s+2}{s+1}. \qquad \square$$

The Diophantine equations (1.100) are an elegant tool to calculate the partial fraction expansions needed in the Wiener–Hopf approach:

$$\Gamma = \frac{C^* D^* E}{S_C^* S_D^* A} = \frac{Z^*}{S_C^* S_D^*} + \frac{Y}{A}.$$

One obtains

$$\begin{aligned} \frac{Y}{A} &= \Lambda\, \frac{Y_R + W_{opt} A_R}{A_R} \Omega \\ &= \Lambda\, \frac{Y_R}{A_R}\Omega + \Lambda W_{opt}\Omega \\ &= \{\Gamma\}_+ + \{\Lambda\, \frac{Y_R}{A_R}\, \Omega\}_- \end{aligned}$$

in view of (1.86). Also

$$\frac{\Lambda}{A_R} = \frac{S_D}{A}, \qquad \frac{\Omega}{A_L} = \frac{S_C}{A}$$

so that

$$T_{K_{opt}} = -\frac{A}{S_C}\frac{Y}{A}\frac{A}{S_D}.$$

Since

$$AX + BY = S_C S_D$$

one recovers the formula

$$K_{opt} = -\frac{Y}{X}.$$

Example 1.7 Consider the plant

$$\begin{aligned} \dot{x} &= x + u \\ z &= u \\ y &= x - v \end{aligned}$$

with the transfer functions

$$M_{11}(s) = 0, \qquad M_{12}(s) = 1$$

$$M_{21}(s) = -1, \quad M_{22}(s) = \frac{1}{s-1}$$

It is straightforward to find the H_2 optimal controller using the Diophantine equations. We put

$$A(s) = \frac{s-1}{s+1}, \qquad E(s) = 0, \qquad D(s) = \frac{s-1}{s+1}$$

$$C(s) = \frac{1-s}{s+1}, \qquad B(s) = \frac{1}{s+1}.$$

and calculate

$$S_C(s) = 1, \qquad S_D(s) = 1, \qquad F(s) = \frac{1-s}{s+1}.$$

Then equations (1.100) become

$$X - Z^* \frac{1}{s+1} = \frac{s+1}{s-1}$$

$$Y + Z^* \frac{s-1}{s+1} = 0$$

and yield the solution triple

$$X(s) = \frac{s+3}{s+1}, \qquad Y(s) = \frac{4}{s+1}, \qquad Z(s) = \frac{4}{s+1}$$

for which $Z/S_C S_D$ is strictly proper.

The conditions of theorems 1.11, 1.14 and 1.15 are all satisfied. The set $\mathcal{S}_M$ is generated by the controllers

$$K(s) = -\frac{\frac{4}{s+1} + \frac{s-1}{s+1} W(s)}{\frac{s+3}{s+1} - \frac{1}{s+1} W(s)}$$

where W is any proper and stable rational function while $\mathcal{S}_{MT}$ is generated by the same formula with W restricted to be strictly proper. The optimal controller corresponds to the choice $W = 0$:

$$K_{opt} = -\frac{4}{s+3}.$$

Assumption (SS.1) fails to hold in this case. This means that the algebraic Riccati equations (1.28) and (1.38) may not have the requisite stabilizing solutions. However, in this particular case,

$$2P_0 - P_0^2 = 0, \qquad 2P_1 - P_1^2 = 0$$

do have the stabilizing solutions

$$P_0 = 2, \qquad P_1 = 2$$

and the state-space approach is applicable.

There results

$$L_0 = 2, \qquad L_1 = 2$$

and the optimal controller is given by (1.48) as $(-3, 2, -2, 0)$, or equivalently

$$K_{opt}(s) = -\frac{4}{s+3},$$

confirming the previous result. □

1.5 References

Anderson, B.D.O. and J.B. Moore (1990). Optimal Control: Linear Quadratic Methods. Prentice Hall, Englewood Cliffs, NJ.

Åström, K.J. (1970). Introduction to Stochastic Control. Academic Press, New York.

Doyle, J.C., K. Glover, P.P. Khargonekar, and B.A. Francis (1989). State space solutions to standard H_2 and H_∞ control problems. IEEE Trans. Automat. Contr., AC-34, 831 - 847.

Kalman, R.E. and R.S. Bucy (1961). New results in linear filtering and prediction theory. Trans. ASME, Ser. D, J. Basic Eng., 83, 95 - 107.

Kučera, V. (1986). Stationary LQG control of singular systems. IEEE Trans. Automat. Contr., AC-31, 31 - 39.

Kučera, V. (1991). Analysis and Design of Discrete Linear Control Systems. Prentice Hall, London.

Kučera, V. (1992). The rational H_2 control problem. Proc. 31st IEEE Conf. Decision and Control, Tucson, AZ, 3610 - 3613.

Park, K. and J.J. Bongiorno (1989). A general theory for the Wiener-Hopf design of multivariable control systems. IEEE Trans. Automat. Contr., AC-34, 619 - 626.

Safonov, M.G., E.A. Jonckheere, M. Verma, and D.J.N. Limebeer (1987). Synthesis of positive real multivariable feedback systems. Int. J. Contr., 45, 817 - 842.

Tse, E. (1971). On the optimal control of stochastic linear systems. IEEE Trans. Automat. Contr., AC-16, 776 - 785.

Vidyasagar, M. (1985). Control System Synthesis: A Factorization Approach. MIT Press, Cambridge, MA.

Wonham, W.M. (1968). On the separation theorem of control. SIAM J. Contr., 6, 312 - 326.

Youla, D.C., H. Jabr, and J.J. Bongiorno (1976). Modern Wiener-Hopf design of optimal controllers - Part II: The multivariable case. IEEE Trans. Automat. Contr., AC-21, 319 - 338.

2

Frequency Domain Solution of the Standard $\mathcal{H}_\infty$ Problem

Huibert Kwakernaak

2.1 Introduction

2.1.1 Introduction

The importance of the famous standard $\mathcal{H}_\infty$ optimal regulator problem [5] as a tool for practical control system design is widely accepted and well-understood [20]. This contribution presents a frequency domain solution of the standard $\mathcal{H}_\infty$ problem that is based on polynomial matrix techniques.

Various solutions of the standard problem exist. The most important is the so-called "two Riccati equation" solution [9], [7]. It relies on a state space representation of the problem and requires the solution of two indefinite algebraic Riccati equations. Implementations on various computational platforms ([1] and [4] document two instances) are available.

There are various reasons to consider the frequency domain solution as an alternative to the state space solution:

- The state space formulation implicitly requires the "generalized plant" to have a proper transfer matrix. Practical design methods based on $\mathcal{H}_\infty$ typically involve nonproper frequency dependent weighting functions, however, resulting in a nonproper generalized plant.

- The state space solution involves a number of technical assumptions on the generalized plant that impose rank conditions, including certain rank conditions on the imaginary axis. These conditions are irrelevant for control system design, and limit the applicability of the standard problem.

- The state space solution does not appear to be a very good tool for studying *optimal* rather than the usual suboptimal — or *sublevel* as they are referred to in this contribution — solutions. As a matter of fact very little has been published on optimal solutions — [10] is an exception.

The frequency domain solution presented in this article bypasses the properness and rank condition problems, and provides access to the study of optimal solutions. It moreover makes it very clear that the solution of the $\mathcal{H}_\infty$ problem — no matter which approach is used — is based on spectral factorization.

The paper also offers a contribution to the understanding of optimal solutions. Two types of optimal solutions are distinguished. It is well known that near-optimal solutions are often accompanied by numerical ill-conditioning. The cause is that at the optimum the spectral factorization often is what is termed here *noncanonical.* Once this phenomenon is understood it is easy to devise algorithms that avoid the numerical difficulties and allow accurate computation of optimal solutions.

The paper is a much expanded version of a conference paper [21]. Several of the results may be found elsewhere [18], Meinsma [25]. New in this paper are the generality under which the results are claimed to hold, the clarification of the role of the canonicity of the spectral factorization, the spectral factorization algorithm of Section 2.7, and the representation of optimal solutions in section 2.8.

Other work on the solution of the $\mathcal{H}_\infty$ problem in the frequency domain relying on J-lossless transfer matrices has been done by [14], [12] and [33].

The work reported here made much of its progress during the period 1989–1993 when Gjerrit Meinsma worked on his dissertation at Twente. In fact it was Gjerrit who recognized the basic characterization of all suboptimal solutions summarized in theorem 2.6. His contributions are most gratefully acknowledged [28], [25], [26].

2.1.2 Problem formulation

We study the frequency domain version of the continuous-time $\mathcal{H}_\infty$ optimal regulator problem. Figure 2.1 depicts the situation. Given the generalized plant

$$\begin{bmatrix} z \\ y \end{bmatrix} = \begin{bmatrix} G_{11} & G_{12} \\ G_{21} & G_{22} \end{bmatrix} \begin{bmatrix} w \\ u \end{bmatrix} \tag{2.1}$$

interconnected with the feedback compensator $u = Ky$ we wish to minimize the ∞-norm of the closed-loop transfer matrix

$$H = G_{11} + G_{12}(I - KG_{22})^{-1}KG_{21} \tag{2.2}$$

with respect to all compensators K that stabilize the closed-loop system.

In Fig. 2.1, the signal w represents external signals, such as driving signals for disturbances or the reference inputs. The output signal z ideally is zero and is referred to as the error signal. The input signal u is available for controlling the system and is called the control input. The output signal y, finally, may be used for feedback, and is called the measured or observed output. The closed-loop transfer matrix H is the transfer matrix from the external signal w to the error signal z.

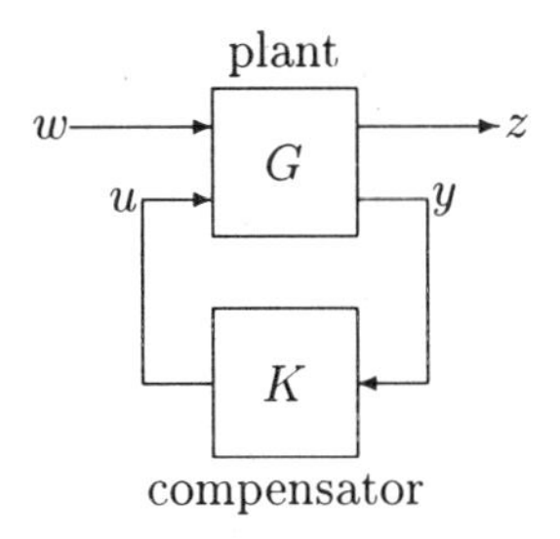

Figure 2.1: The standard problem.

In section 2.2 we discuss the definitions of the ∞-norm and closed-loop stability.

We consider the solution of the standard problem under the following assumptions on the generalized plant G:

1. G is rational but not necessarily proper. G_{11} has dimensions $m_1 \times k_1$, G_{12} has dimensions $m_1 \times k_2$, G_{21} has dimensions $m_2 \times k_1$, and G_{22} has dimensions $m_2 \times k_2$.
2. G_{12} has full normal column rank and G_{21} has full normal row rank. This implies that $k_1 \geq m_2$ and $m_1 \geq k_2$.

These assumptions are much less restrictive than those usually found in the $\mathcal{H}_\infty$ literature [9], [7], [31]). They allow nonproper weighting matrices in the mixed sensitivity problem, which is important for obtaining high frequency roll-off [20]. The fact that G_{12} and G_{21} do no need to have full rank at ∞, as usually required, allows to study the celebrated minimum-sensitivity problem [35]. In subsection 2.5.3 we deal with the situation of "zeros on the imaginary axis," which in the usual state space solution is excluded by suitable rank conditions.

Assumption 2 causes little loss of generality. If the stated rank conditions are not satisfied then often the plant input u or the observed output y or both may be transformed so that the rank conditions hold.

We present two typical examples of $\mathcal{H}_\infty$ optimization problems.

Example 2.1. **The minimum sensitivity problem.** The SISO version of the minimum sensitivity problem historically is the first $\mathcal{H}_\infty$ problem studied [35]. In the block diagram of Fig. 2.2 the block P represents a plant, and K a compensator to be designed. The signal w is an external signal that drives the shaping filter V for the disturbance v. W is a weighting filter that weights the control

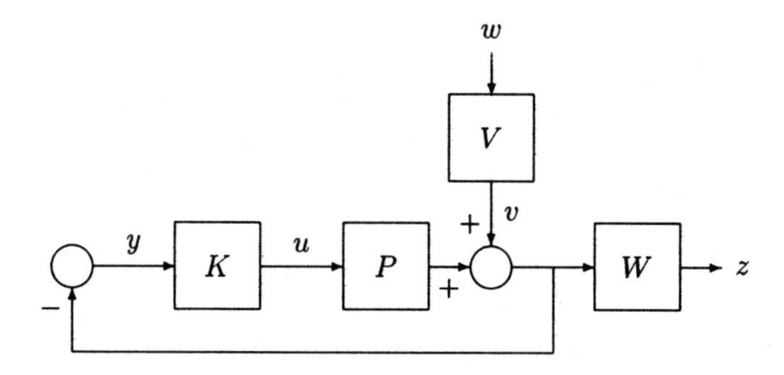

Figure 2.2: The minimum sensitivity problem.

system output. The transfer function from the external signal w to the weighted control system output z is $H = WSV$, where

$$S = (I - PK)^{-1} \tag{2.3}$$

is the sensitivity matrix of the closed-loop system. The minimum sensitivity problem is the problem of minimizing $\|H\|_\infty = \|WSV\|_\infty$. Inspection of Fig. 2.2 shows that this is a standard $\mathcal{H}_\infty$ problem with generalized plant

$$G = \left[\begin{array}{c|c} WV & WP \\ \hline -V & -P \end{array}\right]. \tag{2.4}$$

In the simplest unweighted SISO case, where $W = V = 1$, we have $G_{12} = P$. If the plant transfer function P is strictly proper then $G_{12}(\infty) = 0$ so that the problem does not satisfy the usual assumptions for the standard problem. •

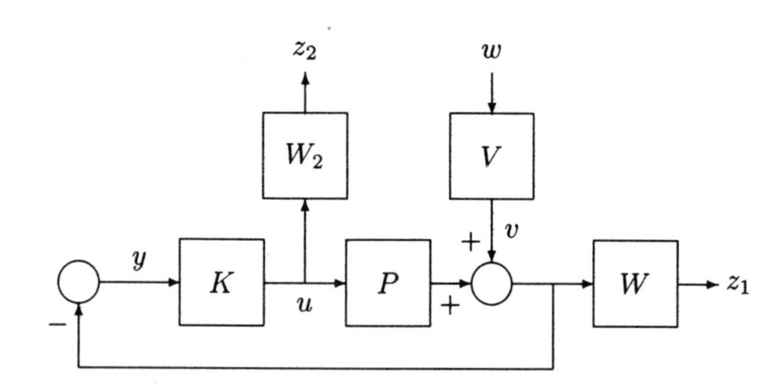

Figure 2.3: The mixed sensitivity problem.

Example 2.2. **The mixed sensitivity problem.** The mixed sensitivity problem is an extension of the minimum sensitivity problem. It is represented by the block diagram of Fig. 2.3. The "error signal" z has as its first component z_1 the weighted control system output, and as its second output z_2 the weighted plant input. The purpose of the second component is to control the closed-loop bandwidth of the system. The transfer matrix from the external signal w to the control error z may easily be found to be

$$H = \begin{bmatrix} W_1 SV \\ -W_2 UV \end{bmatrix}, \tag{2.5}$$

with

$$U = K(I - PK)^{-1} \tag{2.6}$$

what is known as the *input sensitivity matrix.* Inspection of Fig. 2.3 shows that the mixed sensivitity problem is a standard problem with generalized plant

$$G = \left[\begin{array}{c|c} W_1 V & W_1 P \\ 0 & W_2 \\ \hline -V & -P \end{array}\right]. \tag{2.7}$$

To obtain a desired high-frequency roll-off of the input sensitivity U it often is necessary to select the weighting function W_2 to be nonproper. This makes G nonproper and, hence, violates the standard condition that G be proper. •

2.1.3 Polynomial matrix fraction representations

We assume throughout that the generalized plant G has left and right coprime polynomial matrix fraction representations

$$G = D^{-1}N = \bar{N}\bar{D}^{-1}, \tag{2.8}$$

respectively, with the partitionings

$$D = \overset{\quad m_1 \; m_2}{[\, D_1 \; D_2 \,]}, \qquad N = \overset{\quad k_1 \; k_2}{[\, N_1 \; N_2 \,]}, \tag{2.9}$$

$$\bar{D} = \begin{bmatrix} \bar{D}_1 \\ \bar{D}_2 \end{bmatrix} \begin{matrix} k_1 \\ k_2 \end{matrix}, \qquad \bar{N} = \begin{bmatrix} \bar{N}_1 \\ \bar{N}_2 \end{bmatrix} \begin{matrix} m_1 \\ m_2 \end{matrix}. \tag{2.10}$$

Without loss of generality the polynomial matrices

$$[N_1 \;\; D_1] \quad \text{and} \quad \begin{bmatrix} \bar{N}_1 \\ \bar{D}_1 \end{bmatrix} \tag{2.11}$$

may be arranged to be row reduced and column reduced, respectively.

Throughout this paper we assume that basic notions about polynomial matrices, such as row and column reducedness, are understood. We consider the material of chapter 8 of [13] to be known. Another famous source on the use of polynomial matrices in systems and control theory is [15].

2.1.4 Outline

Section 2.2 deals with well-posedness and closed-loop stability of "standard" feedback systems. A well-known lower bound for the minimal ∞-norm, known as the Parrott lower bound, is discussed and derived in section 2.3.

In section 2.4 a characterization of all sublevel solutions for a given level γ is given. This characterization is determined by a rational spectral factorization, which in turn may be reduced to two polynomial spectral factorizations. In section 2.5 canonical factorizations are defined.

The sublevel solutions of section 2.4 are not necessarily stable. In section 2.6 we characterize all stable sublevel solutions, if any exist. This characterization implies a well-known search procedure for optimal solutions. Two types of optimal solutions are introduced: Type A solutions achieve the Parrott lower bound, while type B solutions do not.

In section 2.7 we present a new state space algorithm for the spectral factorization of para-Hermitian polynomial matrices. It is included to clarify the complications that arise when the factorization is noncanonical, and leads to a characterization of noncanonical factorizations.

This characterization is used in section 2.8 to describe all type B optimal solutions. Several properties of the optimal solution are listed.

Section 2.9, finally, presents the conclusions.

Technical proofs are avoided in the main text. They are collected in Appendices 2.10–2.13.

2.2 Well-posedness and closed-loop stability

2.2.1 Introduction

The standard problem concerns the minimization of the ∞-norm of the closed-loop transfer matrix with respect to all stabilizing compensators. After introducing well-posedness of standard feedback systems we explain in this section what we mean by closed-loop stability and the ∞-norm.

2.2.2 Well-posedness

We begin with discussing *well-posedness* of feedback systems as in Fig. 2.1. Let

$$K = X^{-1}Y, \tag{2.12}$$

with X and Y left coprime stable rational matrices (possibly polynomial). Then the closed-loop transfer matrix is

$$\begin{aligned} H &= G_{11} + G_{12}(I - KG_{22})^{-1}KG_{21} \\ &= G_{11} + G_{12}(X - YG_{22})^{-1}YG_{21}. \end{aligned} \tag{2.13}$$

The feedback system is said to be *well-posed* if $X - YG_{22}$ is nonsingular. A feedback system may be well-posed even if X is singular. This formalism allows considering infinite-gain feedback systems, even though such systems cannot be realized.

Write

$$G_{22} = N_o D_o^{-1}, \tag{2.14}$$

with N_o and D_o right coprime polynomial or stable rational matrices. Then the closed-loop system is well-posed iff $XD_o - YN_o$ is nonsingular.

Example 2.3. **Infinite gain.** Consider an unweighted SISO minimum sensitivity problem as in example 2.1. If the plant $P = N_o/D_o$ has no right-half plane zeros then the minimum sensitivity compensator turns out to have infinite gain. The sensitivity problem is solved by the compensator $K = Y/X$ (which actually does not exist) with $X = 0$ and $Y = -1$. Since for the unweighted minimum sensitivity problem we have $G_{11} = 1$, $G_{12} = P$ and $G_{21} = -1$ it follows from (2.13) that $S = H = 0$, so that $\|H\|_\infty = 0$. •

2.2.3 Closed-loop stability

Next we define what it means if the compensator *stabilizes* the closed-loop system. The closed-loop system is said to be stable if the compensator K stabilizes G_{22}, that is, the loop of Fig. 2.4 including K and G_{22} is stable.

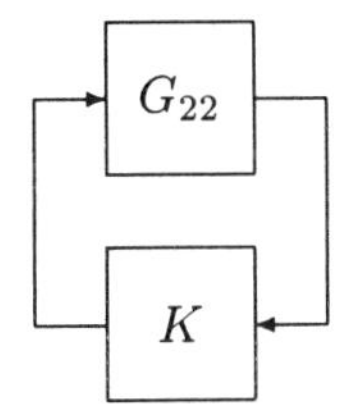

Figure 2.4: Loop with G_{22}.

This definition is less restrictive than the usual requirement that the entire feedback system be stable [7]. It allows to include unstable weighting filters (including such that have poles on the imaginary axis) in the mixed sensitivity problem [20].

By this definition of stability, the compensator K is stabilizing iff the square stable rational or polynomial matrix $XD_o - YN_o$ is strictly Hurwitz, that is, has all its zeros in the open left-half plane.

Example 2.4. **Infinite gain.** In example 2.3 "stability" follows by the fact that $XD_o - YN_o = N_o$ and N_o by assumption has all its zeros in the left-half complex plane. •

2.2.4 Redefinition of the standard problem

Because of our view of closed-loop stability it is necessary to reconsider what is meant by the $\mathcal{H}_\infty$ norm. Normally this norm is defined in the following way. Let

H be the transfer matrix of a linear time-invariant system that is BIBO stable in the sense of the 2-norm on the input and output signals [24]. Then the operator norm induced by these signal norms is the well-known $\mathcal{H}_\infty$ norm

$$\|H\|_\infty = \sup_{\omega \in \mathsf{R}} \sigma_{\max}(H(j\omega)), \tag{2.15}$$

where $\sigma_{\max}$ denotes the largest singular value (that is, the spectral norm). R is the set of real numbers.

For systems that are not BIBO stable the right-hand side of (2.15) is well-defined as long as H has no poles on the imaginary axis (including ∞). Abusing terminology and notation we refer to the right-hand side of (2.15) as the $\mathcal{H}_\infty$ norm of the system and denote it as $\|H\|_\infty$, even if the system is not BIBO stable.

On the basis of these notions of well-posedness, closed-loop stability and the $\mathcal{H}_\infty$ norm we reformulate the $\mathcal{H}_\infty$ problem as the problem of minimizing the ∞-norm as defined by (2.15) of $H = G_{11} + G_{12}(X - YG_{22})^{-1}YG_{21}$ with respect to all stable rational X and Y such that $XD_o - YN_o$ is nonsingular and strictly Hurwitz.

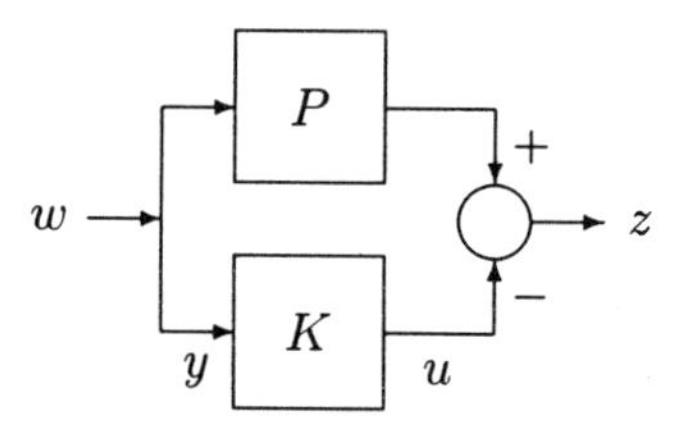

Figure 2.5: The model matching problem.

Example 2.5. **The model matching problem.** What we call the model matching problem does not make much sense as a control or system theoretic problem but helps to illustrate a number of issues. We refer to it several times in this paper.

In its simplest form, the model matching problem is the problem of finding a stable transfer matrix K that minimizes

$$\|P - K\|_\infty, \tag{2.16}$$

with P a given (unstable) transfer matrix. $K = X^{-1}Y$ is the "best" stable approximation of P in the sense of the $\mathcal{H}_\infty$ norm.

Figure 2.5 shows that the model matching problem is a special case of the standard problem. The generalized plant has the transfer matrix

$$G = \left[\begin{array}{c|c} P & -I \\ \hline I & 0 \end{array}\right]. \tag{2.17}$$

By our definition closed-loop stability is obtained if $XD_o - YN_o = X$ has left-half plane zeros only, that is, if K is stable, exactly what is required. If P is unstable then the overall system, with transfer matrix $H = P - K$, is never stable when K is stable. •

2.3 Lower bound

2.3.1 Introduction

In this section we derive what is known as the *Parrott lower bound* [29] for the solution of the $\mathcal{H}_\infty$ problem. While doing this we establish a number of useful preliminary facts.

Theorem 2.1. Bound on the ∞-norm. *Let γ be a nonnegative number and H a rational transfer matrix. Then $\|H\|_\infty \leq \gamma$ is equivalent to either of the following statements:*

(a) $H^\sim H \leq \gamma^2 I$ on the imaginary axis;

(b) $HH^\sim \leq \gamma^2 I$ on the imaginary axis.

If H is a rational matrix function then $H^\sim$ is defined as $H^\sim(s) = H^{\mathrm{T}}(-s)$. The proof of this theorem is given in Appendix 2.10.

Theorem 2.1 makes it clear why it is attractive to study *sublevel* solutions, that is, compensators that make $\|H\|_\infty \leq \gamma$. The theorem shows that sublevel solutions are characterized by nonnegative-definitess requirements.

2.3.2 Lower bound

Consider the closed-loop transfer matrix

$$H = G_{11} + G_{12}K(I - G_{22}K)^{-1}G_{21} = G_{11} + LG_{21}, \tag{2.18}$$

where $L = G_{12}K(I - G_{22}K)^{-1}$. We study under what conditions there exist L that make $\|H\|_\infty \leq \gamma$. Clearly, such conditions are necessary for the existence of compensators K that make $\|H\|_\infty \leq \gamma$.

We have

$$\begin{aligned} \gamma^2 I - HH^\sim &= \gamma^2 I - G_{11}G_{11}^\sim - G_{11}G_{21}^\sim L^\sim - LG_{21}G_{11}^\sim - LG_{21}^\sim G_{21}L^\sim \\ &= [I \; L] \underbrace{\begin{bmatrix} \gamma^2 I - G_{11}G_{11}^\sim & -G_{11}G_{21}^\sim \\ -G_{21}G_{11}^\sim & -G_{21}G_{21}^\sim \end{bmatrix}}_{\Pi_{1,\gamma}} \begin{bmatrix} I \\ L^\sim \end{bmatrix}. \end{aligned} \tag{2.19}$$

The question now is whether there exist L such that

$$[I \; L]\, \Pi_{1,\gamma} \begin{bmatrix} I \\ L^\sim \end{bmatrix} \geq 0 \quad \text{on the imaginary axis.} \tag{2.20}$$

Quadratic forms are closely related to factorization.

Theorem 2.2. Definiteness of quadratic form. *Consider the inequality*

$$[I \quad L]\underbrace{\begin{bmatrix}\Pi_{11} & \Pi_{12}\\ \Pi_{21} & -\Pi_{22}\end{bmatrix}}_{\Pi}\begin{bmatrix}I\\ L^{\sim}\end{bmatrix} \geq 0 \quad \text{on the imaginary axis,} \tag{2.21}$$

with Π a nonsingular para-Hermitian rational matrix such that the $n\times n$ matrix Π_{22} is nonnegative-definite on the imaginary axis. Then the following three facts are equivalent.

(a) *There exists a rational $n\times m$ matrix L such that the inequality (2.21) holds.*

(b) *The Schur complement $\Pi_{11}+\Pi_{12}\Pi_{22}^{-1}\Pi_{21}$ of the (2, 2)-block of Π is nonnegative-definite on the imaginary axis.*

(c) *Π may be factored as $\Pi = ZJZ^{\sim}$, with Z square rational and*

$$J = \begin{bmatrix} I_n & 0\\ 0 & -I_m\end{bmatrix} \tag{2.22}$$

the signature matrix of Π.

The proof of this theorem may be found in Appendix 2.10.

By application of theorem 2.2 to (2.19–2.20) it follows that there exist L that make $\|H\|_\infty \leq \gamma$ if and only if the Schur complement of the (2,2)-block of $\Pi_{1,\gamma}$ is nonnegative-definite on the imaginary axis. It is easy to see that there exists a $\gamma_1 \geq 0$ (possibly ∞) such that the Schur complement is nonnegative-definite on the imaginary axis for all $\gamma \geq \gamma_1$. It is argued in Appendix 2.10 that γ_1 is in fact the first value of γ as γ decreases from ∞ for which $\Pi_{1,\gamma}$ has a zero on the imaginary axis. Zeros of G_{21} need to be excluded.

A similar argument follows by writing $H = G_{11} + G_{12}L$, where now $L = K(I - G_{22}K)^{-1}G_{12}$, and considering $H^{\sim}H \leq \gamma^2 I$. The pertinent coefficient matrix is

$$\Pi_{2,\gamma} = \begin{bmatrix}\gamma^2 I - G_{11}^{\sim}G_{11} & -G_{11}^{\sim}G_{12}\\ -G_{12}^{\sim}G_{11} & -G_{12}^{\sim}G_{12}\end{bmatrix}. \tag{2.23}$$

This leads to a second lower bound γ_2.

The greater of the two numbers γ_1 and γ_2 is the lower bound we are looking for.

Theorem 2.3. Lower bound. *Define γ_1 as the first value of γ as γ decreases from ∞ at which a zero of $\det \Pi_{1,\gamma}$ — excluding the zeros of G_{21} and their mirror images with respect to the imaginary axis— reaches the imaginary axis.*

Likewise, let γ_2 be the first value of γ as γ decreases from ∞ at which a zero of $\det \Pi_{2,\gamma}$ — excluding the zeros of G_{12} and their mirror images with respect to the imaginary axis— reaches the imaginary axis. Then $\|H\|_\infty \geq \gamma_o$, with $\gamma_o = \max(\gamma_1, \gamma_2)$.

In section 2.4 it turns out that the lower bound may actually be achieved (but not necessarily by a stabilizing compensator).

2.3.3 Examples

The following two examples illustrate that no compensator may exist that makes $\|H\|_\infty$ finite, and that if there exist compensators that make $\|H\|_\infty$ finite none of them may stabilize the feedback system.

Example 2.6. **No compensator exists that makes $\|H\|_\infty$ finite.** Consider the generalized plant

$$G(s) = \left[\begin{array}{cc|c} s & 0 & 1 \\ 0 & s^2 & 0 \\ \hline 1 & 0 & 0 \end{array}\right]. \tag{2.24}$$

Inspection of

$$H(s) = \underbrace{\begin{bmatrix} s & 0 \\ 0 & s^2 \end{bmatrix}}_{G_{11}(s)} + \underbrace{\begin{bmatrix} 1 \\ 0 \end{bmatrix}}_{G_{12}(s)} K(s) \underbrace{\begin{bmatrix} 1 & 0 \end{bmatrix}}_{G_{21}(s)} = \begin{bmatrix} s+K(s) & 0 \\ 0 & s^2 \end{bmatrix} \tag{2.25}$$

shows that $\|H\|_\infty = \infty$, no matter how K is chosen. We have

$$\Pi_{1,\gamma} = \begin{bmatrix} \gamma^2 + s^2 & 0 & -s \\ 0 & \gamma^2 - s^4 & 0 \\ s & 0 & -1 \end{bmatrix}. \tag{2.26}$$

The determinant $\det \Pi_{1,\gamma} = \gamma^2(s^4 - \gamma^2)$ has the zeros $\pm\sqrt{\gamma}$ and $\pm j\sqrt{\gamma}$. As none of these zeros may be excluded and two of them are always on the imaginary axis by theorem 2.3 we have $\gamma_1 = \infty$. It follows that $\gamma_o = \infty$, which confirms that there exists no K that makes $\|H\|_\infty$ finite. •

Example 2.7. **No stabilizing compensator exists that makes $\|H\|_\infty$ finite.** For the generalized plant

$$G(s) = \left[\begin{array}{c|c} \frac{1}{s} & 1 \\ \hline 1 & 0 \end{array}\right] \tag{2.27}$$

we have

$$H(s) = \frac{1}{s} + K(s). \tag{2.28}$$

Since $G_{22} = 0$, the compensator K stabilizes the feedback system iff K itself is stable. Inspection of H shows that any compensator that makes $\|H\|_\infty$ finite needs to have a pole at 0. Hence, no compensator that makes $\|H\|_\infty$ finite stabilizes the system.

By simple substitution it follows that

$$\Pi_{1,\gamma} = \begin{bmatrix} \gamma^2 + \frac{1}{s^2} & -\frac{1}{s} \\ \frac{1}{s} & -1 \end{bmatrix}. \tag{2.29}$$

Because $\det \Pi_{1,\gamma} = -\gamma^2$ we have from theorem 2.3 that $\gamma_1 = 0$. It similarly follows that $\gamma_2 = 0$, so that the lower bound is $\gamma_o = 0$. The lower bound is achieved by the compensator $K(s) = -1/s$, of course. •

Example 2.8. **Model matching problem.** Consider the model matching problem of example 2.5 for the "plant" with transfer function

$$P(s) = \frac{1}{1-s}. \tag{2.30}$$

We pursue this example throughout this paper. The corresponding generalized plant is

$$G(s) = \left[\begin{array}{c|c} \frac{1}{1-s} & -1 \\ \hline 1 & 0 \end{array}\right] \tag{2.31}$$

By straightforward computation we find that $\det \Pi_{1,\gamma}$ and $\det \Pi_{2,\gamma}$ are both equal to $-\gamma^2$. Hence, $\gamma_1 = \gamma_2 = \gamma_o = 0$. Indeed, the compensator $K(s) = 1/(1-s)$ achieves $\|P - K\|_\infty = 0$. •

2.3.4 Polynomial formulas

Using the left and right polynomial matrix fraction representations (2.8–2.10) it may easily be found that

$$\Pi_{1,\gamma} = \begin{bmatrix} \gamma^2 I - G_{11}G_{11}^{\sim} & -G_{11}G_{21}^{\sim} \\ -G_{21}G_{11}^{\sim} & -G_{21}G_{21}^{\sim} \end{bmatrix} = (D^{-1})^{\sim}(\gamma^2 D_1 D_1^{\sim} - N_1 N_1^{\sim})D^{-1}, \tag{2.32}$$

$$\Pi_{2,\gamma} = \begin{bmatrix} \gamma^2 I - G_{11}^{\sim}G_{11} & -G_{11}^{\sim}G_{12} \\ -G_{12}^{\sim}G_{11} & -G_{12}^{\sim}G_{12} \end{bmatrix} = (\bar{D}^{\sim})^{-1}(\gamma^2 \bar{D}_1^{\sim}\bar{D}_1 - \bar{N}_1^{\sim}\bar{N}_1)\bar{D}^{-1}. \tag{2.33}$$

It follows that

$$\det \Pi_{1,\gamma} = \frac{\det(\gamma^2 D_1 D_1^{\sim} - N_1 N_1^{\sim})}{\det D \det D^{\sim}}, \tag{2.34}$$

$$\det \Pi_{2,\gamma} = \frac{\det(\gamma^2 \bar{D}_1^{\sim}\bar{D}_1 - \bar{N}_1^{\sim}\bar{N}_1)}{\det \bar{D} \det \bar{D}^{\sim}}. \tag{2.35}$$

We immediately have the following result.

Theorem 2.4. Lower bound in polynomial terms. *Define γ_1 as the first value of γ as γ decreases from ∞ at which a root of* $\det(\gamma^2 D_1 D_1^{\sim} - N_1 N_1^{\sim})$ *— excluding the zeros of G_{21}, any zero of* $\det D$ *and their mirror images with respect to the imaginary axis — reaches the imaginary axis.*

Likewise, let γ_2 be the first value of γ as γ decreases from ∞ at which a root of $\det(\gamma^2 \bar{D}_1^{\sim}\bar{D}_1 - \bar{N}_1^{\sim}\bar{N}_1)$ *— excluding the zeros of G_{12}, any zero of* $\det \bar{D}$ *and their mirror images with respect to the imaginary axis — reaches the imaginary axis. Then $\|H\|_\infty \geq \gamma_o$ with $\gamma_o = \max(\gamma_1, \gamma_2)$.*

The roots of $\det D$ and $\det \bar{D}$, which are identical, are of course the open-loop poles of the generalized plant.

Example 2.9. **Mixed sensitivity for integrating action.** We consider a mixed sensitivity problem for the simple SISO plant

$$P(s) = \frac{a}{s+a}, \tag{2.36}$$

with a a positive constant. The shaping filter V is chosen as

$$V(s) = \frac{M(s)}{s(s+a)}, \tag{2.37}$$

where M is a polynomial of degree 2. The factor s in the denominator of V represents a zero frequency disturbance. It is included to obtain integrating action. By partial pole assignment [16] the roots of the polynomial M are closed-loop poles. By choosing $M(s) = (s+a)(s+b)$, with b positive, we leave the pole at $-a$ in place and introduce a dominant closed-loop pole at $-b$. We furthermore let $W_1(s) = 1$ and $W_2(s) = cs$, with c a positive constant. The factor s in W_2 is needed to allow infinite gain at zero frequency.

From example 2.2 we have for the generalized plant

$$G(s) = \left[\begin{array}{c|c} \frac{M(s)}{s(s+a)} & \frac{a}{s+a} \\ 0 & cs \\ \hline -\frac{M(s)}{s(s+a)} & -\frac{a}{s+a} \end{array}\right] = \left[\begin{array}{c|c} \frac{s+b}{s} & \frac{a}{s+a} \\ 0 & cs \\ \hline -\frac{s+b}{s} & -\frac{a}{s+a} \end{array}\right] \tag{2.38}$$

$$= \underbrace{\left[\begin{array}{cc|c} s(s+a) & 0 & 0 \\ 0 & 1 & 0 \\ 1 & 0 & 1 \end{array}\right]^{-1}}_{D(s)} \underbrace{\left[\begin{array}{c|c} (s+a)(s+b) & as \\ 0 & cs \\ 0 & 0 \end{array}\right]}_{N(s)} \tag{2.39}$$

$$= \underbrace{\left[\begin{array}{cc} s+b & a \\ 0 & cs(s+a) \\ \hline -s-b & -a \end{array}\right]}_{\bar{N}(s)} \underbrace{\left[\begin{array}{cc} s & 0 \\ \hline 0 & s+a \end{array}\right]^{-1}}_{\bar{D}(s)}. \tag{2.40}$$

Since $\det D(d) = \det \bar{D}(s) = s(s+a)$ the open-loop poles are 0 and $-a$. G_{12} has no zeros but G_{21} has a zero at $-b$.

By direct computation it follows that

$$\det(\gamma^2 D_1 D_1^\sim - N_1 N_1^\sim) = -\gamma^4(-s^2+a^2)(-s^2+b^2). \tag{2.41}$$

The fixed zero pair at $\pm a$ cancels against the corresponding open-loop pole $-a$ and its mirror image in the right-half plane. The fixed zero pair at $\pm b$ originates from the zero at $-b$ of G_{21}. The remaining factor of the determinant is nonzero for $\gamma \neq 0$ so that $\gamma_1 = 0$.

Again by straighforward computation it may be found that

$$\begin{aligned} &\det(\gamma^2 \bar{D}_1^\sim \bar{D}_1 - \bar{N}_1^\sim \bar{N}_1) \\ &= s^2\left(c^2(\gamma^2-1)s^4 + c^2(a^2+b^2-\gamma^2 a^2)s^2 + a^2(\gamma^2-b^2c^2)\right). \end{aligned} \tag{2.42}$$

The double zero at 0 cancels against the open-loop pole at 0 and its mirror image. The reader is invited to verify that the first value of γ as γ decreases for which the remaining polynomial factor has a zero on the imaginary axis is given by $\gamma_2 = \max(1, bc)$. It follows that the lower bound is

$$\gamma_o = \max(1, bc). \tag{2.43}$$

•

2.4 Sublevel solutions

2.4.1 Introduction

In this section we outline the main steps that lead to an explicit formula for sublevel solutions to the standard H_∞ optimal regulation problem. Sublevel solutions are solutions such that

$$\|H\|_\infty \leq \gamma, \tag{2.44}$$

with γ a given nonnegative number. The method of solution is based on spectral factorization. It applies to the wide class of problems defined in Section 2.1.

2.4.2 The basic inequality

To motivate the next result, we first consider the model matching problem of example 2.5. For this problem we have $H = P - K$, with P a given transfer matrix, and K a stable transfer matrix to be determined. Substituting $H = P - K$ into $HH^\sim \leq \gamma^2 I$ we obtain from theorem 2.1 that $\|H\|_\infty \leq \gamma$ if and only if

$$PP^\sim - PK^\sim - KP^\sim + KK^\sim \leq \gamma^2 I \quad \text{on the imaginary axis.} \tag{2.45}$$

This in turn we may rewrite as

$$[I \;\; K] \underbrace{\begin{bmatrix} \gamma^2 I - PP^\sim & P \\ P^\sim & -I \end{bmatrix}}_{\Pi_\gamma} \begin{bmatrix} I \\ K^\sim \end{bmatrix} \geq 0 \quad \text{on the imaginary axis.} \tag{2.46}$$

This expression defines the rational matrix Π_γ. Writing $K = X^{-1}Y$, with X and Y stable rational, the inequality (2.46) is equivalent to

$$[X \;\; Y]\,\Pi_\gamma \begin{bmatrix} X^\sim \\ Y^\sim \end{bmatrix} \geq 0 \quad \text{on the imaginary axis.} \tag{2.47}$$

The inequality (2.47) characterizes all sublevel compensators K.

A similar inequality, less easily obtained than for the model matching problem, holds for the general standard problem. Recall that we define the standard problem as the problem of minimizing the ∞-norm of

$$H = G_{11} + G_{12}(X - YG_{22})^{-1}YG_{21} \tag{2.48}$$

with respect to X and Y such that $X - YG_{22}$ is strictly Hurwitz.

Theorem 2.5. Basic inequality. *(X, Y) achieves $\|H\|_\infty \leq \gamma$ for the standard problem, with $\gamma \geq \gamma_o$ a given nonnegative number, if and only if*

$$[X \;\; Y]\,\Pi_\gamma \begin{bmatrix} X^\sim \\ Y^\sim \end{bmatrix} \geq 0 \quad \text{on the imaginary axis,} \tag{2.49}$$

where Π_γ is the rational matrix

$$\Pi_\gamma = \begin{bmatrix} 0 & -I \\ G_{21} & G_{22} \end{bmatrix} \begin{bmatrix} G_{11}^\sim G_{11} - \gamma^2 I & G_{11}^\sim G_{12} \\ G_{12}^\sim G_{11} & G_{12}^\sim G_{12} \end{bmatrix}^{-1} \begin{bmatrix} 0 & G_{21}^\sim \\ -I & G_{22}^\sim \end{bmatrix}. \tag{2.50}$$

The proof is given in Appendix 2.11.

It is useful to have a formula for the inverse of Π_γ. At the cost of some not inconsiderable algebra it may be proved that

$$\Pi_\gamma^{-1} = \begin{bmatrix} G_{12}^\sim & G_{22}^\sim \\ 0 & I \end{bmatrix} \begin{bmatrix} I - \frac{1}{\gamma^2} G_{11} G_{11}^\sim & -\frac{1}{\gamma^2} G_{11} G_{21}^\sim \\ -\frac{1}{\gamma^2} G_{21} G_{11}^\sim & -\frac{1}{\gamma^2} G_{21} G_{21}^\sim \end{bmatrix}^{-1} \begin{bmatrix} G_{12} & 0 \\ G_{22} & I \end{bmatrix}. \tag{2.51}$$

2.4.3 Spectral factorization

A much more explicit characterization of all sublevel compensators (stabilizing or not) than that of theorem 2.5 may be obtained by finding a *spectral factorization* of the rational matrix Π_γ.

A nonsingular para-Hermitian[1] rational matrix Π has a spectral factorization if there exists a nonsingular square rational matrix Z such that both Z and its inverse Z^{-1} have all their poles in the closed left-half complex plane, and

$$\Pi = Z^\sim J Z, \tag{2.52}$$

with J a signature matrix[2]. Z is called a *spectral factor.*

On the imaginary axis Π is Hermitian. Except at the poles and zeros of Π on the imaginary axis J is the signature matrix of Π on the imaginary axis[3]. Hence, a necessary condition for the existence of a spectral factorization is that except in finitely many points Π has constant numbers of positive and negative eigenvalues on the imaginary axis (that is, Π has constant inertia on the imaginary axis). This condition is also sufficient for the existence of a spectral factorization [30].

The factorization $\Pi = Z^\sim J Z$, if it exists, is not at all unique. If Z is a spectral factor then so is UZ, where U is any rational or constant matrix such that $U^\sim J U = J$. The matrix U is said to be *J-unitary.* There are many such matrices. If

$$J = \begin{bmatrix} 1 & 0 \\ 0 & -1 \end{bmatrix}, \tag{2.53}$$

---

[1]A rational matrix Π is para-Hermitian if $\Pi^\sim = \Pi$.

[2]A matrix J is a signature matrix if it is of the form $J = \text{diag}\,(I, -I)$, with the two unit matrices not necessarily of the same dimension.

[3]That is, $J = \text{diag}\,(I, -I)$ with the dimension of I equal to the number of positive eigenvalues and that of $-I$ equal to the number of negative eigenvalues.

for instance, then all constant matrices U, with

$$U = \begin{bmatrix} c_1 \cosh\alpha & c_2 \sinh\alpha \\ c_1 \sinh\alpha & c_2 \cosh\alpha \end{bmatrix}, \tag{2.54}$$

$\alpha \in \mathsf{R}$, $c_1 = \pm 1$, $c_2 = \pm 1$, satisfy $U^{\mathrm{T}} J U = J$.

Sometimes it is appropriate to consider the *spectral cofactorization* $\Pi = ZJZ^{\sim}$ with Z again square such that both Z and Z^{-1} have all their poles in the left-half complex plane, and J a signature matrix.

2.4.4 All sublevel solutions

For $\gamma \geq \gamma_o$ the matrix Π_γ has constant signature on the imaginary axis, that is, the numbers of positive and negative eigenvalues remain constant. It may be checked from the definition (2.50) that for γ sufficiently large Π_γ has k_2 positive eigenvalues and m_2 negative eigenvalues on the imaginary axis. Hence, for $\gamma \geq \gamma_o$ the rational matrix Π_γ has a spectral cofactorization

$$\Pi_\gamma = Z_\gamma J Z_\gamma^{\sim}, \tag{2.55}$$

with

$$J = \begin{bmatrix} I_{k_2} & 0 \\ 0 & -I_{m_2} \end{bmatrix}. \tag{2.56}$$

The subscripts on the unit matrices indicate their dimensions.

By the spectral cofactorization (2.55) the inequality (2.49) is equivalent to

$$[X \;\; Y] Z_\gamma \begin{bmatrix} I_{k_2} & 0 \\ 0 & -I_{m_2} \end{bmatrix} Z_\gamma^{\sim} \begin{bmatrix} X^{\sim} \\ Y^{\sim} \end{bmatrix} \geq 0 \quad \text{on the imaginary axis.} \tag{2.57}$$

Denote

$$[X \;\; Y] Z_\gamma = [A \;\; B], \tag{2.58}$$

with A a $k_2 \times k_2$ rational matrix and B a $k_2 \times m_2$ rational matrix, both with all their poles in the closed left-half plane. Then the inequality (2.57) is equivalent to

$$[X \;\; Y] = [A \;\; B] Z_\gamma^{-1}, \qquad AA^{\sim} \geq BB^{\sim} \quad \text{on the imaginary axis.} \tag{2.59}$$

It is proved in Appendix 2.11 that the closed-loop system defined this way is well-posed iff A is nonsingular.

Theorem 2.6. Sublevel solutions. *Let for $\gamma \geq \gamma_o$ the rational matrix Π_γ have the spectral cofactorization $\Pi_\gamma = Z_\gamma J Z_\gamma^{\sim}$, with*

$$J = \begin{bmatrix} I_{k_2} & 0 \\ 0 & -I_{m_2} \end{bmatrix}. \tag{2.60}$$

Then all sublevel compensators are given by

$$[X \;\; Y] = [A \;\; B] Z_\gamma^{-1}, \tag{2.61}$$

with A nonsingular square rational and B rational, both with all their poles in the closed left-half plane and such that

$$AA^\sim \geq BB^\sim \quad \textit{on the imaginary axis.} \tag{2.62}$$

There are many matrices of stable rational functions A and B that satisfy (2.62). An obvious choice is $A = I$, $B = 0$. We call the corresponding compensator the *central solution.* The central solution as defined here is not at all unique, because as seen in subsection 2.4.3 the spectral factor Z_γ is not unique.

Example 2.10. **Sublevel solutions for a model matching problem.** Consider the model matching problem of example 2.5 with

$$P(s) = \frac{1}{1-s}. \tag{2.63}$$

From example 2.8 we have $\gamma_o = 0$. It is straightforward to find that for this problem Π_γ is given by

$$\Pi_\gamma(s) = \begin{bmatrix} 1 - \frac{\frac{1}{\gamma^2}}{1-s^2} & \frac{\frac{1}{\gamma^2}}{1-s} \\ \frac{\frac{1}{\gamma^2}}{1+s} & -\frac{1}{\gamma^2} \end{bmatrix}. \tag{2.64}$$

It may be verified that for $\gamma \neq \frac{1}{2}$ the matrix Π_γ has the spectral cofactor and signature matrix

$$Z_\gamma(s) = \begin{bmatrix} \frac{\frac{\gamma^2-\frac{3}{4}}{\gamma^2-\frac{1}{4}}+s}{1+s} & -\frac{\frac{\frac{1}{4}}{\gamma(\gamma^2-\frac{1}{4})}}{1+s} \\ \frac{\frac{1}{\gamma^2-\frac{1}{4}}}{1+s} & \frac{\frac{1}{\gamma}\left(\frac{\gamma^2+\frac{1}{4}}{\gamma^2-\frac{1}{4}}+s\right)}{1+s} \end{bmatrix}, \qquad J = \begin{bmatrix} 1 & 0 \\ 0 & -1 \end{bmatrix}. \tag{2.65}$$

The inverse of Z_γ is

$$M_\gamma(s) = Z_\gamma^{-1}(s) = \begin{bmatrix} \frac{\frac{\gamma^2+\frac{1}{4}}{\gamma^2-\frac{1}{4}}+s}{1+s} & \frac{\frac{\frac{1}{4}}{\gamma^2-\frac{1}{4}}}{1+s} \\ -\frac{\frac{\gamma}{\gamma^2-\frac{1}{4}}}{1+s} & \frac{\gamma(\frac{\gamma^2-\frac{3}{4}}{\gamma^2-\frac{1}{4}}+s)}{1+s} \end{bmatrix}. \tag{2.66}$$

For $\gamma = \frac{1}{2}$ something special occurs. We discuss this in section 2.5. It follows from theorem 2.6 that for $\gamma \neq \frac{1}{2}$ all sublevel "compensators" $K = X^{-1}Y$ have the transfer function

$$K(s) = \frac{\frac{\frac{1}{4}}{\gamma^2-\frac{1}{4}} + \gamma\left(\frac{\gamma^2-\frac{3}{4}}{\gamma^2-\frac{1}{4}} + s\right)U(s)}{\left(\frac{\gamma^2+\frac{1}{4}}{\gamma^2-\frac{1}{4}} + s\right) - \frac{\gamma}{\gamma^2-\frac{1}{4}}U(s)}, \tag{2.67}$$

with $U = B/A$ such that $AA^\sim \geq BB^\sim$ on the imaginary axis.

We consider two special cases:

(a) $A = 1$, $B = 0$, so that $U = 0$. We then obtain the "central" compensator

$$K(s) = \frac{\frac{\frac{1}{4}}{\gamma^2 - \frac{1}{4}}}{\frac{\gamma^2+\frac{1}{4}}{\gamma^2-\frac{1}{4}} + s}. \tag{2.68}$$

For $\gamma > \frac{1}{2}$ the compensator is stable, for $0 \leq \gamma < \frac{1}{2}$ it is unstable. The resulting "closed-loop transfer function" is

$$H(s) = P(s) - K(s) = \frac{\frac{\gamma^2}{\gamma^2 - \frac{1}{4}}(1+s)}{(\frac{\gamma^2+\frac{1}{4}}{\gamma^2-\frac{1}{4}} + s)(1-s)}. \tag{2.69}$$

The magnitude of the corresponding frequency response function takes its peak value at the frequency 0, and we find

$$\|H\|_\infty = \frac{\gamma^2}{\gamma^2 + \frac{1}{4}}. \tag{2.70}$$

It may be checked that $\|H\|_\infty \leq \gamma$, with equality only for $\gamma = 0$ and $\gamma = \frac{1}{2}$.

(b) $A = 1$, $B = 1$, so that $U = 1$. After simplification we obtain the compensator

$$K(s) = \frac{\gamma\left(s + \frac{\gamma+\frac{1}{2}-\frac{1}{2\gamma}}{\gamma+\frac{1}{2}}\right)}{s + \frac{\gamma-\frac{1}{2}}{\gamma+\frac{1}{2}}}. \tag{2.71}$$

For this compensator we obtain the closed-loop transfer function

$$H(s) = \gamma \frac{(1+s)(s - \frac{\gamma-\frac{1}{2}}{\gamma+\frac{1}{2}})}{(1-s)(s + \frac{\gamma-\frac{1}{2}}{\gamma+\frac{1}{2}})}. \tag{2.72}$$

The magnitude of the corresponding frequency response function is constant and equal to γ. For this reason, the compensator (2.71) is said to be *equalizing*. It follows that $\|H\|_\infty = \gamma$.

Investigation of (2.67) reveals that for $\gamma \to \frac{1}{2}$ the transfer function of all sublevel compensators approaches $K_o(s) = \frac{1}{2}$. This compensator is (trivially) stable. The resulting closed-loop transfer function is

$$H_o(s) = P(s) - K_o(s) = \frac{1}{2}\frac{1+s}{1-s}, \tag{2.73}$$

so that $\|H_o\|_\infty = \frac{1}{2}$. In example 2.20 we establish that this compensator actually is *optimal*, that is, minimizes $\|H\|_\infty$ with respect to all stable compensators. •

2.4.5 Polynomial formulas

With the polynomial matrix fraction representations (2.8–2.10) it follows that

$$\Pi_\gamma = \begin{bmatrix} -\bar{D}_2 \\ \bar{N}_2 \end{bmatrix} (\bar{N}_1^\sim \bar{N}_1 - \gamma^2 \bar{D}_1^\sim \bar{D}_1)^{-1} [\, -\bar{D}_2^\sim \quad \bar{N}_2^\sim \,], \tag{2.74}$$

$$\Pi_\gamma^{-1} = \begin{bmatrix} N_2^\sim \\ D_2^\sim \end{bmatrix} (D_1 D_1^\sim - \frac{1}{\gamma^2} N_1 N_1^\sim)^{-1} [\, N_2 \quad D_2 \,]. \tag{2.75}$$

The spectral factorization of the rational matrix Π_γ or its inverse may be reduced to two successive polynomial factorizations. Various algorithms for polynomial factorization are described by Kwakernaak and Šebek [23], including one that is based on state space methods.

We summarize the computational steps for the factorization of Π_γ, or, equivalently, Π_γ^{-1}, with $\gamma \geq \gamma_o$.

1. First do the polynomial spectral cofactorization

$$D_1 D_1^\sim - \frac{1}{\gamma^2} N_1 N_1^\sim = Q_\gamma J' Q_\gamma^\sim, \tag{2.76}$$

   such that the square polynomial matrix Q_γ is Hurwitz.

2. Next, do the left-to-right conversion

$$Q_\gamma^{-1} [N_2 \;\; D_2] = \Delta_\gamma \Lambda_\gamma^{-1}. \tag{2.77}$$

   This amounts to the solution of the linear polynomial matrix equation

$$Q \Delta_\gamma = [N_2 \;\; D_2] \Lambda_\gamma \tag{2.78}$$

   for the polynomial matrices Δ_γ and Λ_γ.

3. Finally, do the second polynomial spectral factorization

$$\Delta_\gamma^\sim J' \Delta_\gamma = \Gamma_\gamma^\sim J \Gamma_\gamma, \tag{2.79}$$

   with Γ square and Hurwitz.

The desired factorization then is $\Pi_\gamma^{-1} = M_\gamma^\sim J M_\gamma$, with

$$M_\gamma = Z_\gamma^{-1} = \Gamma_\gamma \Lambda_\gamma^{-1}. \tag{2.80}$$

Example 2.11. **Model matching problem.** We illustrate these steps for the model matching problem of example 2.10. Consider the left matrix fraction representation

$$G(s) = \left[\begin{array}{c|c} \frac{1}{1-s} & -1 \\ \hline 1 & 0 \end{array}\right] = \underbrace{\left[\begin{array}{c|c} 1-s & 0 \\ 0 & 1 \end{array}\right]^{-1}}_{D(s)} \underbrace{\left[\begin{array}{c|c} 1 & -1+s \\ 1 & 0 \end{array}\right]}_{N(s)}. \tag{2.81}$$

1. The polynomial matrix

$$D_1 D_1^{\sim} - \frac{1}{\gamma^2} N_1 N_1^{\sim} = \begin{bmatrix} 1 - \frac{1}{\gamma^2} - s^2 & -\frac{1}{\gamma^2} \\ -\frac{1}{\gamma^2} & -\frac{1}{\gamma^2} \end{bmatrix} \tag{2.82}$$

may be factored as $Q_\gamma(s) J' Q_\gamma^{\sim}(s)$, with

$$Q_\gamma(s) = \begin{bmatrix} 1+s & \frac{1}{\gamma} \\ 0 & \frac{1}{\gamma} \end{bmatrix}, \qquad J' = \begin{bmatrix} 1 & 0 \\ 0 & -1 \end{bmatrix}. \tag{2.83}$$

2. Left-to-right conversion results in

$$Q_\gamma^{-1}(s)[\, N_2(s) \quad D_2(s)\,] = \underbrace{\begin{bmatrix} 1 & -1 \\ -2\gamma & \gamma(1+s) \end{bmatrix}}_{\Delta_\gamma(s)} \underbrace{\begin{bmatrix} 1 & 0 \\ -2 & 1+s \end{bmatrix}^{-1}}_{\Lambda_\gamma(s)}. \tag{2.84}$$

3. It may be verified (see also example 2.18) that for $\gamma \neq \frac{1}{2}$

$$\Delta_\gamma^{\sim}(s) J' \Delta_\gamma(s) = \begin{bmatrix} 1 - 4\gamma^2 & 2\gamma^2(1 - \frac{1}{2\gamma^2} + s) \\ 2\gamma^2(1 - \frac{1}{2\gamma^2} - s) & \gamma^2(-1 + \frac{1}{\gamma^2} + s^2) \end{bmatrix} \tag{2.85}$$

may be factored as $\Gamma_\gamma^{\sim}(s) J \Gamma_\gamma(s)$, with

$$\Gamma_\gamma(s) = \begin{bmatrix} 1 & \frac{\frac{1}{4}}{\gamma^2 - \frac{1}{4}} \\ -2\gamma & \gamma\left(\frac{\gamma^2 - \frac{3}{4}}{\gamma^2 - \frac{1}{4}} + s\right) \end{bmatrix}, \qquad J = \begin{bmatrix} 1 & 0 \\ 0 & -1 \end{bmatrix}. \tag{2.86}$$

For $\gamma \geq 0$ the polynomial matrix Γ_γ is well-defined as long as $\gamma \neq \frac{1}{2}$.

For $\gamma \neq \frac{1}{2}$ the desired factorization hence is $\Pi_\gamma^{-1} = M_\gamma^{\sim} J M_\gamma$, with

$$M_\gamma(s) = \Gamma_\gamma(s) \Lambda_\gamma^{-1}(s) = \begin{bmatrix} \frac{\frac{\gamma^2 + \frac{1}{4}}{\gamma^2 - \frac{1}{4}} + s}{1+s} & \frac{\frac{\frac{1}{4}}{\gamma^2 - \frac{1}{4}}}{1+s} \\ -\frac{\frac{\gamma}{\gamma^2 - \frac{1}{4}}}{1+s} & \frac{\gamma(\frac{\gamma^2 - \frac{3}{4}}{\gamma^2 - \frac{1}{4}} + s)}{1+s} \end{bmatrix}. \tag{2.87}$$

2.5 Canonical spectral factorizations

2.5.1 Definition

In the preceding section a formula for all sublevel solutions of the standard problem is obtained. Actually, we are not interested in *all* sublevel solutions. We only need *stabilizing* solutions, that is, compensators that stabilize the feedback system. To investigate this problem the notion of canonical spectral factorization is needed.

We define canonical factorizations for two distinct situations:

(a) The spectral factorization $\Pi = Z^{\sim} J Z$ of a biproper[4] para-Hermitian ra-

[4] Π is biproper if both Π and its inverse Π^{-1} are proper.

tional matrix Π is *canonical* if the spectral factor Z is biproper.

(b) The spectral factorization $\Pi = Z^{\sim} J Z$ of a diagonally reduced[5] para-Hermitian polynomial matrix Π is canonical if Z is column reduced with column degrees equal to the half diagonal degrees of Π.

A spectral cofactorization $\Pi = ZJZ^{\sim}$ is said to be canonical if the factorization $\Pi(-s) = Z(-s)JZ^{\sim}(-s)$ is canonical.

Part (a) of the definition is standard in the mathematical literature. Part (b) of the definition allows to fit in known results for the spectral factorization of polynomial matrices. If the diagonally reduced para-Hermitian polynomial matrix Π is nonnegative-definite on the imaginary axis then a canonical factorization always exists [3]. In the more general case, where Π is is diagonally reduced but not necessarily nonnegative-definite on the imaginary axis, a column reduced spectral factor does not always exist, that is, a canonical factorization does not always exist.

Example 2.12. **No canonical factorization may exist.** Consider the polynomial matrix

$$\Sigma_\gamma = \begin{bmatrix} 1 - 4\gamma^2 & 2\gamma^2(1 - \frac{1}{2\gamma^2} + s) \\ 2\gamma^2(1 - \frac{1}{2\gamma^2} - s) & \gamma^2(-1 + \frac{1}{\gamma^2} + s^2) \end{bmatrix}, \tag{2.88}$$

which occurs in example 2.11. Σ_γ is diagonally reduced with half diagonal degrees 0 and 1.

For $\gamma \neq 1/2$ the matrix Σ_γ may be be factored as $\Sigma_\gamma = \Gamma_\gamma^{\sim}(s)J\Gamma_\gamma(s)$, with

$$\Gamma_\gamma(s) = \begin{bmatrix} 1 & \frac{\frac{1}{4}}{\gamma^2 - \frac{1}{4}} \\ -2\gamma & \gamma\left(\frac{\gamma^2 - \frac{3}{4}}{\gamma^2 - \frac{1}{4}} + s\right) \end{bmatrix}, \qquad J = \begin{bmatrix} 1 & 0 \\ 0 & -1 \end{bmatrix}. \tag{2.89}$$

For $\gamma \neq 1/2$ the spectral factor Γ_γ of (2.89) is column reduced with column degrees 0 and 1. Hence the factorization is canonical.

For $\gamma = 1/2$ the polynomial matrix

$$\Sigma_{1/2}(s) = \begin{bmatrix} 0 & \frac{-1+s}{2} \\ \frac{-1-s}{2} & \frac{3+s^2}{4} \end{bmatrix} \tag{2.90}$$

does not have a canonical factorization. It does have a (nonunique) noncanonical factorization. One spectral factor is

$$\Gamma_{1/2}(s) = \begin{bmatrix} -\frac{1+s}{2} & \frac{7+s^2}{8} \\ \frac{1+s}{2} & \frac{1-s^2}{8} \end{bmatrix}. \tag{2.91}$$

[5] Define $U(s) = \text{diag}\,(1/s^{n_1}, 1/s^{n_2}, \cdots, 1/s^{n_n}$. The para-Hermitian polynomial matrix Π is diagonally reduced [3] if there exist integers n_1, n_2, $\cdots$, n_n such that $U^{\sim}\Pi U$ is biproper. The numbers n_1, n_2, $\cdots$, n_n are the ***half diagonal degrees*** of Π.

The corresponding spectral factor of $\Pi_{1/2}^{-1}$ is

$$M_{1/2}(s) = \Gamma_{1/2}(s)\Lambda_{1/2}^{-1}(s) = \begin{bmatrix} \frac{(5+s)(1-s)}{4(1+s)} & \frac{7+s^2}{8(1+s)} \\ \frac{3+s}{4} & \frac{1-s}{8} \end{bmatrix}. \tag{2.92}$$

$\Pi_{1/2}^{-1}$ is biproper but the spectral factor is not. The spectral factorization of $\Pi_{1/2}^{-1}$ is not canonical. Actually, no canonical spectral factorization exists.

It appears that the spectral factorizations of both Σ_γ and Π_γ have a singularity at $\gamma = 1/2$. •

2.5.2 Polynomial formulation of the rational factorization

Under the general assumptions of section 2.1 the matrices Π_γ and Π_γ^{-1} introduced in section 2.4 are not necessarily biproper. Because it is not clear how to define canonicity of the spectral factorization of rational matrices that are not biproper we define the canonicity of a spectral factorization $\Pi_\gamma^{-1} = M_\gamma^\sim J M_\gamma$ indirectly by means of the two polynomial factorizations to which the factorization of Π_γ^{-1} may be reduced. To this end, we structure the polynomial algorithm of subsection 2.4.5 for the factorization of

$$\Pi_\gamma^{-1} = \begin{bmatrix} N_2^\sim \\ D_2^\sim \end{bmatrix} (D_1 D_1^\sim - \frac{1}{\gamma^2} N_1 N_1^\sim)^{-1} [N_2 \quad D_2] \tag{2.93}$$

in the following manner:

1. Arrange the left coprime polynomial matrix fraction representation

$$G = [D_1 \quad D_2]^{-1} [N_1 \quad N_2] \tag{2.94}$$

such that $[D_1 \quad N_1]$ is row reduced.

2. Find a polynomial spectral cofactorization — if one exists —

$$D_1 D_1^\sim - \frac{1}{\gamma^2} N_1 N_1^\sim = Q_\gamma J' Q_\gamma^\sim, \tag{2.95}$$

such that Q_γ is row reduced with row degrees equal to the row degrees of $[D_1 \quad N_1]$, that is, the polynomial spectral factorization is canonical.

3. Next, do the left-to-right conversion

$$Q_\gamma^{-1} [N_2 \quad D_2] = \Delta_\gamma \Lambda_\gamma^{-1} \tag{2.96}$$

such that Δ_γ and Λ_γ are right coprime and Δ_γ is column reduced.

4. Finally, determine the polynomial spectral factorization — if one exists —

$$\Delta_\gamma^\sim J' \Delta_\gamma = \Gamma_\gamma^\sim J \Gamma_\gamma, \tag{2.97}$$

such that Γ_γ is column reduced with column degrees equal to the column degrees of Δ_γ, that is, the polynomial factorization is canonical.

The desired factorization then is $\Pi_\gamma^{-1} = M_\gamma^\sim J M_\gamma$, with $M_\gamma = Z_\gamma^{-1} = \Gamma_\gamma \Lambda_\gamma^{-1}$.

If the two canonical polynomial factorizations exist then we say that the factorization of Π_γ^{-1} is canonical.

Example 2.13. **Canonical factorization.** Consider the standard plant

$$G(s) = \left[\begin{array}{c|c} \frac{s^2}{s-1} & 1 \\ \hline 1 & 0 \end{array}\right]. \tag{2.98}$$

By direct calculation it follows that

$$\Pi_\gamma^{-1}(s) = \begin{bmatrix} 1 & -\frac{s^2}{-s+1} \\ -\frac{s^2}{s+1} & -\gamma^2 + \frac{s^4}{-s^2+1} \end{bmatrix}. \tag{2.99}$$

Obviously, Π_γ^{-1} is not biproper. We go through the polynomial algorithm. Letting

$$G(s) = \left[\begin{array}{c|c} s-1 & -s^2 \\ 0 & 1 \end{array}\right]^{-1} \left[\begin{array}{c|c} 0 & s-1 \\ 1 & 0 \end{array}\right] \tag{2.100}$$

we find

$$\begin{aligned} D_1 D_1^\sim - \frac{1}{\gamma^2} N_1 N_1^\sim &= \begin{bmatrix} -s^2-1 & 0 \\ 0 & -1/\gamma^2 \end{bmatrix} \\ &= \underbrace{\begin{bmatrix} s+1 & 0 \\ 0 & 1/\gamma \end{bmatrix}}_{Q(s)} \underbrace{\begin{bmatrix} 1 & 0 \\ 0 & -1 \end{bmatrix}}_{J'} \underbrace{\begin{bmatrix} -s+1 & 0 \\ 0 & 1/\gamma \end{bmatrix}}_{Q^\sim(s)}. \end{aligned} \tag{2.101}$$

This factorization is canonical for any γ. We convert

$$Q^{-1}(s)[N_2 \;\; D_2] = \underbrace{\begin{bmatrix} 1 & -1 \\ -2\gamma & \gamma(s+1) \end{bmatrix}}_{\Delta_\gamma} \underbrace{\begin{bmatrix} -2s-1 & (s+1)^2 \\ -2 & s+1 \end{bmatrix}^{-1}}_{\Lambda_\gamma}. \tag{2.102}$$

It follows that

$$\Delta_\gamma^\sim(s) J' \Delta_\gamma(s) = \begin{bmatrix} 1-4\gamma^2 & 2\gamma^2(1-\frac{1}{2\gamma^2}+s) \\ 2\gamma^2(1-\frac{1}{2\gamma^2}-s) & \gamma^2(-1+\frac{1}{\gamma^2}+s^2) \end{bmatrix}, \tag{2.103}$$

which is precisely the polynomial matrix Σ_γ of (2.88). For $\gamma \neq 1/2$ the matrix Σ_γ may be factored as in (2.89), and we obtain the spectral factor

$$M_\gamma(s) = \Gamma_\gamma(s)\Lambda_\gamma^{-1}(s) = \frac{1}{s+1} \begin{bmatrix} s + \frac{\gamma^2+\frac{1}{4}}{\gamma^2-\frac{1}{4}} & -s^2 - s\frac{2\gamma^2}{\gamma^2-\frac{1}{4}} - \frac{\gamma^2}{\gamma^2-\frac{1}{4}} \\ -\frac{\gamma}{\gamma^2-\frac{1}{4}} & s\frac{\gamma(\gamma^2+\frac{3}{4})}{\gamma^2-\frac{1}{4}} + \frac{\gamma(\gamma^2+\frac{1}{4})}{\gamma^2-\frac{1}{4}} \end{bmatrix}. \tag{2.104}$$

The spectral factor M_γ is not proper. For $\gamma \neq 1/2$ both polynomial factorizations are canonical, so that the factorization $\Pi_\gamma^{-1} = M_\gamma^\sim J M_\gamma$ by definition is also canonical. For $\gamma = 1/2$ no canonical factorization of (2.103) exists, and, hence, no canonical factorization of $\Pi_{1/2}^{-1}$ exists. •

2.5.3 Zeros on the imaginary axis

The polynomial factorization algorithms described by Kwakernaak and Šebek [23], and also the algorithm sketched in section 2.7, fail when the matrices to be factored have roots on the imaginary axis. This occurs

- for $\gamma = \gamma_o$, that is, at the lower bound;
- when $[\,D_1 \;\; N_1\,]$ or $[\,N_2 \;\; D_2\,]$ do not have full rank on the imaginary axis.

The first situation may be resolved by avoiding the value γ_o. To handle the second case the algorithm of subsection 2.5.2 may be modified as follows:

1. First find nonsingular square polynomial matrices $\hat{F}_1$ and F_2 that have all their roots in the closed left-half plane such that

$$[\,D_1 \;\; N_1\,] = \hat{F}_1[\,\check{D}_1 \;\; \check{N}_1\,], \tag{2.105}$$

$$[\,N_2 \;\; D_2\,] = [\,\hat{N}_2 \;\; \hat{D}_2\,]F_2, \tag{2.106}$$

   where $[\check{D}_1 \;\; \check{N}_1]$ is row-reduced and $[\hat{N}_2 \;\; \hat{D}_2]$ column-reduced, and both $[\check{D}_1 \;\; \check{N}_1]$ and $[\hat{N}_2 \;\; \hat{D}_2]$ have full rank everywhere in the open left-half plane. These pre-factorizations remove the offending roots.

2. Next do the right-to-left conversion

$$\hat{F}_1^{-1}[\,\hat{N}_2 \;\; \hat{D}_2\,] = [\,\check{N}_2 \;\; \check{D}_2\,]F_1^{-1}, \tag{2.107}$$

   with $[\,\check{N}_2 \;\; \check{D}_2\,]$ column reduced.

3. Finally, apply the factorization algorithm as outlined in subsection 2.5.2 to

$$\check{\Pi}_\gamma^{-1} = \begin{bmatrix} \check{N}_2^\sim \\ \check{D}_2^\sim \end{bmatrix} (\check{D}_1\check{D}_1^\sim - \frac{1}{\gamma^2}\check{N}_1\check{N}_1^\sim)^{-1}[\check{N}_2 \;\; \check{D}_2] \tag{2.108}$$

   to obtain the factorization $\check{\Pi}_\gamma^{-1} = \check{M}_\gamma^\sim J\check{M}_\gamma$.

4. Then $\Pi_\gamma^{-1} = M_\gamma^\sim JM_\gamma$ with $M_\gamma = \check{M}_\gamma F_1^{-1}F_2$.

Example 2.14. **Zeros on the imaginary axis.** Zeros on the imaginary axis typically arise in the mixed sensitivity problem when designing for integral control [27]. One way to achieve integral control is to let the weighting function V have a pole at the origin. This causes the second factorization in the algorithm of subsection 2.5.2 to have a root at the origin.

By way of example we discuss the two-input two-output MIMO mixed sensitivity problem introduced in [16]. The plant P, the shaping filter V and the weighting filters W_1 and W_2 are given by

$$P(s) = \begin{bmatrix} \frac{1}{s^2} & \frac{1}{2+s} \\ 0 & \frac{1}{2+s} \end{bmatrix}, \qquad V(s) = \begin{bmatrix} \frac{1+s\sqrt{2}+s^2}{s^2} & 0 \\ 0 & \frac{1+s}{s} \end{bmatrix}, \tag{2.109}$$

$$W_1(s) = \begin{bmatrix} 1 & 0 \\ 0 & 1 \end{bmatrix}, \qquad W_2(s) = \begin{bmatrix} c_1(1+rs) & 0 \\ 0 & c_2 s \end{bmatrix}. \tag{2.110}$$

The choice of the weighting filters is explained in [16]. The factor s in the denominator of the (2, 2) entry of V serves to introduce integrating action in the second input channel of the feedback system.

The generalized plant may be represented in left coprime matrix fraction form as

$$\left[\begin{array}{c|c} W_1(s)V(s) & W_1(s)P(s) \\ 0 & W_2(s) \\ \hline -V(s) & -P(s) \end{array}\right] = \left[\begin{array}{cc|cc} \frac{1+s\sqrt{2}+s^2}{s^2} & 0 & \frac{1}{s^2} & \frac{1}{2+s} \\ 0 & \frac{1+s}{s} & 0 & \frac{1}{2+s} \\ 0 & 0 & c_1(1+rs) & 0 \\ 0 & 0 & 0 & c_2 s \\ \hline -\frac{1+s\sqrt{2}+s^2}{s^2} & 0 & -\frac{1}{s^2} & -\frac{1}{2+s} \\ 0 & -\frac{1+s}{s} & 0 & -\frac{1}{2+s} \end{array}\right]$$

$$= D^{-1}(s)N(s), \tag{2.111}$$

with

$$D(s) = \left[\begin{array}{cccc|cc} 1 & 0 & 0 & 0 & 1 & 0 \\ 0 & 1 & 0 & 0 & 0 & 1 \\ 0 & 0 & 1 & 0 & 0 & 0 \\ 0 & 0 & 0 & 1 & 0 & 0 \\ 0 & 0 & 0 & 0 & s^2 & -s^2 \\ 0 & 0 & 0 & 0 & 0 & s(2+s) \end{array}\right], \tag{2.112}$$

$$N(s) = \left[\begin{array}{cc|cc} 0 & 0 & 0 & 0 \\ 0 & 0 & 0 & 0 \\ 0 & 0 & c_1(1+rs) & 0 \\ 0 & 0 & 0 & c_2 s \\ -1-s\sqrt{2}-s^2 & s(1+s) & -1 & 0 \\ 0 & -(2+s)(1+s) & 0 & -s \end{array}\right]. \tag{2.113}$$

Correspondingly, we have

$$[\,D_1(s) \;\; N_1(s)\,] = \begin{bmatrix} 1\,0\,0\,0 & 0 & 0 \\ 0\,1\,0\,0 & 0 & 0 \\ 0\,0\,1\,0 & 0 & 0 \\ 0\,0\,0\,1 & 0 & 0 \\ 0\,0\,0\,0 & -1-s\sqrt{2}-s^2 & s(1+s) \\ 0\,0\,0\,0 & 0 & -(2+s)(1+s) \end{bmatrix} \tag{2.114}$$

$$[\,N_2(s) \;\; D_2(s)\,] = \left[\begin{array}{cc|cc} 0 & 0 & 1 & 0 \\ 0 & 0 & 0 & 1 \\ c_1(1+rs) & 0 & 0 & 0 \\ 0 & c_2 s & 0 & 0 \\ -1 & 0 & s^2 & -s^2 \\ 0 & -s & 0 & s(2+s) \end{array}\right] \tag{2.115}$$

$[\,D_1 \;\; N_1\,]$ has full rank on the imaginary axis, but $[\,N_2(s) \;\; D_2(s)\,]$ loses rank at $s = 0$. It follows that $\hat{F}_1 = I$, $\check{D}_1 = D_1$ and $\check{N}_1 = N_1$. We may let

$$[\,N_2(s) \;\; D_2(s)\,] = \underbrace{\left[\begin{array}{cc|cc} 0 & 0 & 1 & 0 \\ 0 & 0 & 0 & 1 \\ c_1\,(1+rs) & 0 & 0 & 0 \\ 0 & c_2 & 0 & 0 \\ -1 & 0 & s^2 & -s^2 \\ 0 & -1 & 0 & s\,(2+s) \end{array}\right]}_{[\,\hat{N}_2(s) \;\; \hat{D}_2(s)\,]} \underbrace{\begin{bmatrix} 1\,0\,0\,0 \\ 0\,s\,0\,0 \\ 0\,0\,1\,0 \\ 0\,0\,0\,1 \end{bmatrix}}_{F_2(s)} . \tag{2.116}$$

Because $\hat{F}_1 = I$ we have $\check{N}_2 = \hat{N}_2$, $\check{D}_2 = \hat{D}_2$ and $F_1 = I$. The factorization of

$$\check{\Pi}_\gamma^{-1} = \begin{bmatrix} \check{N}_2^\sim \\ \check{D}_2^\sim \end{bmatrix} \left(\check{D}_1 \check{D}_1^\sim - \frac{1}{\gamma^2} \check{N}_1 \check{N}_1^\sim \right)^{-1} [\,\check{N}_2 \; \check{D}_2\,] \tag{2.117}$$

no longer is handicapped by zeros on the imaginary axis.

We continue the discussion of this problem in example 2.21. •

2.6 Stability

2.6.1 Introduction

The sublevel solutions of section 2.5 are not necessarily stable. In the present section we characterize all stable sublevel solutions. This characterization leads to a well-known search procedure for optimal solutions. It turns out that there exist two types of optimal solutions.

2.6.2 All stabilizing sublevel compensators

We consider the question under which conditions a sublevel compensator stabilizes the feedback system.

Theorem 2.7. Closed-loop stability. *Suppose that there exists a compensator that makes $\|H\|_\infty$ finite, and that for a given $\gamma > \gamma_o$ the factorization $\Pi_\gamma^{-1} = M_\gamma^\sim J M_\gamma$ is canonical. Consider all sublevel compensators given by*

$$[\,X \;\; Y\,] = [\,A \;\; B\,] M_\gamma \tag{2.118}$$

with A nonsingular stable and B stable such that $AA^\sim \geq BB^\sim$ on the imaginary axis.

(a) Necessary for any sublevel compensator to be stabilizing is that A be strictly Hurwitz.

(b) All sublevel compensators such that A is strictly Hurwitz are stabilizing if and only if any one of them (in particular, the central solution) is stabilizing.

The proof of this theorem involves a number of technicalitites, and is given in Appendix 2.12.

The result is quite satisfactory. It shows that a necessary condition for a compensator to be stabilizing is that A be strictly Hurwitz. If this condition is satisfied and *any* sublevel compensator stabilizes the closed-loop system then *all* sublevel compensators stabilize the system. Otherwise, no sublevel compensator stabilizes the system.

Example 2.15. **Model matching problem.** Again we consider the model matching problem of example 2.10. We know that for a given $\gamma > 0$ with $\gamma \neq \frac{1}{2}$ all sublevel compensators are given by

$$K(s) = \frac{\frac{\frac{1}{4}}{\gamma^2-\frac{1}{4}} + \gamma\left(\frac{\gamma^2-\frac{3}{4}}{\gamma^2-\frac{1}{4}} + s\right)U(s)}{\left(\frac{\gamma^2+\frac{1}{4}}{\gamma^2-\frac{1}{4}} + s\right) - \frac{\gamma}{\gamma^2-\frac{1}{4}}U(s)}. \tag{2.119}$$

$U = B/A$ with $A^{\sim}A \geq B^{\sim}B$ on the imaginary axis. The central compensator follows by setting $U = 0$ and is given by

$$K(s) = \frac{\frac{\frac{1}{4}}{\gamma^2-\frac{1}{4}}}{\frac{\gamma^2+\frac{1}{4}}{\gamma^2-\frac{1}{4}} + s}. \tag{2.120}$$

For $\gamma > \frac{1}{2}$ the central compensator is stable, while for $0 \leq \gamma < \frac{1}{2}$ it is unstable.

For $\gamma \neq \frac{1}{2}$ the factorization is canonical. Hence, by theorem 2.7 *all* sublevel compensators as given by (2.119) are stable, provided U is stable. One of the sublevel compensators besides the central solution is the equalizing compensator

$$K(s) = \frac{\gamma\left(s + \frac{\gamma+\frac{1}{2}-\frac{1}{2\gamma}}{\gamma+\frac{1}{2}}\right)}{s + \frac{\gamma-\frac{1}{2}}{\gamma+\frac{1}{2}}}. \tag{2.121}$$

Inspection confirms that also this compensator is stable for $\gamma > \frac{1}{2}$ and unstable for $0 \leq \gamma < \frac{1}{2}$. ●

2.6.3 Search procedure – Type A and Type B optimal solutions

Obviously, if a stabilizing sublevel exists for any γ then one exists for γ sufficiently large. Two instances may be distinguished:

(a) *Type A solution.* For any γ greater than or equal to the lower bound γ_o stabilizing compensators exist. In this case, the minimum of $\|H\|_\infty$ is γ_o and any stabilizing compensator that achieves $\|H\|_\infty = \gamma_o$ is optimal.

(b) *Type B solution.* There exists a $\gamma_{\text{opt}} > \gamma_o$ such that no stabilizing sublevel compensators exist for $\gamma < \gamma_{\text{opt}}$. In this case, $\|H\|_\infty = \gamma_{\text{opt}}$.

In case (b), the value of $\gamma_{\rm opt}$ may be obtained by a binary search procedure:

1. Choose $\overline{\gamma}$ such that for $\gamma = \overline{\gamma}$ stabilizing compensators exist, and let $\underline{\gamma} := \gamma_o$.
2. Set
$$\gamma = \frac{\overline{\gamma} + \underline{\gamma}}{2} \tag{2.122}$$
and compute the corresponding "central" compensator.
3. If this compensator is stabilizing then let $\overline{\gamma} := \gamma$; otherwise, let $\underline{\gamma} := \gamma$.
4. Return to step 2.

The number $\overline{\gamma}$ is the smallest known value of γ for which a stabilizing compensator exists, while $\underline{\gamma}$ is the largest known value for which no stabilizing compensator exists. The procedure is terminated when γ is sufficiently close to $\gamma_{\rm opt}$. In sections 2.7–2.8 it is discussed how this may be determined.

In the case of a type B solution something peculiar happens: The factorization of $\Pi_{\gamma_{\rm opt}}$ is noncanonical.

We give examples of both a type A and a type B solution.

Example 2.16. **Model matching problem with stable "plant."** Consider the model matching problem of example 2.5. From subsection 2.4.1 we know that
$$\Pi_\gamma = \begin{bmatrix} \gamma^2 I - PP^\sim & P \\ P^\sim & -I \end{bmatrix}. \tag{2.123}$$

If P is stable (that is, has no poles in the closed right-half complex plane) then Π_γ has the spectral factorization
$$\Pi_\gamma = \underbrace{\begin{bmatrix} \gamma I & P \\ 0 & -I \end{bmatrix}}_{Z_\gamma} \begin{bmatrix} I & 0 \\ 0 & -I \end{bmatrix} \begin{bmatrix} \gamma I & 0 \\ P^\sim & -I \end{bmatrix}. \tag{2.124}$$

It follows that
$$M_\gamma = Z_\gamma^{-1} = \begin{bmatrix} \frac{1}{\gamma} I & \frac{1}{\gamma} P \\ 0 & -I \end{bmatrix}. \tag{2.125}$$

If P is proper then the factorization is canonical for all $\gamma > 0$. All sublevel compensators are characterized by
$$[X_\gamma \;\; Y_\gamma] = [A \;\; B] M_\gamma = [\frac{1}{\gamma} A \;\; \frac{1}{\gamma} AP - B], \tag{2.126}$$

with A nonsingular stable such that A^{-1} is stable and $A \geq B$ on the imaginary axis. The sublevel compensators themselves are given by
$$K_\gamma = X_\gamma^{-1} Y_\gamma = P - \gamma A^{-1} B = P - \gamma U, \tag{2.127}$$

where U is stable such that $\|U\|_\infty = 1$.

The result is unsurprising. As $\gamma \downarrow 0$ the sublevel compensators remain stable. Hence, this is an example of a type A solution, where the optimal solution is obtained for $\gamma = \gamma_o = 0$. The optimal solution — obviously — is $K_{\text{opt}} = P$. •

Example 2.17. **Model matching problem.** Example 2.15 shows that the model matching problem example under consideration is of type B. The lower bound is $\gamma_o = 0$, while $\gamma_{\text{opt}} = \frac{1}{2}$. As γ approaches γ_{opt} several of the coefficients of the spectral factor M_γ and also of the sublevel compensators tend to ∞. Inspection of (2.119) shows that in the limit $\gamma \to \frac{1}{2}$ all sublevel compensators approach

$$K_{1/2}(s) = \frac{1}{2}. \tag{2.128}$$

Clearly, this is the optimal compensator. •

2.7 Factorization algorithm

2.7.1 Introduction

The polynomial algorithm of subsection 2.5.2 requires the spectral factorization of two polynomial matrices. We first need a spectral cofactorization $D_1 D_1^{\sim} - \frac{1}{\gamma^2} N_1 N_1^{\sim} = Q_\gamma J' Q_\gamma^{\sim}$, with Q_γ row-reduced with row degrees equal to the row degrees of $[D_1 \; N_1]$ (which is row-reduced). Additionally we require a factorization $\Delta_\gamma^{\sim} J' \Delta_\gamma = \Gamma_\gamma^{\sim} J \Gamma_\gamma$ such that Γ_γ is column reduced with the same column degrees as Δ_γ (which is column-reduced).

Kwakernaak and Šebek [23] treat several algorithms for the spectral factorization of polynomial matrices, based on a number of methods. In the present paper we outline a state space factorization algorithm that is simpler than that presented in [23]. We include it to clarify the complications that arise when the factorization is noncanoncial or close to noncanonical.

Let $\Phi = P^{\sim} J' P$, with J' a signature matrix and P a column-reduced polynomial matrix. Assume that Φ is nonsingular and has no roots on the imaginary axis. We consider the problem of finding a square Hurwitz polynomial Q which is column-reduced with the same column degrees as P such that

$$\Phi = Q^{\sim} J Q, \tag{2.129}$$

with J a signature matrix.

Once a suitable algorithm for finding the spectral factorization of a polynomial matrix Φ is available, application to Φ^{T} and transposition of the resulting spectral factor yields a cofactorization of Φ.

2.7.2 State space algorithm

Under the assumptions stated the rational matrix Φ^{-1} is proper and, hence, has a minimal state space realization

$$\Phi^{-1}(s) = C(sI - A)^{-1} B + D \tag{2.130}$$

corresponding to the state space system

$$\begin{aligned} \dot{x} &= Ax + Bu, \\ y &= Cx + Du. \end{aligned} \tag{2.131}$$

The eigenvalues of the matrix A are the roots of the polynomial matrix Φ and, hence, occur in pairs that are symmetric with respect to the imaginary axis. By the assumption that Φ has no roots on the imaginary axis A has no eigenvalues on the imaginary axis.

By a suitable ordered Schur transformation [11] A may be rendered as

$$A = U^{\mathrm{H}} \begin{bmatrix} A_{11} & A_{12} \\ 0 & A_{22} \end{bmatrix} U, \tag{2.132}$$

with U a unitary matrix, A_{11} a square matrix with all its eigenvalues in the open left-half plane and A_{22} a square matrix (of the same dimensions as A_{11}) with all its eigenvalues in the open righ-half plane. Redefining $x := Ux$, the state space system (2.131) is transformed to

$$\begin{aligned} \dot{x} &= \begin{bmatrix} A_{11} & A_{12} \\ 0 & A_{22} \end{bmatrix} x + \begin{bmatrix} B_1 \\ B_2 \end{bmatrix} u, \\ y &= \begin{bmatrix} C_1 & C_2 \end{bmatrix} x + Du, \end{aligned} \tag{2.133}$$

where

$$\begin{bmatrix} B_1 \\ B_2 \end{bmatrix} = UB, \quad \begin{bmatrix} C_1 & C_2 \end{bmatrix} = CU^{\mathrm{H}}. \tag{2.134}$$

With (2.133) we may write

$$\begin{aligned} \Phi^{-1}(s) &= \begin{bmatrix} C_1 & C_2 \end{bmatrix} \begin{bmatrix} sI - A_{11} & -A_{12} \\ 0 & sI - A_{22} \end{bmatrix}^{-1} \begin{bmatrix} B_1 \\ B_2 \end{bmatrix} + D \\ &= C_1(sI - A_{11})^{-1}B_1 \\ &\quad + C_1(sI - A_{11})^{-1}A_{12}(sI - A_{22})^{-1}B_2 \\ &\quad + C_2(sI - A_{22})^{-1}B_2 + D. \end{aligned} \tag{2.135}$$

We convert

$$C_1(sI - A_{11})^{-1} = R^{-1}(s)E(s), \tag{2.136}$$

with R a square polynomial matrix with column degrees equal to those of P, and E another polynomial matrix.

By inspection of (2.135) we see that the rational matrix Φ^{-1} has a left factor R^{-1}. Since the roots of R are the left-half plane roots of Φ and R has the desired column degrees we have

$$R(s)\Phi^{-1}(s)R^{\sim}(s) = L, \tag{2.137}$$

with L a constant Hermitian matrix. Given the polynomial matrices Φ and R the constant matrix L follows from

$$L = R(0)\Phi^{-1}(0)R^{\sim}(0). \tag{2.138}$$

If R is nonsingular then L is nonsingular and

$$\Phi(s) = R^{\sim}(s)L^{-1}R(s). \tag{2.139}$$

The desired spectral factorization follows by the decomposition $L^{-1} = V^{\mathrm{H}}JV$, with V a constant matrix, which is easily obtained from a Schur decomposition of L^{-1}. We then have

$$\Phi(s) = Q^{\sim}(s)JQ(s), \tag{2.140}$$

with $Q(s) = VR(s)$.

Theorem 2.8. State space algorithm for the factorization of a para-Hermitian polynomial matrix. *Let $\Phi = P^{\sim}J'P$, with J' a signature matrix and P a column-reduced polynomial matrix. Suppose that Φ is nonsingular and has no roots on the imaginary axis.*

1. *Find a minimal state space realization $\dot{x} = Ax + Bu$, $y = Cx + Bu$ of the system with transfer matrix $\Phi^{-1}(s)$.*

2. *Determine an ordered Schur transformation such that*

$$A = U^{\mathrm{H}} \begin{bmatrix} A_{11} & A_{12} \\ 0 & A_{22} \end{bmatrix} U, \tag{2.141}$$

such that A_{11} has all its roots in the open left-half plane and A_{22} has all its roots in the open right-half plane. Define

$$\begin{bmatrix} C_1 & C_2 \end{bmatrix} = CU^{\mathrm{H}}. \tag{2.142}$$

3. *Convert*

$$C_1(sI - A_{11})^{-1} = R^{-1}(s)E(s), \tag{2.143}$$

with R a square polynomial matrix whose column degrees equal those of P, and E another polynomial matrix.

4. *Then*

$$R(s)\Phi^{-1}(s)R^{\sim}(s) = L, \tag{2.144}$$

with $L = R(0)\Phi^{-1}(0)R^{\sim}(0)$ a constant Hermitian matrix.

5. *If R is nonsingular then L is nonsingular and*

$$\Phi(s) = R^{\sim}(s)L^{-1}R(s). \tag{2.145}$$

The desired spectral factorization

$$\Phi(s) = Q^{\sim}(s)JQ(s), \tag{2.146}$$

with $Q(s) = VR(s)$, follows from the decomposition $L^{-1} = V^{\mathrm{H}}JV$. The constant matrix V is easily obtained from the Schur decomposition of L^{-1}.

Various details needed for the execution of this algorithm may be found in [22].

Example 2.18. **Spectral factorization of a polynomial matrix.** We apply the algorithm to the polynomial matrix

$$\Phi(s) = \begin{bmatrix} 1-4\gamma^2 & 2\gamma^2-1+2\gamma^2 s \\ 2\gamma^2-1-2\gamma^2 s & 1-\gamma^2+\gamma^2 \end{bmatrix}, \tag{2.147}$$

which appears in example 2.11. We cannot immediately apply the Kailath's results [13] to find an observable realization of $\Phi^{-1}(d/dt)y(t) = u(t)$ because Φ is not row-reduced. By first transforming Φ to row-reduced form the minimal realization $\dot{x} = Ax + Bu$, $y = Cx + Du$, may be found, with

$$A = \begin{bmatrix} \frac{2\gamma^2-1}{2\gamma^2} & \frac{1}{2\gamma^2} \\ \frac{4\gamma^2-1}{2\gamma^2} & \frac{1-2\gamma^2}{2\gamma^2} \end{bmatrix}, \quad B = \begin{bmatrix} 0 & -2 \\ 2 & 0 \end{bmatrix}, \quad C = \begin{bmatrix} 1 & 0 \\ 0 & 1 \end{bmatrix}, \quad D = \begin{bmatrix} 1 & 0 \\ 0 & 0 \end{bmatrix}. \tag{2.148}$$

A has the eigenvalues -1 and 1. Rather than using ordered Schur transformation we bring the system in modal form to obtain A in upper triangular form as in (2.132). We find

$$A = \begin{bmatrix} -1 & 0 \\ 0 & 1 \end{bmatrix}, \quad B = \begin{bmatrix} -\frac{1}{2\gamma^2} & -\frac{1}{2\gamma^2} \\ \frac{1}{2\gamma^2} & \frac{1-4\gamma^2}{2\gamma^2} \end{bmatrix}, \quad C = \begin{bmatrix} 1 & 1 \\ 1-4\gamma^2 & 1 \end{bmatrix}, \quad D = \begin{bmatrix} 1 & 0 \\ 0 & 0 \end{bmatrix}. \tag{2.149}$$

We next consider the conversion (2.143). R needs to have column degrees 0 and 1, respectively. It may be found that

$$C_1(sI - A_{11})^{-1} = \begin{bmatrix} 1 \\ 1-4\gamma^2 \end{bmatrix} \frac{1}{s+1} = \underbrace{\begin{bmatrix} 1 & \frac{1}{4\gamma^2-1} \\ 0 & s+1 \end{bmatrix}^{-1}}_{R(s)} \underbrace{\begin{bmatrix} 0 \\ 1-4\gamma^2 \end{bmatrix}}_{E(s)}. \tag{2.150}$$

For $\gamma \neq \frac{1}{2}$ the polynomial matrix R is well-defined. L may then be computed as

$$L = R(0)\Phi^{-1}(0)R^{\sim}(0) = \begin{bmatrix} 1 & 2 \\ 2 & \frac{4\gamma^2-1}{2\gamma^2} \end{bmatrix}. \tag{2.151}$$

L^{-1} may be diagonalized as

$$L^{-1} = \underbrace{\begin{bmatrix} 1 & -2\gamma \\ 0 & \gamma \end{bmatrix}}_{V^{\mathrm{T}}} \underbrace{\begin{bmatrix} 1 & 0 \\ 0 & -1 \end{bmatrix}}_{J} \underbrace{\begin{bmatrix} 1 & 0 \\ -2\gamma & \gamma \end{bmatrix}}_{V}. \tag{2.152}$$

It follows that for $\gamma \neq \frac{1}{2}$ the desired spectral factor is

$$P(s) = VR(s) = \begin{bmatrix} 1 & \frac{\frac{1}{4}}{\gamma^2-\frac{1}{4}} \\ -2\gamma & \gamma\left(s + \frac{\gamma^2-\frac{3}{4}}{\gamma^2-\frac{1}{4}}\right) \end{bmatrix}. \tag{2.153}$$

•

2.7.3 Noncanonical factorizations

A diagonally reduced para-Hermitian polynomial matrix Φ may have no canonical factorization, that is, it may have no spectral factor whose columns have the required degrees. In the algorithm of theorem 2.8 this deficiency manifests itself by the fact that the polynomial matrix R obtained in step 3 is singular. As a result, also the matrix L is singular and the expressions (2.145) and (2.146) are not defined. The expression (2.144)

$$R(s)\Phi^{-1}(s)R^{\sim}(s) = L \tag{2.154}$$

is still appropriate, however.

In section 2.8 we show how (2.154) may be used to characterize optimal solutions of type B problems. It is useful to arrange (2.154) such that all coefficients of R are finite and L is diagonal and finite with its diagonal entries in order of increasing value. Also other factorization algorithms [23] may be used to arrange the factorization in this way.

Example 2.19. **Noncanonical factorization.** For $\gamma = \frac{1}{2}$ the polynomial matrix Φ of example 2.18 has no canonical factorization. By multiplying the first row of both R and E as obtained in (2.150) by $4\gamma^2 - 1$ we modify R to

$$R(s) = \begin{bmatrix} 4\gamma^2 - 1 & 1 \\ 0 & s+1 \end{bmatrix}. \tag{2.155}$$

Accordingly, L needs to be changed to

$$L = \begin{bmatrix} (1-4\gamma^2)^2 & 2(1-4\gamma^2) \\ 2(1-4\gamma^2) & \frac{4\gamma^2-1}{\gamma^2} \end{bmatrix}. \tag{2.156}$$

Both R and L are singular at $\gamma = \frac{1}{2}$. L may be diagonalized as

$$L = \begin{bmatrix} 1 & -2\gamma^2 \\ 0 & 1 \end{bmatrix} \begin{bmatrix} (1-4\gamma^2)(1+4\gamma^2) & 0 \\ 0 & \frac{4\gamma^2-1}{\gamma^2} \end{bmatrix} \begin{bmatrix} 1 & 0 \\ -2\gamma^2 & 1 \end{bmatrix}. \tag{2.157}$$

Accordingly, we redefine L and R as

$$L_\gamma := \begin{bmatrix} (1-4\gamma^2)(1+4\gamma^2) & 0 \\ 0 & \frac{4\gamma^2-1}{\gamma^2} \end{bmatrix}, \tag{2.158}$$

$$R_\gamma(s) := \begin{bmatrix} 1 & -2\gamma^2 \\ 0 & 1 \end{bmatrix}^{-1} R(s) = \begin{bmatrix} 4\gamma^2 - 1 & 1 + 2\gamma^2 + 2\gamma^2 s \\ 0 & s+1 \end{bmatrix}. \tag{2.159}$$

•

2.8 Optimal solutions

2.8.1 Introduction

In this section we discuss optimal solutions of type B.

For $\gamma > \gamma_o$ sublevel solutions corresponding to canonical factorizations of the rational matrix Π_γ cannot be optimal. The reason is that among the sublevel solutions there is always one that is *strictly* sublevel. Hence, optimal solutions, if they exist, correspond to values of γ such that the factorization is noncanonical.

A reasonable conjecture is that — possibly under mild assumptions — noncanonical factorizations of Π_γ occur at a finite number of values of γ only. In the general model matching problem of example 2.5 problem for instance, these values are the Hankel singular values of the plant [19].

If Π_γ has no canonical factorization then one or both of the polynomial factorizations in the algorithm of subsection 2.5.2 is noncanonical. Suppose that at $\gamma_{\rm opt}$ the factorization of Π_γ is noncanonical because the second polynomial factorization (2.97) is noncanonical, while the first factorization (2.95) is canonical.

No theory is available for the situation that both factorizations are noncanonical. In the dual problem, obtained by transposing the transfer matrix G of the generalized plant, the roles of the factorizations interchange. Hence, if the first factorization turns out to be noncanonical then switching to the dual problem may be a way out.

2.8.2 All optimal compensators

Under the assumption that the first factorization is canonical but the second is not, consider finding compensators $K_\gamma = X_\gamma^{-1} Y_\gamma$ that satisfy the inequality

$$[\, X_\gamma \;\; Y_\gamma \,]\, \Pi_\gamma \begin{bmatrix} X_\gamma^\sim \\ Y_\gamma^\sim \end{bmatrix} \geq 0 \quad \text{on } j\mathsf{R} \tag{2.160}$$

at $\gamma = \gamma_{\rm opt}$. With (2.95) and (2.96) we have $\Pi_\gamma = \Lambda_\gamma Z_\gamma^{-1} \Lambda_\gamma^\sim$, where $Z_\gamma = \Delta_\gamma^\sim J' \Delta_\gamma$. The inequality (2.160) is satisfied if

$$[\, X \;\; Y \,]\, \Lambda_\gamma Z_\gamma^{-1} \Lambda_\gamma^\sim \begin{bmatrix} X^\sim \\ Y^\sim \end{bmatrix} \geq 0 \quad \text{on } j\mathsf{R}. \tag{2.161}$$

Arrange the spectral factorization of Z_γ according to subsection 2.7.3 as

$$\Gamma_\gamma Z_\gamma^{-1} \Gamma_\gamma = L_\gamma \tag{2.162}$$

with the coefficients of Γ_γ normalized in a suitable sense and L_γ diagonal with the diagonal entries arranged in ascending order. Then we may write

$$L_\gamma = \text{diag}\,(L_{\gamma,1},\; -L_{\gamma,2}), \tag{2.163}$$

with $L_{\gamma,1}$ and $L_{\gamma,2}$ diagonal with nonnegative entries, and the partitioning consistent with the natural partitioning for $\gamma \neq \gamma_{\rm opt}$. Let

$$[\, X_\gamma \;\; Y_\gamma \,]\, \Lambda_\gamma = [\, A \;\; B \,]\, \Gamma_\gamma, \tag{2.164}$$

with A and B stable rational matrices to be determined. A is square. Then (2.161) holds if

$$AL_{\gamma,1}A^{\sim} \geq BL_{\gamma,2}B^{\sim} \quad \text{on } j\mathbb{R}. \tag{2.165}$$

The simplest way to accomplish this is to let $A = I$ and $B = 0$.

Equations (2.164–2.165) characterize all suboptimal stabilizing compensators, provided A and B are stable with A strictly Hurwitz.

If the factorization of Π_γ is canonical then Γ_γ is nonsingular and the characterization (2.164–2.165) is trivially equivalent to (2.61). The difference is that (2.164–2.165) also hold at $\gamma = \gamma_{\text{opt}}$.

At a type B optimum several peculiarities occur:

1. Because the factorization is noncanonical, both L_γ and Γ_γ are singular at the optimum.

   L_γ manifests this by the fact that $L_{\gamma,1}$ and $L_{\gamma,2}$ have one or several zero diagonal entries. By conservation of inertia $L_{\gamma,1}$ and $L_{\gamma,2}$ have the same numbers of zero diagonal entries. Generically, $L_{\gamma,1}$ and $L_{\gamma,2}$ each have one zero diagonal entry.

   Because of the singularity of L_γ at the optimum there is an associated loss of dimensionality in the set of all A and B that satisfy (2.165).

2. There can be no strictly sublevel solutions at the optimum. All optimal compensators satisfy $\|H\|_\infty = \gamma_{\text{opt}}$. All optimal compensators have the property that the largest singular value of H is constant on the imaginary axis. This is called the *equalizing property* and is proved in Appendix 2.13.

3. At the optimum one or several closed-loop poles "cross over the imaginary axis" from one half of the complex plane to the other. Because the closed-loop transfer matrix H cannot have any poles on the imaginary axis — otherwise its ∞-norm would be ∞ — the poles that cross over necessarily cancel. They cancel within the compensator transfer matrix $K_\gamma = X_\gamma^{-1}Y_\gamma$.

   The canceling poles may be removed by making $[\,A \quad B\,]\Gamma_\gamma$ on the right-hand side of (2.164) left prime after having chosen A and B. The cancellation results in order reduction of the compensator, generically by 1. Because of the indeterminacy introduced in the factorization by the zero diagonal entries of L_γ these canceling poles actually may not lie on the imaginary axis.

2.8.3 Examples

We discuss two examples.

Example 2.20. **Model matching problem.** We conclude the sequence of examples dealing with the model matching problem of example 2.5 by determining the optimal solution. According to example 2.11 we have

$$\Lambda_\gamma(s) = \begin{bmatrix} 1 & 0 \\ -2 & 1+s \end{bmatrix}. \tag{2.166}$$

From example 2.19 we have the modified spectral factorization

$$\Delta_\gamma^\sim J' \Delta_\gamma = \Gamma_\gamma^\sim L_\gamma^{-1} \Gamma_\gamma, \tag{2.167}$$

with

$$L_\gamma = \begin{bmatrix} (1-4\gamma^2)(1+4\gamma^2) & 0 \\ 0 & \frac{4\gamma^2-1}{\gamma^2} \end{bmatrix}, \tag{2.168}$$

$$\Gamma_\gamma(s) = \begin{bmatrix} 4\gamma^2 - 1 & 1 + 2\gamma^2 + 2\gamma^2 s \\ 0 & s+1 \end{bmatrix}. \tag{2.169}$$

For $\gamma = \frac{1}{2}$ the factorization is noncanonical, and

$$\Lambda_{\frac{1}{2}}(s) = \begin{bmatrix} 1 & 0 \\ -2 & 1+s \end{bmatrix}, \quad L_{\frac{1}{2}} = 0, \quad \Gamma_{\frac{1}{2}}(s) = \begin{bmatrix} 0 & \frac{3}{2} + \frac{1}{2}s \\ 0 & s+1 \end{bmatrix}. \tag{2.170}$$

We use (2.164-2.165) to determine all optimal compensators. Because $L_{\frac{1}{2}} = 0$ the relation (2.165) imposes no constraints on A and B. The right-hand side of (2.164) yields

$$[A \;\; B]\Gamma_{\frac{1}{2}}(s) = \begin{bmatrix} 0 & A(\frac{3}{2} + \frac{1}{2}s) + B(s+1) \end{bmatrix}. \tag{2.171}$$

By canceling the common left factor we may simplify the right-hand side to $[0 \;\; 1]$. The compensator now follows from

$$[X_{\frac{1}{2}}(s) \;\; Y_{\frac{1}{2}}(s)] \underbrace{\begin{bmatrix} 1 & 0 \\ -2 & 1+s \end{bmatrix}}_{\Lambda_{\frac{1}{2}}(s)} = [0 \;\; 1], \tag{2.172}$$

or

$$[X_{\frac{1}{2}}(s) \;\; Y_{\frac{1}{2}}(s)] = [0 \;\; 1] \begin{bmatrix} 1 & 0 \\ -2 & 1+s \end{bmatrix}^{-1} = \frac{1}{1+s}[2 \;\; 1]. \tag{2.173}$$

It follows that

$$K_{\text{opt}}(s) = \frac{Y_{\frac{1}{2}}(s)}{X_{\frac{1}{2}}(s)} = \frac{1}{2}. \tag{2.174}$$

Note that for $\gamma \neq \frac{1}{2}$ the set of all sublevel compensators has dimensionality 1, but that there is a single optimal compensator. •

Example 2.21. **MIMO mixed sensitivity problem.** As a second example we continue the MIMO mixed sensitivity problem of example 2.14. We go through the steps of the algorithm with slight modifications. First, because

$$[\,D_1(s)\quad N_1(s)\,] = \begin{bmatrix} 1\,0\,0\,0 & 0 & 0 \\ 0\,1\,0\,0 & 0 & 0 \\ 0\,0\,1\,0 & 0 & 0 \\ 0\,0\,0\,1 & 0 & 0 \\ 0\,0\,0\,0 & -1-s\sqrt{2}-s^2 & s(1+s) \\ 0\,0\,0\,0 & 0 & -(2+s)(1+s) \end{bmatrix} \tag{2.175}$$

is square and Hurwitz, we consider the spectral cofactorization

$$D_1 D_1^{\sim} - \frac{1}{\gamma_2} N_1 N_1^{\sim} = Q J_\gamma Q^{\sim}, \tag{2.176}$$

with $Q = [\,D_1 \quad N_1\,]$ and

$$J_\gamma = \begin{bmatrix} 1\,0\,0\,0 & 0 & 0 \\ 0\,1\,0\,0 & 0 & 0 \\ 0\,0\,1\,0 & 0 & 0 \\ 0\,0\,0\,1 & 0 & 0 \\ 0\,0\,0\,0 & -\frac{1}{\gamma^2} & 0 \\ 0\,0\,0\,0 & 0 & -\frac{1}{\gamma^2} \end{bmatrix}. \tag{2.177}$$

Next we obtain the left-to-right conversion

$$Q^{-1}[\,\hat{N}_2 \quad \hat{D}_2] = \Delta\Lambda^{-1} \tag{2.178}$$

with

$$\Delta(s) = \begin{bmatrix} -1 & -\frac{1}{2}\sqrt{2} & \frac{1}{2}(1+s\sqrt{2}) & 0 \\ 0 & -1 & 0 & 1+s \\ c_1(1+s\sqrt{2})(1+rs) & \frac{1}{2}\sqrt{2}c_1(1+rs) & \frac{1}{2}c_1(1+rs) & 0 \\ 0 & c_2(2+s) & 0 & 0 \\ 1 & \frac{1}{2}\sqrt{2} & \frac{1}{2}(1-s\sqrt{2}) & 0 \\ 0 & 1 & 0 & -s \end{bmatrix},$$

$$\Lambda(s) = \begin{bmatrix} 1+s\sqrt{2} & \frac{1}{2}\sqrt{2} & \frac{1}{2} & 0 \\ 0 & 2+s & 0 & 0 \\ -1 & -\frac{1}{2}\sqrt{2} & \frac{1}{2}(1+s\sqrt{2}) & 0 \\ 0 & -1 & 0 & 1+s \end{bmatrix}. \tag{2.179}$$

Δ is column reduced. We now need to consider the spectral factorization

$$\Delta^{\sim} J_\gamma^{-1} \Delta = \Gamma_\gamma^{\sim} L_\gamma^{-1} \Gamma_\gamma, \tag{2.180}$$

with the coefficients of the polynomial matrix Γ_γ suitably scaled and L_γ diagonal.

At this point symbolic calculation has to be abandoned for numerical computation. As in [16] we let $c_1 = c_2 = \frac{1}{2}$ and $r = 1$. In the earlier work γ_{opt} was found to be 1.80680. Application of one of the algorithms of [23] leads for this value of γ to a spectral factorization with

$$L_\gamma = \begin{bmatrix} 1.0 & 0 & 0 & 0 \\ 0 & .0000020690 & 0 & 0 \\ 0 & 0 & -.0000017567 & 0 \\ 0 & 0 & 0 & -1.0 \end{bmatrix}, \tag{2.181}$$

$$\Gamma_\gamma(s) = \left[\begin{matrix} -.17763 - 0.058098s - 0.026677s^2 & -0.28428 + 0.098383s \\ 0.34932 - 0.51644s - 0.12727s^2 & 0.050235 - 0.085079s \\ -0.32187 - 0.61435s - 0.11727s^2 & -0.046292 - 0.0783941s \\ -0.39573 - 0.0049083s - 0.016880s^2 & -1.7909 + 0.042922s \end{matrix}\right.$$

$$\left.\begin{matrix} 0.61149 + 0.27446s & 0.86500 + 0.69380s \\ 0.56585 - 0.95829s & -0.30672 + 0.51945s \\ -0.052139 - 0.88299s & 0.28262 + 0.47863s \\ -0.011534 + 0.44581s & -0.76645 + 1.5767s \end{matrix}\right]. \tag{2.182}$$

The first and last diagonal entries of L have been scaled to 1. Both L_γ and Γ_γ are singular, or at least close to it. Choosing $A = I$ and $B = 0$ we obtain

$$[\,A \;\; B\,]\Gamma_\gamma(s) = \left[\begin{matrix} -.17763 - 0.058098s - 0.026677s^2 & -0.28428 + 0.098383s \\ 0.34932 - 0.51644s - 0.12727s^2 & 0.050235 - 0.085079s \end{matrix}\right.$$

$$\left.\begin{matrix} 0.61149 + 0.27446s & 0.86500 + 0.69380s \\ 0.56585 - 0.95829s & -0.30672 + 0.51945s \end{matrix}\right]. \tag{2.183}$$

The polynomials in the second row have a common root 0.59048, which may be canceled. This results in

$$[\,X_\gamma(s) \;\; Y_\gamma(s)\,]\Lambda_\gamma(s) = \left[\begin{matrix} -.17763 - 0.058098s - 0.026677s^2 & -.28428 + 0.098383s \\ -0.12727s - 0.59159 & -0.085079 \end{matrix}\right.$$

$$\left.\begin{matrix} 0.61149 + 0.27446s & 0.865 + 0.6938s \\ -0.95829 & 0.51945 \end{matrix}\right]. \tag{2.184}$$

Right-to-left conversion of

$$\left[\begin{matrix} -.17763 - 0.058098s - 0.026677s^2 & -.28428 + 0.098383s \\ -0.12727s - 0.59159 & -0.085079 \end{matrix}\right.$$

$$\left.\begin{matrix} 0.61149 + 0.27446s & 0.865 + 0.6938s \\ -0.95829 & 0.51945 \end{matrix}\right] \Lambda_\gamma^{-1}(s) \tag{2.185}$$

and cancelation of the denominator results in $[\,X_\gamma \;\; Y_\gamma\,]$. •

2.9 Conclusions

The frequency domain solution of the $\mathcal{H}_\infty$-optimal regulation problem presented in this paper is altogether transparent. It clearly exhibits the central role of spectral factorization. The results may be specialized to the well-known state space solution [9], [6] with the appropriate algebraic Riccati equations using the Kalman-Jacubovič-Popov equality [2].

The paper clarifies how loss of canonicity often accompanies optimality. Nearly noncanonical factorizations may be dealt with by replacing the signature matrix by its inverse and scaling it. This slight but significant modification resolves the problem of large coefficients near the optimum that plagues the conventional state space algorithm. It also allows the determination of all optimum solutions.

The solution method applies under very mild assumptions. Nonstabilizable and nondetectable problems may be considered. The generalized plant does not need to be proper. Zeros on the imaginary axis (including those at ∞) present no great difficulty.

The solution method is amenable to numerical computation [17]. A number of routines for polynomial matrix computations are needed. The most important algorithms are for polynomial matrix fraction conversion and polynomial spectral factorization. State space algorithms for these problems [22], such as the factorization algorithm outlined in section 2.7, appear quite attractive. Work on a MATLAB-based "polynomial toolbox" with a comprehensive set of routines for polynomial matrix operations [32] is in progress.

An unresolved problem is to find a fast method to compute the minimal norm γ_{opt} for type B solutions. A method that in principle offers quadratic convergence is to apply a root finding procedure to reduce the smallest diagonal entry of the matrix L_γ in subsection 2.8.2 equal to zero. The nonuniqueness of the spectral factorization causes an indeterminacy that may well make the algorithm fail, however. The only safe method that is currently available to compute γ_{opt} is a binary search on whether or not stabilizing solutions exist, as outlined in subsection 2.6.3. The convergence of this method is linear and, hence, slow.

Another open problem is the reliable factorization of para-Hermitian polynomial matrices with zeros on the imaginary axis, such as needed for the precise determination of type A optimal solutions.

2.10 Appendix: Proofs for section 2.3

Proof. **Theorem 2.1: ∞-norm.** By our definition of the ∞-norm the inequality $\|H\|_\infty \leq \gamma$ is equivalent to $\lambda_{\max}(H^{\text{H}}(j\omega)H(j\omega)) \leq \gamma^2$ for $\omega \in \mathsf{R}$, with $\lambda_{\max}$ denoting the largest eigenvalue. This in turn is equivalent to $H^{\text{H}}(j\omega)H(j\omega) \leq \gamma^2 I$ for $\omega \in \mathsf{R}$, which proves (a). The proof of (b) is similar. •

Proof. **Theorem 2.2: Definiteness of quadratic form.** *(a) $\Rightarrow$ (b).* The inequality

$$[I \quad L] \underbrace{\begin{bmatrix} \Pi_{11} & \Pi_{12} \\ \Pi_{21} & -\Pi_{22} \end{bmatrix}}_{\Pi} \begin{bmatrix} I \\ L^\sim \end{bmatrix} \geq 0 \quad \text{on the imaginary axis,} \tag{2.186}$$

is equivalent to

$$(L - \Pi_{12}\Pi_{22}^{-1})\Pi_{22}(L - \Pi_{12}\Pi_{22}^{-1})^\sim \leq \Pi_{11} + \Pi_{12}\Pi_{22}^{-1}\Pi_{21} \tag{2.187}$$

on the imaginary axis. Hence, if L exists such that (2.186) holds then necessarily $\Pi_{11} + \Pi_{12}\Pi_{22}^{-1}\Pi_{21}$ is nonnegative-definite on the imaginary axis.

(b) ⇒ (a). Conversely, suppose that $\Pi_{11}+\Pi_{12}\Pi_{22}^{-1}\Pi_{21}$ is nonnegative-definite on the imaginary axis. Then by a well-known result [34] there exists a rational matrix function V_1 such that

$$\Pi_{11} + \Pi_{12}\Pi_{22}^{-1}\Pi_{21} = V_1 V_1^\sim. \tag{2.188}$$

By the same theorem, there exists a rational matrix function V_2 such that

$$\Pi_{22} = V_2 V_2^\sim. \tag{2.189}$$

It is easily seen that the inequality $(L - \Pi_{12}\Pi_{22}^{-1})V_2V_2^\sim(L - \Pi_{12}\Pi_{22}^{-1})^\sim \leq V_1V_1^\sim$ is satisfied (with equality) by $L = V_1V_2^{-1} + \Pi_{12}\Pi_{22}$.

(b) ⇒ (c). Since by assumption Π_{22} is invertible we have

$$\Pi = \begin{bmatrix} I & -\Pi_{12}\Pi_{22}^{-1} \\ 0 & I \end{bmatrix} \begin{bmatrix} \Pi_{11} + \Pi_{12}\Pi_{22}^{-1}\Pi_{21} & 0 \\ 0 & -\Pi_{22} \end{bmatrix} \begin{bmatrix} I & 0 \\ -\Pi_{22}^{-1}\Pi_{21} & I \end{bmatrix}. \tag{2.190}$$

If $\Pi_{11}+\Pi_{12}\Pi_{22}^{-1}\Pi_{21}$ is nonnegative-definite on the imaginary axis then it follows with (2.188) and (2.189) that

$$\Pi = \begin{bmatrix} I & -\Pi_{12}\Pi_{22}^{-1} \\ 0 & I \end{bmatrix} \begin{bmatrix} V_1V_1^\sim & 0 \\ 0 & -V_2V_2^\sim \end{bmatrix} \begin{bmatrix} I & 0 \\ -\Pi_{22}^{-1}\Pi_{21} & I \end{bmatrix} = ZJZ^\sim, \tag{2.191}$$

where

$$Z = \begin{bmatrix} I & -\Pi_{12}\Pi_{22}^{-1} \\ 0 & I \end{bmatrix} \begin{bmatrix} V_1 & 0 \\ 0 & V_2 \end{bmatrix}. \tag{2.192}$$

(c) ⇒ (b). If the Schur complement $\Pi_{11} + \Pi_{12}\Pi_{22}^{-1}\Pi_{21}$ is not nonnegative-definite on the imaginary axis then (2.190) shows that the number of positive eigenvalues of Π on the imaginary axis is less than n and the number of negative eigenvalues greater than m. This contradicts the existence of the required factorization and, hence, proves that (b) ⇐ (c). •

Proof. **Theorem 2.3: Lower bound.** By application of Theorem 2.2 to (2.19–2.20) it follows that there exist L that make $\|H\|_\infty \leq \gamma$ if and only if the Schur complement

$$\begin{aligned}&(\gamma^2 I - G_{11}G_{11}^\sim) + G_{11}G_{21}^\sim\left(G_{21}G_{21}^\sim\right)^{-1}G_{21}G_{11}^\sim \\ &= \gamma^2 I - G_{11}\left(I - G_{21}^\sim\left(G_{21}G_{21}^\sim\right)^{-1}G_{21}\right)G_{11}^\sim\end{aligned} \tag{2.193}$$

of the (2, 2) block of $\Pi_{1,\gamma}$ is nonnegative-definite on the imaginary axis. If there at all exists a value of γ such that (2.193) is nonnegative-definite on the imaginary axis then (2.193) is nonnegative-definite on the imaginary axis for all γ greater than or equal to this value. In this case, there exists a number γ_1 such that (2.193) is nonnegative-definite on the imaginary axis for all $\gamma \geq \gamma_1$. This number γ_1 is a lower bound for the existence of an L such that (2.20) holds, and, hence, for the smallest value that $\|H\|_\infty$ may assume.

We carry the analysis a little further. Consider $\Pi_{1,\gamma}$, and suppose first that $G_{21}G_{21}^{\sim}$ is positive-definite on the imaginary axis. Then if γ_1 is finite the Schur complement (2.193) of the (2,2)-block of $\Pi_{1,\gamma}$ is positive-definite on the imaginary axis for $\gamma > \gamma_1$. Hence, for $\gamma > \gamma_1$ the matrix $\Pi_{1,\gamma}$ is nonsingular on the imaginary axis, that is, $\det \Pi_{1,\gamma}$ has no zeros on the imaginary axis. If γ decreases then for $\gamma = \gamma_1$ the matrix $\Pi_{1,\gamma}$ becomes singular in one or several points on the imaginary axis. This means that γ_1 is the first value as γ decreases from ∞ such that $\det \Pi_{1,\gamma}$ has a zero on the imaginary axis.

If $G_{21}G_{21}^{\sim}$ is nonnegative-definite but not positive-definite on the imaginary axis then $\det G_{21}G_{21}^{\sim}$ has one or several zeros on the imaginary axis. These zeros are also zeros of $\det \Pi_{1,\gamma}$ and need to be ignored when determining γ_1. Note that the zeros of $\det G_{21}G_{21}^{\sim}$ are those points where G_{21} — which by assumption is wide — loses row rank, together with the mirror images of these points with respect to the imaginary axis. That is, the zeros of $\det G_{21}G_{21}^{\sim}$ are the zeros of G_{21} and their mirror images with respect to the imaginary axis. •

2.11 Appendix: Proofs for section 2.4

The lemma that follows presents, in polished form, a result originally obtained by Meinsma [18].

Lemma 2.9. Meinsma's lemma. *Let Π be a nonsingular para-Hermitian rational matrix function that admits a factorization $\Pi = Z^{\sim}JZ$, with*

$$J = \begin{bmatrix} I_n & 0 \\ 0 & -I_m \end{bmatrix} \tag{2.194}$$

and Z square rational. Suppose that V is an $(n+m)\times n$ rational matrix of full column rank and W an $m \times (n+m)$ rational matrix of full row rank such that $W^{\sim}V = 0$. Then

(a) $V^{\sim}\Pi V \geq 0$ on the imaginary axis if and only if $W^{\sim}\Pi^{-1}W \leq 0$ on the imaginary axis.

(b) $V^{\sim}\Pi V > 0$ on the imaginary axis if and only if $W^{\sim}\Pi^{-1}W < 0$ on the imaginary axis.

Proof. **Meinsma's lemma.** (a) $V^{\sim}\Pi V \geq 0$ on the imaginary axis is equivalent to

$$V = Z^{-1}\begin{bmatrix} A \\ B \end{bmatrix}, \tag{2.195}$$

with A an $n \times n$ nonsingular rational matrix and B an $n \times m$ rational matrix such that $A^\sim A \geq B^\sim B$ on the imaginary axis. Define an $m \times m$ nonsingular rational matrix $\bar{A}$ and an $m \times n$ rational matrix $\bar{B}$ such that

$$[-\bar{B} \;\; \bar{A}] \begin{bmatrix} A \\ B \end{bmatrix} = 0. \tag{2.196}$$

It follows that $\bar{B}A = \bar{A}B$, and, hence, $\bar{A}^{-1}\bar{B} = BA^{-1}$. From $A^\sim A \geq B^\sim B$ on the imaginary axis it follows that $\|BA^{-1}\|_\infty \leq 1$, so that also $\|\bar{A}^{-1}\bar{B}\|_\infty \leq 1$, and, hence, $\bar{A}\bar{A}^\sim \geq \bar{B}\bar{B}^\sim$ on the imaginary axis. Since

$$W^\sim V = W^\sim Z^{-1} \begin{bmatrix} A \\ B \end{bmatrix} = 0, \tag{2.197}$$

we have from (2.196) that

$$W^\sim Z^{-1} = U[-\bar{B} \;\; \bar{A}], \tag{2.198}$$

with U some square nonsingular rational matrix. It follows that $W^\sim = U[-\bar{B} \;\; \bar{A}]Z$, and, hence,

$$W^\sim \Pi^{-1} W = U(\bar{B}\bar{B}^\sim - \bar{A}\bar{A}^\sim)U^\sim. \tag{2.199}$$

This proves that $W^\sim \Pi^{-1} W \leq 0$ on the imaginary axis. The proof of the converse and that of (b) are similar. •

Proof. **Theorem 2.5: Basic inequality.** We rewrite the expression $H = G_{11} + G_{12}(X - YG_{22})^{-1}YG_{21}$ as $H = G_{11} + G_{12}L$, with $L = (X - YG_{22})^{-1}YG_{21}$. Since $\|H\|_\infty \leq \gamma$ is equivalent to $H^\sim H \leq \gamma^2 I$ on $j\mathsf{R}$, it easily follows that $\|H\|_\infty \leq \gamma$ is equivalent to

$$[I \;\; L^\sim] \begin{bmatrix} \gamma^2 I - G_{11}^\sim G_{11} & -G_{11}^\sim G_{12} \\ -G_{12}^\sim G_{11} & -G_{12}^\sim G_{12} \end{bmatrix} \begin{bmatrix} I \\ L \end{bmatrix} \geq 0 \quad \text{on } j\mathsf{R}. \tag{2.200}$$

It follows from theorems 2.2 and 2.3 that for $\gamma \geq \gamma_o$ the coefficient matrix has a factorization. Then by Meinsma's lemma (2.200) is equivalent to

$$[-L \;\; I] \begin{bmatrix} \gamma^2 I - G_{11}^\sim G_{11} & -G_{11}^\sim G_{12} \\ -G_{12}^\sim G_{11} & -G_{12}^\sim G_{12} \end{bmatrix}^{-1} \begin{bmatrix} -L^\sim \\ I \end{bmatrix} \leq 0 \quad \text{on } j\mathsf{R}. \tag{2.201}$$

After multiplying through by -1, multiplication on the left by $X - YG_{22}$ and on the right by its adjoint, the inequality takes the form

$$[YG_{21} \quad YG_{22} - X] \begin{bmatrix} G_{11}^\sim G_{11} - \gamma^2 I & G_{11}^\sim G_{12} \\ G_{12}^\sim G_{11} & G_{12}^\sim G_{12} \end{bmatrix}^{-1} \begin{bmatrix} G_{21}^\sim Y^\sim \\ G_{22}^\sim Y^\sim - X^\sim \end{bmatrix} \geq 0 \tag{2.202}$$

on $j\mathsf{R}$. Since

$$[YG_{21} \quad YG_{22} - X] = [X \quad Y] \begin{bmatrix} 0 & -I \\ G_{21} & G_{22} \end{bmatrix} \tag{2.203}$$

we see that $\|H\|_\infty \leq \gamma$ is equivalent to (2.49–2.50). •

Proof. **Theorem 2.6: Sublevel solutions.** We discuss the well-posedness of the feedback system defined by the compensator $[X \;\; Y] = [A \;\; B]Z_\gamma^{-1}$, with A nonsingular and $AA^\sim \geq BB^\sim$ on the imaginary axis. Denote $M_\gamma = Z_\gamma^{-1}$. It is easily seen that

$$[\,X - YG_{22} \quad Y\,] = [\,A \quad B\,]M_\gamma \begin{bmatrix} I & 0 \\ -G_{22} & I \end{bmatrix}. \tag{2.204}$$

If $X - YG_{22}$ is nonsingular, that is, the feedback system is well-posed, then the left-hand side has full normal row rank, and, hence, so has $[A \;\; B]$. If at the same time $AA^\sim \geq BB^\sim$ on $j\mathsf{R}$, then necessarily A needs to be nonsingular.

It remains to prove that, conversely, if A is nonsingular then the feedback system is well-posed. Partition

$$M_\gamma = \begin{bmatrix} M_{\gamma,11} & M_{\gamma,12} \\ M_{\gamma,21} & M_{\gamma,22} \end{bmatrix} \tag{2.205}$$

consistently with the partitioning of Π_γ. Then if $M_{\gamma,11} - M_{\gamma,12}G_{22}$ is nonsingular we have

$$\begin{aligned} X - YG_{22} &= A(M_{\gamma,11} - M_{\gamma,12}G_{22}) + B(M_{\gamma,21} - M_{\gamma,22}G_{22}) \\ &= A(I + TC)(M_{\gamma,11} - M_{\gamma,12}G_{22}) \end{aligned} \tag{2.206}$$

with $C = (M_{\gamma,21} - M_{\gamma,22}G_{22})(M_{\gamma,11} - M_{\gamma,12}G_{22})^{-1}$ and $T = A^{-1}B$. We prove in a moment that there exists a point on the imaginary axis such that

(a) $M_{\gamma,11} - M_{\gamma,21}G_{22}$ is nonsingular, and

(b) $\|C\|_2 < 1$, with $\|\cdot\|_2$ the spectral norm (that is, the largest singular value).

Since $\|T\|_2 \leq 1$ on any point on the imaginary axis, it follows by (b) that $I + TC$ is nonsingular. Hence, $X - YG_{22}$ is nonsingular, and the feedback system is well-posed.

We finish by proving the claims (a) and (b). Postmultiplication of $M_\gamma^\sim JM_\gamma = \Pi_\gamma^{-1}$ as given in (2.51) by

$$\begin{bmatrix} I & 0 \\ -G_{22} & I \end{bmatrix} \tag{2.207}$$

and premultiplication by its adjoint yields

$$\begin{aligned} &\begin{bmatrix} M_{\gamma,11} - M_{\gamma,12}G_{22} & M_{\gamma,12} \\ M_{\gamma,21} - M_{\gamma,22}G_{22} & M_{\gamma,22} \end{bmatrix}^\sim J \begin{bmatrix} M_{\gamma,11} - M_{\gamma,12}G_{22} & M_{\gamma,12} \\ M_{\gamma,21} - M_{\gamma,22}G_{22} & M_{\gamma,22} \end{bmatrix} \\ &= \begin{bmatrix} G_{12}^\sim & 0 \\ 0 & I \end{bmatrix} \begin{bmatrix} I - \frac{1}{\gamma^2}G_{11}G_{11}^\sim & -\frac{1}{\gamma^2}G_{11}G_{21}^\sim \\ -\frac{1}{\gamma^2}G_{21}G_{11}^\sim & -\frac{1}{\gamma^2}G_{21}G_{21}^\sim \end{bmatrix}^{-1} \begin{bmatrix} G_{12} & 0 \\ 0 & I \end{bmatrix}. \end{aligned} \tag{2.208}$$

Evaluation of the 11-block results in

$$\begin{aligned} &(M_{\gamma,11} - M_{\gamma,12}G_{22})^\sim(M_{\gamma,11} - M_{\gamma,21}G_{22}) \\ &\quad - (M_{\gamma,21} - M_{\gamma,22}G_{22})^\sim(M_{\gamma,21} - M_{\gamma,22}G_{22}) \\ &= [\,G_{12}^\sim \;\; 0\,] \begin{bmatrix} I - \frac{1}{\gamma^2}G_{11}G_{11}^\sim & -\frac{1}{\gamma^2}G_{11}G_{21}^\sim \\ -\frac{1}{\gamma^2}G_{21}G_{11}^\sim & -\frac{1}{\gamma^2}G_{21}G_{21}^\sim \end{bmatrix}^{-1} \begin{bmatrix} G_{12} \\ 0 \end{bmatrix}. \end{aligned} \tag{2.209}$$

By the assumption that G_{12} has full normal column rank and G_{21} full normal row rank there exists a point on the imaginary axis where both G_{12} has full column rank and G_{21} has full row rank. In this point, by Meinsma's lemma the right-hand side of (2.209) is positive-definite because

$$[0\ I]\begin{bmatrix} I-\frac{1}{\gamma^2}G_{11}G_{11}^{\sim} & -\frac{1}{\gamma^2}G_{11}G_{21}^{\sim} \\ -\frac{1}{\gamma^2}G_{21}G_{11}^{\sim} & -\frac{1}{\gamma^2}G_{21}G_{21}^{\sim}\end{bmatrix}\begin{bmatrix}0\\ I\end{bmatrix} = -\frac{1}{\gamma^2}G_{21}G_{21}^{\sim} < 0. \tag{2.210}$$

Hence, the matrix $(M_{\gamma,11}-M_{\gamma,12}G_{22})^{\sim}(M_{\gamma,11}-M_{\gamma,12}G_{22})$ is positive-definite, so that $M_{\gamma,11}-M_{\gamma,12}G_{22}$ is nonsingular. Moreover, by postmultiplication of (2.209) by $(M_{\gamma,11}-M_{\gamma,12}G_{22})^{-1}$ and premultiplication by its adjoint we obtain $I-C^{\sim}C>0$, so that $\|C\|_2<1$. •

2.12 Appendix: Proof of theorem 2.7

This appendix is devoted to the proof of theorem 2.7. The hypotheses of theorem 2.7 apply throughout. First a number of lemmas are presented. The appendix ends with the actual proof of theorem 2.7.

Lemma 2.10. Zeros of $A+\varepsilon B$. *Let A and B be square rational matrices of the same dimensions such that (1) A is nonsingular, (2) A has no poles or zeros on the imaginary axis, and (3) $\|A^{-1}B\|_\infty<1$. Then as the real variable ε varies from 0 to 1 no zeros of $A+\varepsilon B$ cross the imaginary axis.*

Proof. **Zeros of $A+\varepsilon B$.** Write $A+\varepsilon B = A(I+\varepsilon T)$, with $T=A^{-1}B$. Then $\|T\|_\infty<1$, and we need to prove that $I+\varepsilon T$ has no zeros on the imaginary axis for $\varepsilon\in[0,1]$. Since by assumption A has no zero on the imaginary axis we may take $\varepsilon\in(0,1]$.

The proof is by contradiction. Suppose that $I+\varepsilon T$ has a zero on the imaginary axis, say at $s=j\omega$. Then (by the definition of a zero) $I+\varepsilon T(j\omega)$ is singular, and there exists a (complex-valued) vector e such that $[I+\varepsilon T(j\omega)]e=0$. Since $T(j\omega)e=-\frac{1}{\varepsilon}e$ we have

$$\frac{\|T(j\omega)e\|_2}{\|e\|_2} = \frac{e^{\mathrm{H}}T^{\mathrm{H}}(j\omega)T(j\omega)e}{e^{\mathrm{H}}e} = \frac{1}{\varepsilon^2}\geq 1. \tag{2.211}$$

This contradicts the assumption that $\|T\|_\infty<1$. •

Lemma 2.11. Transformation of a rational matrix. *Suppose that G is a tall rational matrix with full normal column rank. Then there exists a nonsingular square rational matrix V such that GV is proper and has no zeros on the imaginary axis, including at ∞ (that is, has full rank at ∞).*

Proof. The matrix V may be constructed as follows.

(a) Find a right coprime polynomial faction representation $G=ND^{-1}$. Factor $N=\hat{N}N_o$, where $\hat{N}$ is column reduced and has no zeros on the imaginary axis, and N_o square. N_o contains any zeros of N on the imaginary axis.

(b) Find a square nonsingular polynomial matrix Q such that $\hat{N}Q^{-1}$ is proper with full rank at ∞. The matrix Q may for instance be chosen as $Q(s) = \text{diag}\,(s^{-n_1}, s^{-n_2}, \cdots, s^{-n_n})$, with n_1, n_2, $\cdots$, n_n the column degrees of $\hat{N}$.

Then

$$\underbrace{ND^{-1}}_{G} \cdot \underbrace{DN_o^{-1}Q^{-1}}_{V} = \hat{N}Q^{-1} \tag{2.212}$$

is proper and has no zeros on the imaginary axis (including at ∞), and V is nonsingular square rational. •

By the hypothesis of theorem 2.7 that there exists a compensator that makes $\|H\|_\infty$ finite there exists a compensator K_o such that

$$H = G_{11} + G_{12}\underbrace{(I - K_oG_{22})^{-1}K_o}_{L_o}G_{21} \tag{2.213}$$

is proper. Denote $G_{11} + G_{12}L_oG_{21} = \hat{G}_{11}$.

By the assumption that G_{12} has full normal column rank there exists according to lemma 2.11 a nonsingular rational matrix V_1 such that $\hat{G}_{12} = G_{12}V_1$ is proper without zeros on the imaginary axis with full column rank at ∞. Likewise, since G_{21} has full normal row rank there exists a bistable rational matrix V_2 such that $\hat{G}_{21} = V_2G_{21}$ is proper without zeros on the imaginary axis and with full row rank at ∞.

Define

$$T = \begin{bmatrix} I & 0 \\ -G_{22} & I \end{bmatrix} \begin{bmatrix} I & -L_o \\ 0 & I \end{bmatrix} \begin{bmatrix} V_1 & 0 \\ 0 & V_2^{-1} \end{bmatrix}. \tag{2.214}$$

Lemma 2.12. Biproperness of $M_\gamma T$. *If the spectral factorization $\Pi_\gamma^{-1} = M_\gamma^{\sim} J M_\gamma$ is canonical then $M_\gamma T$ is biproper.*

Proof. From

$$G = \begin{bmatrix} G_{11} & G_{12} \\ G_{21} & G_{22} \end{bmatrix} = [D_1 \;\; D_2]^{-1}[N_1 \;\; N_2] \tag{2.215}$$

it follows that

$$N_1 = D_1G_{11} + D_2G_{21}, \qquad N_2 = D_1G_{12} + D_2G_{22}. \tag{2.216}$$

Rearrangement yields

$$[N_2 \;\; D_2]\begin{bmatrix} 0 & I \\ G_{21} & -G_{22} \end{bmatrix} = [D_1 \;\; N_1]\begin{bmatrix} -G_{11} & G_{12} \\ I & 0 \end{bmatrix}. \tag{2.217}$$

Rewriting this as

$$[N_2 \;\; D_2]TT^{-1}\begin{bmatrix} 0 & I \\ G_{21} & -G_{22} \end{bmatrix} = [D_1 \;\; N_1]\begin{bmatrix} -G_{11} & G_{12} \\ I & 0 \end{bmatrix} \tag{2.218}$$

and postmultiplying both sides by

$$\begin{bmatrix} I & 0 \\ -L_o G_{21} & 0 \end{bmatrix} \begin{bmatrix} I & 0 \\ 0 & V_1 \end{bmatrix} \tag{2.219}$$

we obtain

$$[N_2 \;\; D_2] T \begin{bmatrix} 0 & I \\ \hat{G}_{21} & 0 \end{bmatrix} = [D_1 \;\; N_1] \begin{bmatrix} -\hat{G}_{11} & \hat{G}_{12} \\ I & 0 \end{bmatrix}. \tag{2.220}$$

Further postmultiplication by the matrix

$$\begin{bmatrix} 0 & \hat{G}_{21}^{\sim}(\hat{G}_{21}\hat{G}_{21}^{\sim})^{-1} \\ I & 0 \end{bmatrix} \tag{2.221}$$

results in

$$[N_2 \;\; D_2] T = [D_1 \;\; N_1] S, \tag{2.222}$$

where

$$S = \begin{bmatrix} \hat{G}_{12} & -\hat{G}_{11}\hat{G}_{21}^{\sim}(\hat{G}_{21}\hat{G}_{21}^{\sim})^{-1} \\ 0 & \hat{G}_{21}^{\sim}(\hat{G}_{21}\hat{G}_{21}^{\sim})^{-1} \end{bmatrix}. \tag{2.223}$$

S is tall and proper and has full rank at ∞.

Canonicity of the spectral factorization $\Pi_\gamma^{-1} = M_\gamma^{\sim} J M_\gamma$ is defined in polynomial matrix terms in subsection 2.5.2. Premultiplication of (2.222) by Q_γ^{-1}, with Q_γ the spectral cofactor defined in subsection 2.5.2, yields

$$\underbrace{Q_\gamma^{-1}[N_2 \;\; D_2]}_{\Delta_\gamma \Lambda_\gamma^{-1}} T = Q_\gamma^{-1}[N_1 \;\; D_1] S. \tag{2.224}$$

From this we obtain

$$\Delta_\gamma \Gamma_\gamma^{-1} \cdot \underbrace{\Gamma_\gamma \Lambda_\gamma^{-1}}_{M_\gamma} T = Q_\gamma^{-1}[D_1 \;\; N_1] S. \tag{2.225}$$

Since by assumption the factorization is canonical, both $\Delta_\gamma \Gamma_\gamma^{-1}$ and $Q_\gamma^{-1}[N_1 \;\; D_1]$ are proper with full rank at ∞. It follows from (2.225) that $M_\gamma T$ is proper. By multiplying both sides of (2.225) on the right by $(M_\gamma T)^{-1}$ it follows that also $(M_\gamma T)^{-1}$ is proper. Hence, $M_\gamma T$ is biproper. •

Lemma 2.13. $\|C\|_\infty < 1$. *Partition*

$$M_\gamma = \begin{bmatrix} M_{\gamma,11} & M_{\gamma,12} \\ M_{\gamma,21} & M_{\gamma,22} \end{bmatrix} \tag{2.226}$$

and define

$$C = (M_{\gamma,21} - M_{\gamma,22} G_{22})(M_{\gamma,11} - M_{\gamma,12} G_{22})^{-1}. \tag{2.227}$$

Then $\|C\|_\infty < 1$.

Proof. From the proof of theorem 2.6 in Appendix 2.11 we have

$$\begin{aligned}&(M_{\gamma,11} - M_{\gamma,12}G_{22})^{\sim}(M_{\gamma,11} - M_{\gamma,12}G_{22})\\&\quad - (M_{\gamma,21} - M_{\gamma,22}G_{22})^{\sim}(M_{\gamma,21} - M_{\gamma,22}G_{22})\\&\quad = \begin{bmatrix} G_{12}^{\sim} & 0 \end{bmatrix} \begin{bmatrix} I - \frac{1}{\gamma^2}G_{11}G_{11}^{\sim} & -\frac{1}{\gamma^2}G_{11}G_{21}^{\sim} \\ -\frac{1}{\gamma^2}G_{21}G_{11}^{\sim} & -\frac{1}{\gamma^2}G_{21}G_{21}^{\sim} \end{bmatrix}^{-1} \begin{bmatrix} G_{12} \\ 0 \end{bmatrix}.\end{aligned} \tag{2.228}$$

Postmultiplication of both sides by V_1, premultiplication of both sides by $V_1^{\sim}$ and premultiplication of

$$\begin{bmatrix} I - \frac{1}{\gamma^2}G_{11}G_{11}^{\sim} & -\frac{1}{\gamma^2}G_{11}G_{21}^{\sim} \\ -\frac{1}{\gamma^2}G_{21}G_{11}^{\sim} & -\frac{1}{\gamma^2}G_{21}G_{21}^{\sim} \end{bmatrix} \quad \text{and} \quad \begin{bmatrix} G_{12} \\ 0 \end{bmatrix} \tag{2.229}$$

by

$$\begin{bmatrix} I & 0 \\ 0 & V_2 \end{bmatrix} \begin{bmatrix} I & G_{12}L_o \\ 0 & I \end{bmatrix} \tag{2.230}$$

yields

$$\begin{aligned}&V_1^{\sim}(M_{\gamma,11} - M_{\gamma,12}G_{22})^{\sim}(M_{\gamma,11} - M_{\gamma,12}G_{22})V_1\\&\quad - V_1^{\sim}(M_{\gamma,21} - M_{\gamma,22}G_{22})^{\sim}(M_{\gamma,21} - M_{\gamma,22}G_{22})V_1\\&\quad = \begin{bmatrix} \hat{G}_{12}^{\sim} & 0 \end{bmatrix} \begin{bmatrix} I - \frac{1}{\gamma^2}\hat{G}_{11}\hat{G}_{11}^{\sim} & -\frac{1}{\gamma^2}\hat{G}_{11}\hat{G}_{21}^{\sim} \\ -\frac{1}{\gamma^2}\hat{G}_{21}\hat{G}_{11}^{\sim} & -\frac{1}{\gamma^2}\hat{G}_{21}\hat{G}_{21}^{\sim} \end{bmatrix}^{-1} \begin{bmatrix} \hat{G}_{12} \\ 0 \end{bmatrix}.\end{aligned} \tag{2.231}$$

$\hat{G}_{12}$ and $\hat{G}_{21}$ have full rank everywhere on the imaginary axis, including at ∞. Hence, by Meinsma's lemma (lemma 2.9) the right-hand side of (2.231) is positive-definite on the imaginary axis (including at ∞) because

$$\begin{bmatrix} 0 & I \end{bmatrix} \begin{bmatrix} I - \frac{1}{\gamma^2}\hat{G}_{11}\hat{G}_{11}^{\sim} & -\frac{1}{\gamma^2}\hat{G}_{11}\hat{G}_{21}^{\sim} \\ -\frac{1}{\gamma^2}\hat{G}_{21}\hat{G}_{11}^{\sim} & -\frac{1}{\gamma^2}\hat{G}_{21}\hat{G}_{21}^{\sim} \end{bmatrix} \begin{bmatrix} 0 \\ I \end{bmatrix} = -\frac{1}{\gamma^2}\hat{G}_{21}\hat{G}_{21}^{\sim} < 0 \tag{2.232}$$

on the imaginary axis. Because by lemma 2.12

$$M_{\gamma}T = \begin{bmatrix} (M_{\gamma,11} - M_{\gamma,12}G_{22})V_1 & [-(M_{\gamma,11} - M_{\gamma,12}G_{22})L_o + M_{\gamma,12}]V_2^{-1} \\ (M_{\gamma,21} - M_{\gamma,22}G_{22})V_1 & [-(M_{\gamma,21} - M_{\gamma,22}G_{22})L_o + M_{\gamma,22}]V_2^{-1} \end{bmatrix} \tag{2.233}$$

is proper both $(M_{\gamma,11} - M_{\gamma,12}G_{22})V_1$ and $(M_{\gamma,21} - M_{\gamma,22}G_{22})V_1$ are proper. It then follows from (2.231) that $(M_{\gamma,11} - M_{\gamma,21}G_{22})V_1$ is positive-definite and, hence, nonsingular on the imaginary axis, including at ∞. Moreover, postmultiplying (2.231) by $[(M_{\gamma,11} - M_{\gamma,21}G_{22})V_1]^{-1}$ and premultiplying by its adjoint we find that $I - C^{\sim}C > 0$ on the imaginary axis (including at ∞). Hence, $\|C\|_{\infty} < 1$. •

We are now fully prepared to prove theorem 2.7.

Proof. **theorem 2.7: Closed-loop stability.** The compensator $[X \;\; Y] = [A \;\; B]M_\gamma$ stabilizes the closed-loop system iff $XD_o - YN_o$ is Hurwitz, where $G_{22} = N_oD_o^{-1}$. By the proof of theorem 2.6 in Appendix 2.11 we have

$$XD_o - YN_o = (X - YG_{22})D_o = A(I + TC)(M_{\gamma,11}D_o - M_{\gamma,12}N_o), \quad (2.234)$$

where $C = (M_{\gamma,21} - M_{\gamma,22}G_{22})(M_{\gamma,11} - M_{\gamma,12}G_{22})^{-1}$ and $T = A^{-1}B$. By the way A and B are chosen $\|T\|_\infty \leq 1$. By lemma 2.13, $\|C\|_\infty < 1$, so that $\|TC\|_\infty \leq \|T\|_\infty\|C\|_\infty < 1$. By lemma 2.10 no zeros of $A(I + \varepsilon TC)(M_{\gamma,11}D_o - M_{\gamma,12}N_o)$ cross the imaginary axis as ε varies from 0 to 1. It follows that the compensator stabilizes the closed-loop system iff both A and $M_{\gamma,11}D_o - M_{\gamma,12}N_o$ have all their zeros in the open left-half complex plane.

Hence, a necessary condition for stability is that A have all its zeros in the open left-half complex plane. If this condition is satisfied then the condition that $M_{\gamma,11}D_o - M_{\gamma,12}N_o$ have all its zeros in the open left-half plane is equivalent to the condition that the central compensator stabilizes the closed-loop system. Furthermore, if A has all its zeros in the open left-half plane then the central compensator stabilizes the closed-loop system iff *any* sublevel compensator stabilizes the closed-loop system. •

2.13 Appendix: Proof of the equalizing property

Proof. **Equalizing property.** We prove the equalizing property of type B optimal compensators stated in subsection 2.8.2. By adapting the proof of theorem 2.5 and that of Meinsma's lemma in Appendix 2.11 it may be shown that for any sublevel compensator $[X \;\; Y]$ corresponding to the level γ the matrix $\gamma^2 I - H^\sim H$ is singular iff

$$[X \;\; Y]\Pi_\gamma \begin{bmatrix} X \\ Y \end{bmatrix} \quad (2.235)$$

is singular.

Let $\gamma = \gamma_{\text{opt}}$. Write $\Pi_\gamma = \Lambda_\gamma Z_\gamma^{-1}\Lambda_\gamma^\sim$ as in subsection 2.8.2. Then for any type B optimal compensator defined by (2.164–2.165) we have

$$[X_\gamma \;\; Y_\gamma]\Pi_\gamma \begin{bmatrix} X_\gamma \\ Y_\gamma \end{bmatrix} = AL_{\gamma,1}A^\sim - BL_{\gamma,2}B^\sim \geq 0. \quad (2.236)$$

Consider any point on the imaginary axis. Since $L_{\gamma,1}$ is singular there exists a nontrivial constant vector x such that $x^{\text{H}}AL_{\gamma,1}A^\sim x = 0$ in this point. Because $x^{\text{H}}(AL_{\gamma,1}A^\sim - BL_{\gamma,2}B^\sim)x \geq 0$ it follows that also $x^{\text{H}}BL_{\gamma,2}B^\sim x = 0$. We conclude that $AL_{\gamma,1}A^\sim - BL_{\gamma,2}B^\sim$ is singular in this point, and, hence, also $\gamma^2 I - H^\sim H$.

Thus, H has a singular value $\gamma = \gamma_{\text{opt}}$ everywhere on the imaginary axis. Since by construction this is the largest singular value it follows that the largest singular value of H is constant and equal to γ_{opt}. •

2.14 References

[1] G. J. Balas, J. C. Doyle, K. Glover, A. Packard, and R. Smith. *User's Guide, μ-Analysis and Synthesis Toolbox.* The MathWorks, Natick, Mass., 1991.

[2] O. H. Bosgra and H. Kwakernaak. Design methods for control systems. Course for the Dutch Graduate Network on Systems and Control, 1994. Available from the authors.

[3] F. M. Callier. On polynomial matrix spectral factorization by symmetric extraction. *IEEE Trans. Aut. Control*, 30(5):453–464, 1985.

[4] R. Y. Chiang and M. G. Safonov. *User's Guide, Robust Control Toolbox.* The MathWorks, Natick, Mass., USA, 1992.

[5] J. C. Doyle. Lecture Notes, ONR/Honeywell Workshop on Advances in Multivariable Control, Minneapolis, Minn., 1984.

[6] J. C. Doyle, G. J. Balas, A. Packard, and K. Glover. μ and H_∞: A short course. Cambridge University/Delft University, June 1990, Cambridge Control, Cambridge, UK, 1990.

[7] J. C. Doyle, K. Glover, P. P. Khargonekar, and B. A. Francis. State-space solutions to standard $\mathcal{H}_2$ and $\mathcal{H}_\infty$ control problems. *IEEE Trans. Aut. Control*, 34:831–847, 1989.

[8] B. A. Francis. *A Course in H_∞ Control Theory*, volume 88 of *Lecture Notes in Control and Information Sciences.* Springer-Verlag, Berlin, etc, 1987. Corrected first printing.

[9] K. Glover and J. C. Doyle. State-space formulae for all stabilizing controllers that satisfy an H_∞-norm bound and relations to risk sensitivity. *Systems & Control Letters*, 11:167–172, 1988.

[10] K. Glover, D. J. N. Limebeer, J. C. Doyle, E. M. Kasenally, and M. G. Safonov. A characterization of all solutions to the four block general distance problem. *SIAM J. Control and Optimization*, 29:283–324, 1991.

[11] G. M. Golub and C. Van Loan. *Matrix Computations.* The Johns Hopkins University Press, Baltimore, Maryland, 1983.

[12] M. Green. H^∞ controller synthesis by J-lossless coprime factorization. *SIAM J. Control and Optim.*, 30:522–547, 1992.

[13] T. Kailath. *Linear Systems.* Prentice Hall, Englewood Cliffs, N. J., 1980.

[14] M. Kimura, Y. Lu, and R. Kawatani. On the structure of H^∞ control systems and related extensions. *IEEE Trans. Aut. Control*, 36:653–667, 1990.

[15] V. Kucera. *Discrete Linear Control.* Wiley, Chichester, 1979.

[16] H. Kwakernaak. A polynomial approach to minimax frequency domain optimization of multivariable systems. *Int. J. Control*, 44:117–156, 1986.

[17] H. Kwakernaak. MATLAB macros for polynomial $\mathcal{H}_\infty$ control system optimization. Memorandum 881, Department of Applied Mathematics, University of Twente, September 1990.

[18] H. Kwakernaak. The polynomial approach to $\mathcal{H}_\infty$-optimal regulation. In E. Mosca and L. Pandolfi, editors, *H_∞-Control Theory*, volume 1496 of *Lecture Notes in Mathematics*, pages 141–221. Springer-Verlag, Heidelberg, etc., 1991.

[19] H. Kwakernaak. Polynomial computation of Hankel singular values. In *Proc. 31st IEEE Decision and Control Conference, Tucson, AZ*, pages 3595–3599, December 1992.

[20] H. Kwakernaak. Robust control and H^∞-optimization. *Automatica*, 29:255–273, 1993.

[21] H. Kwakernaak. Frequency domain solution of the standard $\mathcal{H}_\infty$ problem. In *Preprints IFAC Symposium on Robust Control Design*, Rio de Janeiro, September 14–16, 1994.

[22] H. Kwakernaak. State space algorithms for polynomial matrix computations. In *Proceedings EURACO Workshop on Recent Results in Robust and Adaptive Control*, Florence, Italy, September 1995.

[23] H. Kwakernaak and M. Sebek. Polynomial J-spectral factorization. *IEEE Trans. Auto. Control*, 39(2):315–328, 1994.

[24] H. Kwakernaak and R. Sivan. *Modern Signals and Systems.* Prentice Hall, Englewood Cliffs, NJ, 1991.

[25] G. Meinsma. *Frequency domain methods in $\mathcal{H}_\infty$ control.* PhD thesis, University of Twente, 1993.

[26] G. Meinsma. Polynomial solutions to H_∞ problems. *Int. J. of Robust and Nonlinear Control*, 4:323–351, 1994.

[27] G. Meinsma. Unstable and nonproper weights in H_∞ control. Technical Report EE9420, Department of Electrical and Computer Engineering, University of Newcastle, Newcastle, Australia, 1994.

[28] G. Meinsma and H. Kwakernaak. H_∞-optimal control and behaviors. In *Proceedings, First European Control Conference*, pages 1741–1746, Grenoble, France, 1991.

[29] S. Parrott. On a quotient norm and the Sz. Nagy-Foia lifting theorem. *J. Functional Analysis*, 30:311–328, 1978.

[30] A. C. M. Ran and L. Rodman. On symmetric factorizations of rational matrix functions. *Linear and Multilinear Algebra*, 29:243–261, 1991.

[31] A. A. Stoorvogel. *The H-Infinity Control Problem: A State Space Approach.* Prentice Hall, Englewood Cliffs, NJ, 1992.

[32] R. C. W. Strijbos. A polynomial toolbox: Implementation aspects. Memorandum, Department of Applied Mathematics, University of Twente, 1995.

[33] M. Tsai and C. Tsai. A transfer matrix framework approach to the synthesis of H^∞ controllers. *Int. J. of Robust and Nonlinear Control*, 5(3):155–173, 1995.

[34] D. C. Youla. On the factorization of rational matrices. *IRE Trans. Inform. Theory*, IT-7(3):172–189, 1961.

[35] G. Zames. Feedback and optimal sensitivity: Model reference transformations, multiplicative seminorms, and approximate inverses. *IEEE Trans. Aut. Control*, 26:301–320, 1981.

3

LQG Multivariable Regulation and Tracking Problems for General System Configurations

Alessandro Casavola and Edoardo Mosca

3.1 Introduction

This chapter deals with the polynomial equation approach to the Linear Quadratic Gaussian (LQG) regulation and tracking problem for a general system structure that comprises most control system configurations of practical interest. Within this structure, the problem is usually referred to as the "standard" H_2 control problem [5].

Recently, a transfer-matrix Wiener-Hopf approach to this problem was considered in [11] for the discrete-time case and in [21] for the continuous-time one. These solutions, however, suffer from the fact that the controller, not being in irreducible form, need not be free of unstable hidden modes, if the plant is open-loop unstable. In such a case, because of numerical inaccuracy, the stability of the closed-loop system may be lost [11]. The same problem was also approached via polynomial techniques in [10] and [15]. However, both solutions have their own drawbacks. The solution in [10] was determined under special restrictions on the control system structure so that the resulting problem is equivalent to a two-block problem. In contrast, the present solution addresses the full four-block problem. The solution presented in [15] requires some non standard polynomial operations such as a non-square spectral factorizations. Further, in both the above solutions the denominators of the rational system transfer matrices are forced to be block diagonal, introducing unnecessary degree inflation in the system description.

The limits of the above solutions testify that there are difficulties with the direct application of the polynomial equation approach to general system configurations without requiring - like the Stochastic Dynamic Programming approach - the explicit solution of an intermediate filtering problem. A possible remedy is to work with an innovations representation, obtainable from the physical description of the system by solving a mimimum mean-square error filtering (MMSE) problem. This, in turn, can be solved via polynomial equations as well [3], [4]. Consequently, the whole polynomial LQG design for general system configurations explicitly involves a two stage procedure which is reminiscent of the Certainty-Equivalence property of the Stochastic Dynamic Programming solution for the LQG regulation.

All the results of this chapter are given for the discrete-time case. The reader is referred to [1], whose notation is used as much as possible, for the basic facts on polynomial equations and their use in control system theory.

The outline of the chapter is as follows. In section 3.2, the polynomial equation approach to the general LQG problem is formulated and conditions are given for its solvability. In section 3.2.1, the polynomial solution is derived. In section 3.2.2, the connections with the Wiener-Hopf approach are briefly explored and in section 3.2.3 a polynomial procedure is given to obtain the innovations representation of interest for a quite general class of cascade systems. Next, in section 3.2.4, the relationships with other known polynomial solutions are investigated in order to exemplify the theory. In section 3.3, LQG tracking and servo problems are addressed. In particular, it is shown how the regulation LQG solution is modified by the existence of references and accessible disturbances. Finally, in section 3.4, some concluding remarks are reported.

3.2 Regulation problem

Consider the discrete-time, linear, time-invariant multivariable stochastic system of figure 3.1, represented as

$$\begin{bmatrix} z(t) \\ y(t) \end{bmatrix} = \Sigma_p(d) \begin{bmatrix} \nu(t) \\ u(t) \end{bmatrix} = \begin{bmatrix} \hat{Q}(d) & \hat{P}(d) \\ Q(d) & P(d) \end{bmatrix} \begin{bmatrix} \nu(t) \\ u(t) \end{bmatrix}, \tag{3.1}$$

where d is the unit backward shift operator, viz. $dy(t) := y(t-1)$, and $\Sigma_p(d)$ the interconnection matrix that embeds all signal flows between the inputs and the outputs of the system. Equation (3.1), will be referred to as the *physical* system description. In (3.1): $y(t) \in \text{Re}^{\,l}$ denotes the measured variables; $z(t) \in \text{Re}^{\,p}$ accounts for the regulated variables; $u(t) \in \text{Re}^{\,m}$ for the controlled inputs; and $\nu(t) \in \text{Re}^{\,n}$ collects all exogenous signals. $\hat{P}(d) \in \text{Re}^{\,pm}(d)$, $P(d) \in \text{Re}^{\,lm}(d)$, $\hat{Q}(d) \in \text{Re}^{\,pn}(d)$ and $Q(d) \in \text{Re}^{\,ln}(d)$ are causal transfer-matrices, $\text{Re}^{\,pm}(d)$ denoting the set of the $p \times m$ matrices with elements in Re (d), the set of polynomial fractions in the indeterminate d. Moreover, the following assumptions are adopted:

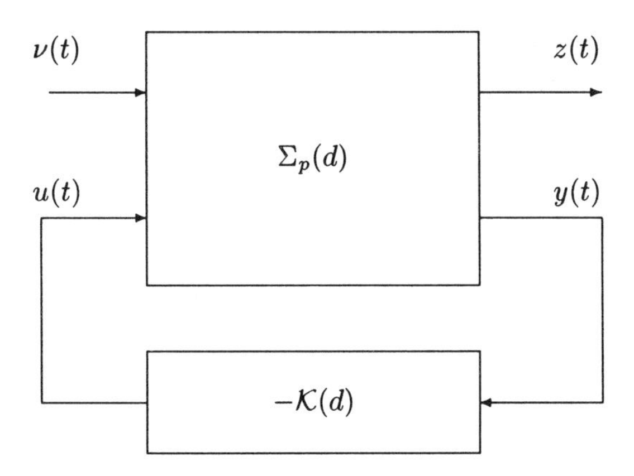

Figure 3.1: Physical system structure.

(A.3.1)
- $\{\nu(t)\} \in \mathrm{Re}^{\,n}$ is a zero-mean white Gaussian sequence with covariance $\Phi_{\nu\nu} > 0$.
- $\hat{P}(d)$ and $P(d)$ are strictly causal transfer matrices.

The problem addressed hereafter is to find, amongst the *linear non anticipative* control laws

$$u(t) = -\mathcal{K}(d)y(t), \tag{3.2}$$

$\mathcal{K} \in \mathrm{Re}^{\,ml}(d)$, the one that, in stochastic steady-state (s.s.s.), stabilizes the system (3.1) and minimizes the unconditional quadratic cost

$$J = \varepsilon\{\|z(t)\|^2_{\Psi_z} + \|u(t)\|^2_{\Psi_u}\}. \tag{3.3}$$

In (3.3): $\Psi_z = \Psi_z' \geq 0$, $\Psi_u = \Psi_u' \geq 0$, $\|\nu\|^2_\Psi := \nu'\Psi\nu$; the prime denotes transpose and ε stands for expectation.

Because of technical difficulties, the above LQG problem (3.1)-(3.3) is hard to be directly solved via polynomial equation approach unless $\dim y = \dim z$ or $\dim y = \dim \nu$. In fact, the case $\dim y = \dim z$ is representative of the typical two-blocks LQG problems solved in the literature. E.g., polynomial LQG solutions are discussed in [9] for cascade configurations under the assumption $\dim z = \dim y$ (two-block problem). The other case, amenable to be solved via polynomial equations is the one where $\dim y = \dim \nu$, irrespectively of $\dim z$. This situation, is also quite limitative if ν denotes the set of physical exogenous signals. A possible way to make the latter situation general, is to substitute the physical representation (3.1) with its *innovations representation* depicted in figure 3.2,

$$\begin{bmatrix} \hat{z}(t) \\ y(t) \end{bmatrix} = \Sigma_i(d) \begin{bmatrix} e(t) \\ u(t) \end{bmatrix} = \begin{bmatrix} \mathcal{H}_{\hat{z}e}(d) & \hat{P}(d) \\ \mathcal{H}_{ye}(d) & P(d) \end{bmatrix} \begin{bmatrix} e(t) \\ u(t) \end{bmatrix}, \tag{3.4}$$

where $\dim e = \dim y$, irrespectively of $\dim \hat{z}$. In (3.4): $e(t) \in \mathrm{Re}^{\,l}$ represents the innovations process of $y(t)$, $e(t) := y(t) - \varepsilon\{y(t) \mid I^{t-1}\}$, where $I^t := \{y^t, u^{t-1}\}$, with $y^t := \{y(t), y(t-1), ...\}$, is the available information up to time t and $\varepsilon\{\cdot \mid I^{t-1}\}$ is the conditional expectation operator based on I^{t-1}; $\hat{z}(t) \in \mathrm{Re}^{\,p}$ is the MMSE estimate of $z(t)$ based on I^t, viz. $\hat{z}(t) := \varepsilon\{z(t) \mid I^t\}$; $\mathcal{H}_{\hat{z}e}(d) \in \mathrm{Re}^{\,pl}(d)$ and $\mathcal{H}_{ye}(d) \in \mathrm{Re}^{\,ll}(d)$ are causal transfer matrices. Therefore, description (3.4) can be drawn out thanks to the linear dependence of $\hat{z}$ and y on I^t and for the strictly causality of $\hat{P}$ and P.

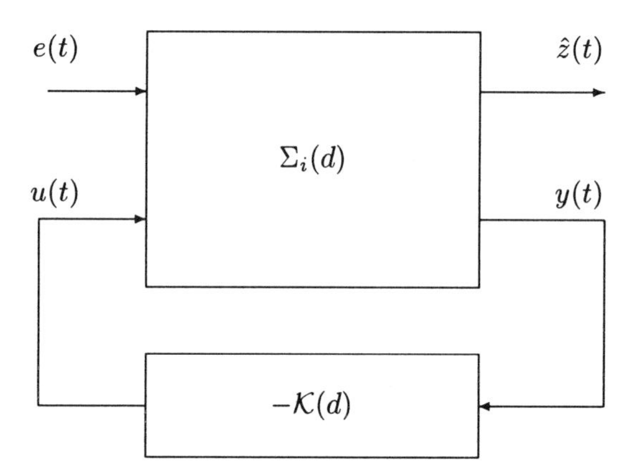

Figure 3.2: Innovations system structure.

The rationale for adopting (3.4) instead of (3.1) is the following. Consider the decomposition $z(t) = \hat{z}(t) + \tilde{z}(t)$, with $\tilde{z}(t)$ independent of I^t, viz. $\varepsilon\{v\tilde{z}'\} = 0$, $\forall v \in \mathcal{L}^m\{I^{t-1}\}$, $\mathcal{L}^m\{z\}$ denoting the subspace of $\mathrm{Re}^{\,m}$-valued random vectors with finite covariance spanned by z. Then, the cost (3.3) can be equivalently rewritten as

$$J = \varepsilon\{\varepsilon\{\|z(t)\|^2_{\Psi_z} + \|u(t)\|^2_{\Psi_u} \mid I^t\}\} = \varepsilon\{\|\hat{z}(t)\|^2_{\Psi_z} + \|u(t)\|^2_{\Psi_u}\} + \varepsilon\{\|\tilde{z}(t)\|^2_{\Psi_z}\},$$

from which, it is seen that the problem of finding the control law (3.2) which minimizes (3.3) is equivalent to minimizing the following quadratic cost

$$\hat{J} = \varepsilon\{\|\hat{z}(t)\|^2_{\Psi_z} + \|u(t)\|^2_{\Psi_u}\}. \tag{3.5}$$

Previous available polynomial solutions can be easily encompassed in this general setting.

Next step is to delineate the class of systems (3.4) for which a stabilizing controller exists. Let (A_Σ, B_Σ) be a right coprime matrix fraction description (r.c.m.f.d.) of $\Sigma_i(d)$, $\Sigma_i(d) = B_\Sigma A_\Sigma^{-1}$, with $A_\Sigma \in \mathrm{Re}^{\,(l+m)(l+m)}[d]$ and

$B_\Sigma \in \mathrm{Re}^{(p+l)(l+m)}[d]$ polynomial matrices with elements in $R[d]$, the ring of polynomials in d. Further, let (A_p, B_p) be a left coprime matrix fraction description (l.c.m.f.d.) of $P(d)$, $P = A_p^{-1}B_p$, with $A_p \in \mathrm{Re}^{ll}[d]$ and $B_p \in \mathrm{Re}^{lm}[d]$.

The following are assumed.

(A.3.2)
- The system to be controlled is a realization of Σ_i free of unstable hidden modes.
- Every unstable root of the characteristic polynomial ($\det A_\Sigma$) of the irreducible m.f.d. $B_\Sigma A_\Sigma^{-1}$ of Σ_i is also a root with the same multiplicity of the characteristic polynomial ($\det A_p$) of the irreducible m.f.d. $A_p^{-1}B_p$ of P.

The second condition of (A.3.2) states that all the unhidden unstable modes of the overall system are the ones due to the realization of P. In [7], it is shown that (A.3.2) is a necessary and sufficient condition for the stabilizability of Σ_i via a dynamic compensator acting on u only and having input y. In such a case, it is proved that $\mathcal{K}$ stabilizes Σ_i if and only if it stabilizes P.

The stabilizability conditions (A.3.2) have a tight connection with the structure of any left and right coprime matrix fraction description of all stabilizable systems. To this end, consider the following lemma, whose proof is reported in the appendix.

Lemma 3.1. *Let the system (3.4) be stabilizable by a dynamic control law of the form $u = -\mathcal{K}y$, $\mathcal{K} \in R^{ml}(d)$, viz. (A.3.2) be fulfilled. Then, there exist a r.c.m.f.d. of Σ_i of the form*

$$\Sigma_i = \begin{bmatrix} \hat{C} & \hat{B}_2 \\ C & B_2 \end{bmatrix} \begin{bmatrix} A_{2s} & 0 \\ A_{lm} & A_2 \end{bmatrix}^{-1} = \begin{bmatrix} \hat{C}A_{2s}^{-1} - \hat{B}_2 A_2^{-1} A_{lm} A_{2s}^{-1} & \hat{B}_2 A_2^{-1} \\ CA_{2s}^{-1} - B_2 A_2^{-1} A_{lm} A_{2s}^{-1} & B_2 A_2^{-1} \end{bmatrix} \tag{3.6}$$

with $A_{2s} \in \mathrm{Re}^{ll}[d]$ strictly Schur, and a l.c.m.f.d. of Σ_i of the form

$$\Sigma_i = \begin{bmatrix} A_{1s} & A_{pl} \\ 0 & A_1 \end{bmatrix}^{-1} \begin{bmatrix} \hat{D} & \hat{B}_1 \\ D & B_1 \end{bmatrix} = \begin{bmatrix} A_{1s}^{-1}\hat{D} - A_{1s}^{-1}A_{pl}A_1^{-1}D & A_{1s}^{-1}\hat{B}_1 - A_{1s}^{-1}A_{pl}A_1^{-1}B_1 \\ A_1^{-1}D & A_1^{-1}B_1 \end{bmatrix}, \tag{3.7}$$

with $A_{1s} \in \mathrm{Re}^{pp}[d]$ strictly Schur. (A_2, B_2) and (A_1, B_1) are respectively a right and left m.f.d.'s of P, whose only *possible right and, respectively, left divisors are* strictly Schur. □

It is further assumed that

(A.3.3) { • D is Schur.

Assumption (A.3.3) is equivalent to assuming that the innovations process of $y(t)$ is nonsingular. In fact, D can be thought as a Schur spectral factor, solution of the spectral factorization problem involved in the determination of the innovations process of $y(t)$.

3.2.1 Problem solution

Consider the system (3.4), described by left and right matrix fraction descriptions (3.6) and (3.7). Assume that (A.3.1)-(A.3.3) are fulfilled. By the sake of simplicity, the argument d will be dropt hereafter unless required to avoid possible confusion.

Further, in order to solve the problem, the quadratic cost (3.5) can be rewritten as follows,

$$\hat{J} = tr\{\prec \Psi_z \mathcal{W}_{\hat{z}} \succ + \prec \Psi_u \mathcal{W}_u \succ\} \tag{3.8}$$

where: tr denotes *trace*; $\mathcal{W}_{\hat{z}}(d)$ and $\mathcal{W}_u(d)$ stand for the closed-loop power density functions of $\hat{z}(t)$ and $u(t)$, $\mathcal{W}_s(d) := \sum_{j=-\infty}^{\infty} \varepsilon\{s(t+j)s'(t)\}d^j$; and the operator $\prec \mathcal{W}(d) \succ = \prec \sum_{j=-\infty}^{\infty} w(j)d^j \succ := w(0)$ denotes extraction of the zero-power coefficient from the two-sided matrix sequence $\mathcal{W}(d)$.

Let, respectively, $(\mathcal{N}_1, \mathcal{M}_1)$ and $(\mathcal{N}_2, \mathcal{M}_2)$ be left and right fractional representations of the stabilizing controllers $\mathcal{K}$, viz.

$$\mathcal{K} = \mathcal{M}_1^{-1}\mathcal{N}_1 = \mathcal{N}_2\mathcal{M}_2^{-1}, \tag{3.9}$$

with $\mathcal{N}_1 \in \text{Re}^{\,ml}(d)$, $\mathcal{M}_1 \in \text{Re}^{\,mm}(d)$, $\mathcal{N}_2 \in \text{Re}^{\,ml}(d)$ and $\mathcal{M}_2 \in \text{Re}^{\,ll}(d)$ causal and stable transfer matrices fulfilling the identities

$$A_1\mathcal{M}_2 + B_1\mathcal{N}_2 = I_l \tag{3.10}$$

$$\mathcal{M}_1 A_2 + \mathcal{N}_1 B_2 = I_m. \tag{3.11}$$

By simple algebraic manipulations, the following closed-loop expressions can be obtained by considering (3.6), (3.7), (3.10) and (3.11):

$$\begin{aligned}
\begin{bmatrix} \hat{z}(t) \\ y(t) \end{bmatrix} &= \left[I_{p+l} + \begin{bmatrix} \hat{P} \\ P \end{bmatrix} \begin{bmatrix} 0 \; \mathcal{N}_2 \end{bmatrix} \begin{bmatrix} \mathcal{M}_o & 0 \\ 0 & \mathcal{M}_2 \end{bmatrix}^{-1} \right]^{-1} \begin{bmatrix} \mathcal{H}_{\hat{z}e} \\ \mathcal{H}_{ye} \end{bmatrix} e(t) \\
&= \begin{bmatrix} I_p & -T\mathcal{N}_2\mathcal{M}_2^{-1}(I_l + P\mathcal{N}_2\mathcal{M}_2^{-1})^{-1} \\ 0 & (I_l + P\mathcal{N}_2\mathcal{M}_2^{-1})^{-1} \end{bmatrix} \begin{bmatrix} \mathcal{H}_{\hat{z}e} \\ \mathcal{H}_{ye} \end{bmatrix} e(t) \\
&= \begin{bmatrix} I_p & -\hat{B}_2\mathcal{N}_1 \\ 0 & \mathcal{M}_2 A_1 \end{bmatrix} \begin{bmatrix} \mathcal{H}_{\hat{z}e} \\ \mathcal{H}_{ye} \end{bmatrix} e(t)
\end{aligned} \tag{3.12}$$

$$\begin{aligned}
u(t) &= -\begin{bmatrix} 0 \; \mathcal{N}_2 \end{bmatrix} \begin{bmatrix} \mathcal{M}_o & 0 \\ 0 & \mathcal{M}_2 \end{bmatrix}^{-1} \begin{bmatrix} I_p & -\hat{B}_2\mathcal{N}_1 \\ 0 & \mathcal{M}_2 A_1 \end{bmatrix} \begin{bmatrix} \mathcal{H}_{\hat{z}e} \\ \mathcal{H}_{ye} \end{bmatrix} e(t) \\
&= \begin{bmatrix} 0 \; -A_2\mathcal{N}_1 \end{bmatrix} \begin{bmatrix} \mathcal{H}_{\hat{z}e} \\ \mathcal{H}_{ye} \end{bmatrix} e(t)
\end{aligned} \tag{3.13}$$

Therefore, the power density functions in (3.8) result to be given by

$$\begin{aligned}
\mathcal{W}_{\hat{z}} &= \mathcal{V}_{\hat{z}}\mathcal{V}_{\hat{z}}^* \\
&= \begin{bmatrix} I_p \; -\hat{B}_2\mathcal{N}_1 \end{bmatrix} \begin{bmatrix} A_{1s} & A_{pl} \\ 0 & A_1 \end{bmatrix}^{-1} \begin{bmatrix} \hat{D} \\ D \end{bmatrix} \\
&\quad \begin{bmatrix} \hat{D}^* \; D^* \end{bmatrix} \begin{bmatrix} A_{1s} & A_{pl} \\ 0 & A_1 \end{bmatrix}^{-*} \begin{bmatrix} I_p \\ -\mathcal{N}_1^*\hat{B}_2^* \end{bmatrix}
\end{aligned} \tag{3.14}$$

$$\mathcal{W}_u = \mathcal{V}_u \mathcal{V}_u^*$$
$$= \begin{bmatrix} 0 & -A_2\mathcal{N}_1 \end{bmatrix} \begin{bmatrix} A_{1s} & A_{pl} \\ 0 & A_1 \end{bmatrix}^{-1} \begin{bmatrix} \hat{D} \\ D \end{bmatrix}$$
$$\begin{bmatrix} \hat{D}^* & D^* \end{bmatrix} \begin{bmatrix} A_{1s} & A_{pl} \\ 0 & A_1 \end{bmatrix}^{-*} \begin{bmatrix} 0 \\ -\mathcal{N}_1^* A_2^* \end{bmatrix} \tag{3.15}$$

where the star denotes the conjugate operator, viz. $\mathcal{H}^*(d) := \mathcal{H}'(d^{-1})$; and $A^{-*} := \left(A^{-1}\right)^*$. Further, denote for simplicity

$$\mathcal{Y}_d := \begin{bmatrix} A_{1s} & A_{pl} \\ 0 & A_1 \end{bmatrix}^{-1} \begin{bmatrix} \hat{D} \\ D \end{bmatrix} = \begin{bmatrix} A_{1s}^{-1} & -A_{1s}^{-1}A_{pl}A_1^{-1} \\ 0 & A_1^{-1} \end{bmatrix} \begin{bmatrix} \hat{D} \\ D \end{bmatrix}. \tag{3.16}$$

Then, by using (3.14) and (3.15) in (3.8) and taking into account (3.16), one gets:

$$\hat{J} = tr \prec \mathcal{V}_{\hat{z}}^* \Psi_{\hat{z}} \mathcal{V}_z + \mathcal{V}_u^* \Psi_u \mathcal{V}_u \succ$$
$$= tr \prec \mathcal{Y}_d^* \left(\begin{bmatrix} \Psi_z & -\Psi_z \hat{B}_2 \mathcal{N}_1 \\ -\mathcal{N}_1^* \hat{B}_2^* \Psi_z & \mathcal{N}_1^* \hat{B}_2^* \Psi_z \hat{B}_2 \mathcal{N}_1 \end{bmatrix} + \begin{bmatrix} 0 & 0 \\ 0 & \mathcal{N}_1^* A_2^* \Psi_u A_2 \mathcal{N}_1 \end{bmatrix} \right) \mathcal{Y}_d \succ$$
$$= tr \prec \mathcal{Y}_d^* \begin{bmatrix} \Psi_z & -\Psi_z \hat{B}_2 \mathcal{N}_1 \\ -\mathcal{N}_1^* \hat{B}_2^* \Psi_z & \mathcal{N}_1^* E^* E \mathcal{N}_1 \end{bmatrix} \mathcal{Y}_d \succ . \tag{3.17}$$

In the last expression, E is a Schur right-spectral factor, $E \in \text{Re}^{\,mm}[d]$, solving the following right-spectral factorization problem

$$\boxed{E^* E = A_2^* \Psi_u A_2 + \hat{B}_2^* \Psi_z \hat{B}_2} \tag{3.18}$$

The following is also assumed for the existence of E [1]:

$$(\mathbf{A.3.4}) \left\{ \bullet \text{ rank} \begin{bmatrix} \Psi_u A_2 \\ \Psi_z \hat{B}_2 \end{bmatrix} (e^{j}\omega) = m, \qquad \forall \omega \in [0, 2\pi] \right.$$

e.g. (A.3.4) is fulfilled whenever $\Psi_u > 0$. The cost expression (3.17) can be conveniently split as follows

$$J = J_1 + J_2$$
$$J_1 := tr \prec L^* L \succ \tag{3.19}$$
$$J_2 := tr \prec \mathcal{Y}_d^* \begin{bmatrix} \Psi_z - \Psi_z \hat{B}_2 E^{-1} E^{-*} \hat{B}_2^* \Psi_z & 0 \\ 0 & 0 \end{bmatrix} \mathcal{Y}_d \succ \tag{3.20}$$
$$L := \begin{bmatrix} -E^{-*} \hat{B}_2^* \Psi_z & E \mathcal{N}_1 \end{bmatrix} \mathcal{Y}_d. \tag{3.21}$$

Since J_2 does not depend on $\mathcal{N}_1$, minimizing J is equivalent to minimizing J_1. Define

$$p := \max\{\partial A_2, \partial \hat{B}_2, \partial E\},$$

∂A_2 denoting the degree of A_2. Let

$$\bar{A}_2 := d^p A_2^*, \quad \bar{\hat{B}}_2 := d^p \hat{B}_2^*, \quad \bar{E} := d^p E^* \tag{3.22}$$

Further, define the following r.c.m.f.d.'s $D_2 A_3^{-1}$ and $[\hat{D}A_3 - A_{pl}D_2]_2 A_{3s}^{-1}$ of respectively $A_1^{-1}D$ and $A_{1s}^{-1}(\hat{D}A_3 - A_{pl}D_2)$, such that

$$D_2 A_3^{-1} = A_1^{-1}D, \quad [\hat{D}A_3 - A_{pl}D_2]_2 A_{3s}^{-1} = A_{1s}^{-1}(\hat{D}A_3 - A_{pl}D_2), \tag{3.23}$$

with A_{3s} strictly Schur since A_1 is such. Then, considering that $E^{-*}\hat{B}_2^* = \bar{E}^{-1}\bar{\hat{B}}_2$, and taking into account (3.16), (3.22) and (3.23), (3.21) can be rewritten as follows

$$\begin{aligned} L &= -\bar{E}^{-1}\bar{\hat{B}}_2\Psi_z A_{1s}^{-1}(\hat{D}A_3 - A_{pl}D_2)A_3^{-1} + E\mathcal{N}_1 D_2 A_3^{-1} \\ &= -\bar{E}^{-1}\bar{\hat{B}}_2\Psi_z[\hat{D}A_3 - A_{pl}D_2]_2 A_{3s}^{-1}A_3^{-1} + E\mathcal{N}_1 D_2 A_3^{-1}. \end{aligned} \tag{3.24}$$

Then, in order to minimize J_1 is convenient to reduce the two-sided sequence matrix L to a sum of a causal plus a strictly-anticausal Schur sequence matrix. This causal/strictly-anticausal decomposition can be accomplished by considering the following Diophantine equation

$$\boxed{\bar{E}Y + ZA_3A_{3s} = \bar{\hat{B}}_2\Psi_z[\hat{D}A_3 - A_{pl}D_2]_2} \tag{3.25}$$

to be solved, provided that a solution exists, for the polynomial matrix pair (Y, Z) with the degree condition $\partial Z < \partial \bar{E}$. Based on (3.25), (3.24) can be reduced to

$$L = -YA_{3s}^{-1}A_3^{-1} - \bar{E}^{-1}Z + E\mathcal{N}_1 D_2 A_3^{-1}. \tag{3.26}$$

Consequently, the cost term J_1 becomes

$$J_1 = J_3 + J_4 + J_5, \tag{3.27}$$

with

$$J_3 := tr \prec Z^*\bar{E}^{-*}\bar{E}^{-1}Z \succ \tag{3.28}$$

$$J_4 := tr \prec A_3^{-*}A_{3s}^{-*}(Y - E\mathcal{N}_1 D_2 A_{3s})^*(Y - E\mathcal{N}_1 D_2 A_{3s})A_{3s}^{-1}A_3^{-1} \succ \tag{3.29}$$

$$\begin{aligned} J_5 := tr\{&\prec Z^*\bar{E}^{-*}(Y - E\mathcal{N}_1 D_2 A_{3s})A_{3s}^{-1}A_3^{-1} \succ \\ &+ \prec A_3^{-*}A_{3s}^{-*}(Y - E\mathcal{N}_1 D_2 A_{3s})^*\bar{E}^{-1}Z \succ\} = 0 \end{aligned} \tag{3.30}$$

The cost term $J_5 = tr \prec Z^*\bar{E}^{-*}(Y - E\mathcal{N}_1 D_2 A_{3s})A_{3s}^{-1}A_3^{-1} \succ = tr \prec A_3^{-*}A_{3s}^{-*}(Y - E\mathcal{N}_1 D_2 A_{3s})^*\bar{E}^{-1}Z \succ = 0$, since $\bar{E}^{-1}Z$ is a strictly anticausal sequence matrix. This follows from the fact that $\bar{E}$ is *regular* [1], viz. the coefficient matrix of the higher power of d in $\bar{E}$ is nonsingular, and thanks to the degree condition $\partial Z < \partial \bar{E}$ of the required solution of (3.25).

Therefore, the minimum is attained for $Y - E\mathcal{N}_1 D_2 A_{3s} = 0$, i.e.

$$\boxed{\mathcal{N}_1 = E^{-1}YA_{3s}^{-1}D_2^{-1}} \tag{3.31}$$

that is a causal and Schur transfer matrix.

By simple algebraic manipulations, the corresponding minimal cost turns out to be

$$J_{min} = tr \prec A_3^{-*} A_{3s}^{-*} ([\hat{D}A_3 - A_{pl}D_2]_2^* \Psi_z [\hat{D}A_3 - A_{pl}D_2]_2 - Y^*Y) A_{3s}^{-1} A_3^{-1} \succ .$$

A second Diophantine equation, to be satisfied along (3.25), must be deduced from stability requirements. This second equation allows one to uniquely determine $\mathcal{M}_1$. To this end, by considering (3.7), (3.11), (3.18) and (3.25), after few straightforward steps, one finds

$$\mathcal{M}_1 A_2 = I_m - \mathcal{N}_1 B_2 \quad \Longrightarrow \bar{E}E\mathcal{M}_1 = \bar{E}EA_2^{-1} - \bar{E}E\mathcal{N}_1 B_2 A_2^{-1}$$

$$\begin{aligned}
\bar{E}E\mathcal{M}_1 &= \bar{A}_2\Psi_u + \bar{\hat{B}}_2\Psi_z\hat{B}_2A_2^{-1} - \bar{E}YA_{3s}^{-1}D_2^{-1}B_2A_2^{-1} \\
&= \bar{A}_2\Psi_u + \bar{\hat{B}}_2\Psi_z\hat{B}_2A_2^{-1} \\
&\quad -(\bar{\hat{B}}_2\Psi_z[\hat{D}A_3 - A_{pl}D_2]_2 - ZA_3A_{3s})A_{3s}^{-1}D_2^{-1}B_2A_2^{-1} \\
&= \bar{A}_2\Psi_u + ZA_3D_2^{-1}B_2A_2^{-1} \\
&\quad +\bar{\hat{B}}_2\Psi_zA_{1s}^{-1}\left[A_{1s} \; -(\hat{D}A_3 - A_{pl}D_2)D_2^{-1}\right]\begin{bmatrix}\hat{B}_2\\ B_2\end{bmatrix}A_2^{-1} \\
&= \bar{A}_2\Psi_u + ZA_3D_2^{-1}A_1^{-1}B_1 \\
&\quad +\bar{\hat{B}}_2\Psi_zA_{1s}^{-1}\left(\hat{B}_1 - \hat{D}A_3D_2^{-1}A_1^{-1}B_1\right).
\end{aligned} \tag{3.32}$$

Further, define the following r.c.m.f.d.'s $B_3D_1^{-1}$ and $[\hat{B}_1D_1 - \hat{D}B_3]_1A_{4s}^{-1}$ for $D^{-1}B_1$ and $A_{1s}^{-1}(\hat{B}_1D_1 - \hat{D}B_3)$ respectively, such that

$$B_3D_1^{-1} = D^{-1}B_1, \quad [\hat{B}_1D_1 - \hat{D}B_3]_1A_{4s}^{-1} = A_{1s}^{-1}(\hat{B}_1D_1 - \hat{D}B_3) \tag{3.33}$$

with A_{4s} strictly Schur since A_{1s} is such. Considering (3.23) and (3.33) in (3.32), one arrives at

$$\bar{E}E\mathcal{M}_1D_1A_{4s} = \bar{A}_2\Psi_uD_1A_{4s} + ZB_3A_{4s} + \bar{\hat{B}}_2\Psi_z[\hat{B}_1D_1 - \hat{D}B_3]_1. \tag{3.34}$$

Then, for closed-loop stability, the R.H.S. of the polynomial equality (3.34) must have $\bar{E}$ as left divisor, i.e. the following bilateral Diophantine equation must be satisfied

$$\boxed{\bar{E}X - ZB_3A_{4s} = \bar{A}_2\Psi_uD_1A_{4s} + \bar{\hat{B}}_2\Psi_z[\hat{B}_1D_1 - \hat{D}B_3]_1} \tag{3.35}$$

for some polynomial matrix X. Finally, by using (3.35) in (3.34) one finds

$$\boxed{\mathcal{M}_1 = E^{-1}XA_{4s}^{-1}D_1^{-1}} \tag{3.36}$$

A condition under which (3.25) and (3.35) are solvable with $\partial Z < \partial \bar{E}$ is given by the next lemma.

Lemma 3.2 ([1]). *Let the greatest common left divisor of (A_1, B_1), with (A_1, B_1) l.m.f.d. of P, $P = A_1^{-1}B_1$ be strictly Schur. Then, there exists a unique solution (X, Y, Z) of (3.25) and (3.35) with $\partial Z < \partial \bar{E}$.* □

Finally, all the results of this section are summarized in the following theorem.

Theorem 3.3. *Let the assumptions (A.3.1)-(A.3.4) be fulfilled. Let E be a Schur right-spectral factor solution of the right-spectral factorization problem (3.18). Then, the pair of bilateral Diophantine equations (3.25) and (3.35) have a unique solution (X, Y, Z) with $\partial Z < \partial \bar{E}$.*

Further, provided that $\mathcal{M}_1$ and $\mathcal{N}_1$, given by (3.36) and, respectively, (3.31) are stable transfer matrices, the optimal stabilizing LQG regulator is given by

$$\mathcal{K} = \mathcal{M}_1^{-1}\mathcal{N}_1.$$

Proof. A constructive derivation of this result has been previously given. Further, the condition (A.3.2) ensures that P is free of unstable hidden modes. Consequently lemma 3.2 can be exploited so as to guarantee solvability of (3.25) and (3.35). □

It is to be pointed out that stability of the closed-loop system is guaranteed from the outset if E and D are both strictly Schur. When E or D are Schur in a wide-sense only, stability has to be checked after $\mathcal{N}_1$ and $\mathcal{M}_1$ are evaluated [1].

3.2.2 Connection with the Wiener-Hopf solution

The Wiener-Hopf solution of the general LQG stochastic regulation problem here addressed, has been presented in [11] for the discrete-time case and, more recently, Park and Bongiorno [21] gave analogue solution for the continuous-time one. In both references the solution is fully general but suffers from some disadvantages. The first of them is that the optimal controller is not given in irreducible form and some cancellation procedure has to be used to achieved a controller of minimal degree. Further, because the open-loop modes of the system are common factors of the optimal controller realization, difficulties can arise for open-loop unstable systems ([11] - pag. 113).

The main aim of this section is to show how Wiener-Hopf procedure can be reduced to the polynomial one presented in the previous section, when an innovations model is used to represent the system. In this way, all of the above mentioned difficulties disappeared.

To this end, consider the physical description (3.1). Thus, the above problem can be formulated in the Wiener-Hopf approach as follows:

Determine the optimal control sensitivity matrix

$$\mathcal{M} = \mathcal{K}(I + P\mathcal{K})^{-1}, \tag{3.37}$$

with $\mathcal{K}$ a stabilizing controller as in (3.4), that, in s.s.s, minimizes

$$\begin{aligned} J &= tr\{\prec \Psi_z W_z \succ + \prec \Psi_u W_u \succ\} = \\ &= tr\{\prec (\hat{Q} - \hat{P}\mathcal{M}Q)^* \Psi_z (\hat{Q} - \hat{P}\mathcal{M}Q) \succ + \prec \hat{P}^* \mathcal{M}^* \Psi_u \mathcal{M}\hat{P} \succ\}. \end{aligned} \tag{3.38}$$

The solution is given by [11]

$$\mathcal{M} = \mathcal{Y}_e^{-1} \left\{ \mathcal{Y}_e^{-*} \hat{P}^* \Psi_z \hat{Q} Q^* \mathcal{Y}_f^{-*} \right\}_+ \mathcal{Y}_f^{-1}, \tag{3.39}$$

where $\mathcal{Y}_e$ and $\mathcal{Y}_f$ are the generalized spectral factors [22] defined by

$$\mathcal{Y}_e^* \mathcal{Y}_e := \hat{P}^* \Psi_z \hat{P} + \Psi_u \tag{3.40}$$

$$\mathcal{Y}_f \mathcal{Y}_f^* := Q Q^* \tag{3.41}$$

and $\{\cdot\}_+$ denotes the causal component in the partial fraction expansion, i.e. $\{\sum_{j=-\infty}^{\infty} w(j) d^j\}_+ = \sum_{j=0}^{\infty} w(j) d^j$.

Then, the optimal controller can be computed by $\mathcal{M}$ and it is given by

$$\mathcal{K} = (I - \mathcal{M}P)^{-1} \mathcal{M}. \tag{3.42}$$

In order to show as obtaining a polynomial solution of the Wiener-Hopf problem, we apply the above procedure to the innovations representation (3.4). To this end, let us assume hereafter that (3.6) and (3.7) represent respectively a right and left c.m.f.d's of (3.4). Further, denote $\theta = \mathcal{Y}_e^{-*} \hat{P}^* \Psi_z \mathcal{H}_{\hat{z}e} \mathcal{H}_{ye}^* \mathcal{Y}_f^{-*}$ for notational simplicity.

Based on the above setting, the generalized spectral factors $\mathcal{Y}_e$ and $\mathcal{Y}_f$ and the term θ can be rewritten in polynomial form as follows

$$\begin{aligned}
\mathcal{Y}_e^* \mathcal{Y}_e &:= \hat{P}^* \Psi_z \hat{P} + \Psi_u = A_2^{-*} (A_2^* \Psi_u A_2 + \hat{B}_2^* \Psi_z \hat{B}_2) A_2^{-1} \\
&= A_2^{-*} E^* E A_2^{-1} \qquad \Longrightarrow \mathcal{Y}_e = E A_2^{-1} \\
\mathcal{Y}_f \mathcal{Y}_f^* &= \mathcal{H}_{ye} \mathcal{H}_{ye}^* = A_1^{-1} D D^* A_1^{-*} \qquad \Longrightarrow \mathcal{Y}_f = A_1^{-1} D \quad \text{for (A.3.3)} \\
\theta &= E^{-*} A_2^* A_2^{-*} \hat{B}_2^* \Psi_z A_{1s}^{-1} \left(\hat{D} - A_{pl} A_1^{-1} D \right) D^* A_1^{-*} A_1^* D^{-*} \\
&= \bar{E}^{-1} \bar{\hat{B}}_2 \Psi_z [\hat{D} A_3 - A_{pl} D_2]_2 A_{3s}^{-1} A_3^{-1}
\end{aligned}$$

with E, A_3, A_{3s}, $[\hat{D} A_3 - A_{pl} D_2]_2$, D_2, $\bar{\hat{B}}_2$ and $\bar{E}$ defined in the previous section.

Thus, in order to accomplish the causal extraction $\{\theta\}_+$, the following Diophantine equation can be introduced

$$\bar{E} Y + Z A_3 A_{3s} = \bar{\hat{B}}_2 \Psi_z [\hat{D} A_3 - A_{pl} D_2]_2 \tag{3.43}$$

Provided that (3.43) has an unique solution (Y, Z), with $\partial Z < \partial \bar{E}$, the optimal $\mathcal{M}$ can be uniquely determined, taking into account the following causal extraction

$$\{\theta\}_+ = \{\bar{E}^{-1} Z + Y A_{3s}^{-1} A_3^{-1}\}_+ = Y A_{3s}^{-1} A_3^{-1}, \tag{3.44}$$

based on foregoing arguments.

Consequently,

$$\begin{aligned}
\mathcal{M} &= \mathcal{Y}_e^{-1} Y A_{3s}^{-1} A_3^{-1} \mathcal{Y}_f^{-1} \\
&= A_2 E^{-1} Y A_{3s}^{-1} A_3^{-1} D^{-1} A_1 \\
&= A_2 E^{-1} Y A_{3s}^{-1} D_2^{-1}.
\end{aligned} \tag{3.45}$$

When A_3 is not strictly Schur, however, (3.43) could have more solutions with $\partial Z < \partial \bar{E}$. This happens when $\det(\bar{E})$ and $\det(A_3)$ have common factors [17]. Therefore, according to what done in the previous section, a second Diophantine equation, to be satisfied along with (3.43), must be deduced from stability requirements. In fact, the optimization procedure requires that $\mathcal{Y}_e \mathcal{M} \mathcal{Y}_f - Y A_{3s}^{-1} A_3^{-1}$ is strictly Schur. The second Diophantine equation can be determined as done previously by considering (3.9), (3.11), (3.18), (3.43) and (3.44). It equals, as expected,

$$\bar{E}X - ZB_3A_{4s} = \bar{A}_2 \Psi_u D_1 A_{4s} + \bar{\hat{B}}_2 \Psi_z [\hat{B}_1 D_1 - \hat{D} B_3]_1 \tag{3.46}$$

with A_{4s} and $[\hat{B}_1 D_1 - \hat{D} B_3]_1$ defined in (3.33). Finally, by using (3.18), (3.43) and (3.46), the control law (3.42) can be simplified to

$$\begin{aligned}
\mathcal{K} &= \left(I - A_2 E^{-1} Y A_{3s}^{-1} D_2^{-1} B_2 A_2^{-1} A_2^{-1}\right)^{-1} A_2 E^{-1} Y A_{3s}^{-1} D_2^{-1} \\
&= A_2 \left(\bar{E}E - \bar{E} Y A_{3s}^{-1} D_2^{-1} B_2\right)^{-1} \bar{E} Y A_{3s}^{-1} D_2^{-1} \\
&= (E^{-1} X A_{4s}^{-1} D_1^{-1})^{-1} (E^{-1} Y A_{3s}^{-1} D_2^{-1}),
\end{aligned} \tag{3.47}$$

that equals the expression for $\mathcal{K}$ given in theorem 3.3 by considering (3.31) and (3.36).

3.2.3 Innovations representations

In order to clarify the above arguments a simple and well known regulation problem will be solved with the proposed general procedure. To this end, a polynomial procedure is given that allows one to determine an innovations model for the special class of cascade systems configuration. This class, although it is not the most general one can consider, compries most of the regulation problems treated in literature and it is, anyway, sufficient for the exemplification purposes of this section. Consider the physical cascade system of figure 3.3, that is

$$\begin{aligned}
z(t) &= A_1^{-1} C_d \nu_1(t) + A_1^{-1} B_1 u(t) \\
y(t) &= A_f^{-1} B_f z(t) + A_n^{-1} C_n \nu_2(t).
\end{aligned} \tag{3.48}$$

In (3.48) $A_1 \in \mathrm{Re}^{\,pp}[d]$, $B_1 \in \mathrm{Re}^{\,pm}[d]$, $A_f \in \mathrm{Re}^{\,ll}[d]$, $B_f \in \mathrm{Re}^{\,lp}[d]$, $C_d \in \mathrm{Re}^{\,pp}[d]$, $C_n \in \mathrm{Re}^{\,ll}[d]$ and $A_n \in \mathrm{Re}^{\,ll}[d]$ strictly Schur. Further, let $\nu_1 \in \mathrm{Re}^{\,p}$ and $\nu_2 \in \mathrm{Re}^{\,l}$ be zero-mean jointly Gaussian and mutually independent stationary sequences with covariance $\Phi_{\nu_1} > 0$ and, respectively $\Phi_{\nu_2} > 0$.

The structural specialization (3.48) is quite natural in control applications where output and measurement disturbances act in cascade along the same path. In particular, $A_1^{-1} C_d \nu_1$ and $A_n^{-1} C_n \nu_2$ are coloured noise sources acting on the regulated and, respectively, measured signals and $A_f^{-1} B_f$ represents the feedback dynamic due to the sensors.

A polynomial LQG solution for this class of systems, under the limitative assumption $\dim z = \dim y$, can be found in [9]. In the present context, such a restriction can be removed.

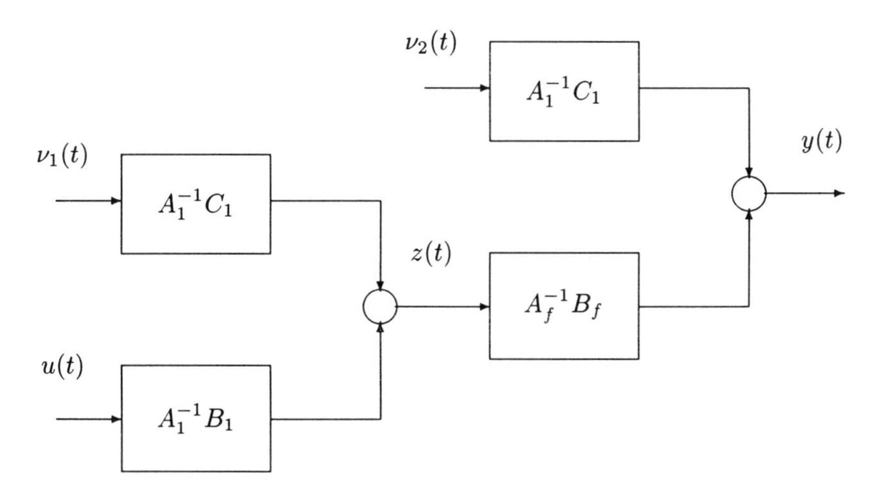

Figure 3.3: Cascade system structure.

Remark 3.4. In order to derive the innovations model of interest, in the sequel $u(t)$ will be set equal to zero w.l.o.g.. The effect of $u(t)$ on $y(t)$ and $z(t)$ will be properly taken into account later in a separate way. Stationary Kalman filtering theory is a relevant example of the above setup. □

Notice that the structural system specialization in (3.48) regards only the way $u(t)$ affects the systems, whereas it is fully general with respect to the exogenous signals. In fact, there is any loss of generality in assuming that the set of the exogenous signals ν can be partitioned in two groups, one of them, ν_2, not affecting z. In fact, all exogenous signals affecting z, but not y, do not change the solution and, hence, can be disregarded in solving the problem. As a consequence, taking into account the considerations of remark 3.4, the exogenous signals representation of (3.48) is fully general and can be rewritten as

$$\begin{aligned} A_1 z(t) &= C_d \nu_1(t) \\ A_{n1} A_f y(t) &= A_{n1} B_f z(t) + A_{f1} C_n \nu_2(t) \end{aligned} \tag{3.49}$$

with A_{n1} and A_{f1} such that

$$A_f A_n^{-1} = A_{n1}^{-1} A_{f1}. \tag{3.50}$$

The control design problem requires the determination of the innovations system representation for (3.48). The latter, for any problems, can be converted in the solution of the following MMSE Multichannel Deconvolution Problem:

MMSE MD Problem - Let $z(t) = \hat{z}(t) + \tilde{z}(t)$ be a suitable orthogonal decomposition of $z(t)$, with $\hat{z}(t) := \varepsilon\{z(t) \mid y^t\}$ the linear filtered estimate of $z(t)$ based on y^t. Thus, the problem is to determine, amongst the stable and causal transfer matrices, the optimal deconvolution filter $\mathcal{H}_{\hat{z}y}$ minimizing

$$J = tr\ \varepsilon\{\tilde{z}(t)\tilde{z}'(t)\}. \tag{3.51}$$

Then, $\hat{z}(t)$ and $y(t)$ can be computed according to

$$\hat{z}(t) = \mathcal{H}_{\hat{z}y}y(t) = \mathcal{H}_{\hat{z}y}\mathcal{H}_{ye}e(t) \tag{3.52}$$

$$y(t) = \mathcal{H}_{ye}e(t), \tag{3.53}$$

where $\mathcal{H}_{ye}$ is the generalized spectral factor [22], solution of

$$A_f^{-1}B_f A_1^{-1}C_d\Phi_{\nu 1}C_d^* A_1^{-*}B_f^* A_f^{-*} + A_n^{-1}C_n\Phi_{\nu 2}C_n^* A_n^{-*} = \mathcal{H}_{ye}\mathcal{H}_{ye}^*$$

and $\{e(t)\}$ is the innovations process of $\{y(t)\}$, $e(t) := y(t) - \varepsilon\{y(t) \mid y^{t-1}\}$. □

A polynomial solution to the above problem has been presented in [3]. All results relevant for the present discussion are summarized in the following theorem 3.5, which proof is based on [3] and is reported in the Appendix for completeness.

Theorem 3.5. *Let (A.1)-(A.3) be fulfilled by (3.48). Let $u(t) \equiv 0$. Further, let D be a Schur left-spectral factor solution of the following left-spectral factorization problem*

$$DD^* = B_o C_d \Phi_{\nu_1} C_d^* B_o^* + A_o A_{f1} C_n \Phi_{\nu_2} C_n^* A_{f1}^* A_o^*, \tag{3.54}$$

with A_o and B_o such that

$$A_{n1}B_f A_1^{-1} = A_o^{-1}B_o, \tag{3.55}$$

and, let the matrix polynomial triplet $(\hat{X}, \hat{Y}, \hat{Z})$ be the minimal solution w.r.t. $\hat{Z}$, viz. $\partial\hat{Z} < \partial\bar{D}$, of the following pair of bilateral Diophantine equations

$$\hat{Y}\bar{D} + A_1\hat{Z} = C_d\Phi_{\nu_1}\overline{C_d B_o} \tag{3.56}$$

$$\hat{X}\bar{D} - A_{n1}B_f\hat{Z} = A_{f1}C_n\Phi_{\nu_2}\overline{C_n A_{f1} A_o}. \tag{3.57}$$

with $\bar{D}$, $\overline{C_d B_o}$ and $\overline{C_n A_{f1} A_o}$ defined by

$$\bar{D} := d^p D^*, \quad \overline{C_d B_o} := d^p C_d^* B_o^*, \quad \overline{C_n A_{f1} A_o} := d^p C_n^* A_{f1}^* A_o^*, \tag{3.58}$$

where p is an integer such that

$$p = \max\{\partial D, \partial(B_o C_d), \partial(A_o A_{f1} C_n)\}. \tag{3.59}$$

Then, the optimal deconvolution filter $\mathcal{H}_{\hat{z}y}$ and the output innovations models $\mathcal{H}_{ye}$ and $\mathcal{H}_{\hat{z}e}$ are given by

$$\mathcal{H}_{\hat{z}y} = A_1^{-1}\hat{Y}D^{-1}A_o A_{n1} A_f \tag{3.60}$$

$$\mathcal{H}_{ye} = A_f^{-1}A_{n1}^{-1}A_o^{-1}D \tag{3.61}$$

$$\mathcal{H}_{\hat{z}e} = A_1^{-1}\hat{Y}.\ \square \tag{3.62}$$

Finally, considering that

$$\varepsilon\{z(t)|u^{t-1}\} = A_1^{-1}B_1u(t) \quad \varepsilon\{y(t)|u^{t-1}\} = A_f^{-1}B_fA_1^{-1}B_1u(t),$$

one finally finds the following innovations system representation for (3.49)

$$\hat{z}(t) = A_1^{-1}\hat{Y}e(t) + A_1^{-1}B_1u(t) \tag{3.63}$$
$$y(t) = A_f^{-1}A_{n1}^{-1}A_o^{-1}De(t) + A_f^{-1}A_{n1}^{-1}A_o^{-1}B_oB_1u(t). \tag{3.64}$$

Last, the above system description has to be converted to the form (3.7), in order to use the controller design procedure given in theorem 3.3.

Example 3.1. In order to clarify the above arguments, consider the system of figure 3.4, that is

$$\begin{aligned} \gamma(t) &= \tfrac{b_1}{a_1}u(t) + \tfrac{c_1}{a_1}\nu_1(t) \\ y(t) &= \tfrac{b_3}{a_3}\gamma(t) + \tfrac{c_3}{a_3}\nu_2(t) \\ z(t) &= \begin{bmatrix} a_2 & 0 \\ 0 & 1 \end{bmatrix}^{-1} \begin{bmatrix} b_2 \\ 1 \end{bmatrix} \gamma(t). \end{aligned} \tag{3.65}$$

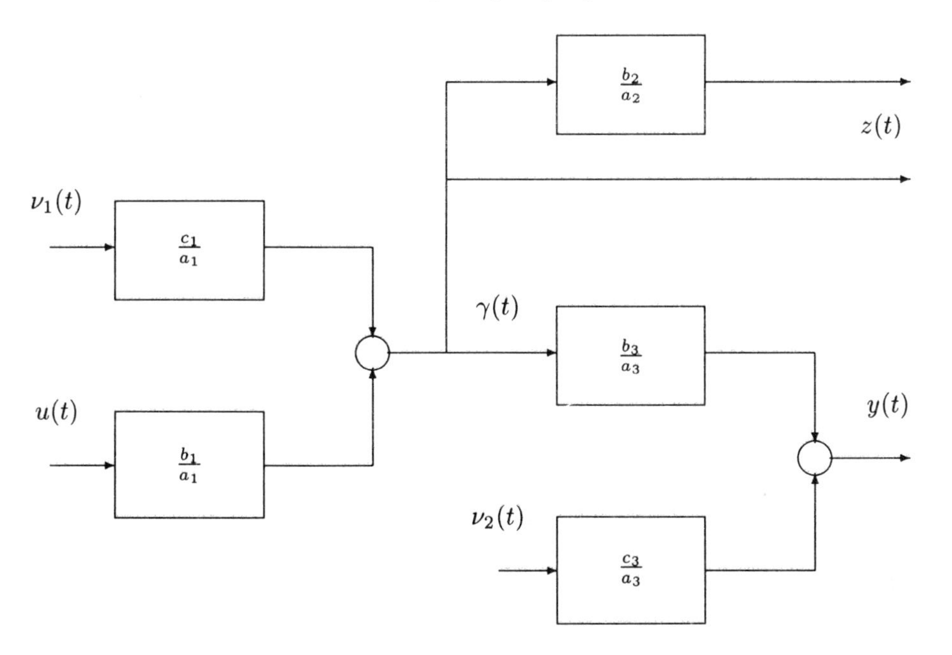

Figure 3.4: A scalar cascade system.

In (3.65), $\gamma(t)$ is an internal variable and a_1, a_2, a_3, b_1, b_2, b_3, c_1 and c_3 are polynomials with a_2 strictly Schur. Further, let ν_1 and ν_2 be scalar zero-mean jointly Gaussian and mutually independent stationary stochastic sequences with covariance $\sigma_{\nu_1}^2$ and, respectively $\sigma_{\nu_2}^2$..

In order to solve the problem, an innovations representation for (3.65) has to be determined. To this end the following polynomial procedure that is based on the results of theorem 3.5 can be adopted.

Determine δ, a Schur spectral factor, solution of following scalar spectral factorization problem

$$\delta\delta^* = b_3 c_1 \sigma_{\nu_1}^2 c_1^* b_3^* + a_1 c_3 \sigma_{\nu_2}^2 a_1^* c_3^*. \tag{3.66}$$

Further, determine $(\hat{y}, \hat{x}, \hat{z})$, the unique scalar polynomial triplet minimal degree solution w.r.t. $\hat{z}$, viz. $\partial\hat{z} < \partial\bar{\delta}$, of the following pair of scalar Diophantine equations

$$\hat{y}\bar{\delta} + \hat{z}a_1 = c_1\sigma_{\nu_1}^2 \overline{b_3 c_1}, \quad \hat{x}\bar{\delta} - \hat{z}b_3 = c_3\sigma_{\nu_2}^2 \overline{a_1 c_3}. \tag{3.67}$$

In (3.67): $\bar{\delta} := d^p\delta^*$, $\overline{b_3 c_1} := d^p b_3^* c_1^*$, $\overline{a_1 c_3} := d^p a_1^* c_3^*$ with $p = \max\{\partial\delta, \partial(c_1 b_3), \partial(a_1 c_3)\}$. It also results, see equation (3.127) in the Appendix, that $\hat{x}$ and $\hat{y}$ fulfill the following equality

$$a_1\hat{x} + b_3\hat{y} = \delta. \tag{3.68}$$

Then, the innovations models $\mathcal{H}_{\hat{z}e} = \begin{bmatrix} \frac{b_2\hat{y}}{a_1 a_2} \\ \frac{\hat{y}}{a_1} \end{bmatrix}$ and $\mathcal{H}_{ye} = \frac{\delta}{a_1 a_3}$ are determined. These give rise to the following innovations representation of (3.65)

$$\begin{aligned} \hat{z}(t) &= \begin{bmatrix} a_2 & 0 \\ 0 & 1 \end{bmatrix}^{-1} \begin{bmatrix} b_2 \\ 1 \end{bmatrix} \frac{\hat{y}}{a_1} e(t) + \begin{bmatrix} a_2 & 0 \\ 0 & 1 \end{bmatrix}^{-1} \begin{bmatrix} b_2 \\ 1 \end{bmatrix} \frac{b_1}{a_1} u(t) \\ y(t) &= \frac{\delta}{a_1 a_3} e(t) + \frac{b_1 b_3}{a_1 a_3} u(t) \end{aligned} \tag{3.69}$$

Next step is to convert (3.69) in the form (3.7) in order to use the controller design procedure given in theorem 3.3. To this end, the following representation is achieved

$$\begin{aligned} \Sigma_i &= \begin{bmatrix} \begin{bmatrix} a_2 & 0 \\ 0 & 1 \end{bmatrix} & \begin{bmatrix} b_2 \\ 1 \end{bmatrix} a_3 t_1 \\ \begin{bmatrix} 0 & 0 \end{bmatrix} & a_1 a_3 \end{bmatrix}^{-1} \begin{bmatrix} \begin{bmatrix} b_2 \\ 1 \end{bmatrix} t_2 & \begin{bmatrix} b_2 \\ 1 \end{bmatrix} b_1 t_3 \\ \delta & b_1 b_3 \end{bmatrix} \\ &= \begin{bmatrix} \begin{bmatrix} \frac{b_2\hat{y}}{a_1 a_2} \\ \frac{\hat{y}}{a_1} \end{bmatrix} & \begin{bmatrix} \frac{b_2 b_1}{a_2 a_1} \\ \frac{b_1}{a_1} \end{bmatrix} \\ \frac{\delta}{a_1 a_3} & \frac{b_1 b_3}{a_1 a_3} \end{bmatrix}, \end{aligned} \tag{3.70}$$

where (t_1, t_2, t_3) is any polynomial triplet satisfying the following set of scalar Diophantine equations

$$a_1 t_2 - \delta t_1 = \hat{y}, \quad a_1 t_3 - b_3 t_1 = 1, \quad b_3 t_2 - \delta t_3 = -\hat{x}. \tag{3.71}$$

In (3.71), the last equation is determined from the first two ones by taking into account (3.68). It is to be pointed out that, if solvable, (3.71) admits a family of solutions for (t_1, t_2, t_3) [1]. The latter could give rise to the erroneous conclusion that different optimal controllers correspond to different triplets, all solutions of

(3.71). However, this is not the case. The rationale is that, as it will be made clear in the sequel, only $\hat{x}$ and $\hat{y}$ and not (t_1, t_2, t_3), are required in applying the control design procedure.

In fact, the relevant quantities for the latter become: the spectral factor e is achieved by solving the following scalar spectral factorization problem

$$e^* e = a_3^2 a_2^* a_1^* \Psi_u a_1 a_2 a_3 + a_3^* b_1^* \begin{bmatrix} b_2^* & a_2^* \end{bmatrix} \Psi_z \begin{bmatrix} b_2 \\ a_2 \end{bmatrix} b_1 a_3, \tag{3.72}$$

with $A_2 = a_1 a_2 a_3$ and $\hat{B}_2 = \begin{bmatrix} b_2 \\ a_2 \end{bmatrix} b_1 a_3$; the polynomial triplet (x, y, z) is the unique minimal degree solution w.r.t. z, viz. $\partial z < \partial \bar{e}$, of the following pair of scalar Diophantine equations

$$\begin{aligned} \bar{e} y + z a_1 a_3 a_2 &= \bar{a}_3 \bar{b}_1 \begin{bmatrix} \bar{b}_2 & \bar{a}_2 \end{bmatrix} \Psi_z \begin{bmatrix} 1 & 0 \\ 0 & a_2 \end{bmatrix} \begin{bmatrix} b_2 \\ 1 \end{bmatrix} a_3 \hat{y} \\ \bar{e} x - z b_1 b_3 a_2 &= \bar{a}_3 \bar{a}_2 \bar{a}_1 \Psi_u \delta a_2 + \bar{a}_3 \bar{b}_1 \begin{bmatrix} \bar{b}_2 & \bar{a}_2 \end{bmatrix} \Psi_z \begin{bmatrix} 1 & 0 \\ 0 & a_2 \end{bmatrix} \begin{bmatrix} b_2 \\ 1 \end{bmatrix} b_1 \hat{x}, \end{aligned} \tag{3.73}$$

where

$$D = D_1 = D_2 = \delta, \quad A_{3s} = A_{4s} = a_2, \quad B_1 = B_3 = b_1 b_3,$$

$$A_1 = A_3 = a_1 a_3, \quad A_{1s} = \begin{bmatrix} a_2 & 0 \\ 0 & 1 \end{bmatrix}$$

$$\left[\hat{D} A_3 - A_{pl} D_2\right]_2 = \begin{bmatrix} 1 & 0 \\ 0 & a_2 \end{bmatrix} \begin{bmatrix} b_2 \\ 1 \end{bmatrix} a_3 \hat{y}, \quad \left[\hat{B}_1 D_1 - \hat{D} B_3\right]_1 = \begin{bmatrix} 1 & 0 \\ 0 & a_2 \end{bmatrix} \begin{bmatrix} b_2 \\ 1 \end{bmatrix} b_1 \hat{x}.$$

Then, the optimal controller can be evaluated according to (3.9) and it is given by

$$\mathcal{K} = \mathcal{M}_1^{-1} \mathcal{N}_1 = \frac{y}{x}. \tag{3.74}$$

3.2.4 Relationships with other polynomial solutions

Another interesting issue is to verify how the above two stage general procedure (MMSE MD plus control design) reduces to a specific one, determined by approaching directly a particular problem. To this end, we discuss two well known examples, that can be derived via a suitable specialization of (3.48). To be specific, for both examples, the following restrictions are assumed

$$\dim y = \dim z \ (l = p), \quad A_f = B_f = I_l.$$

Case 1: $A_n = C_n = I$ - Under the above setting, (3.48) becomes the usual regulation problem structure with white Gaussian measurement noise. A polynomial solution for this problem can be found in [1]. This problem can be easily recast in the above MMSE multichannel deconvolution problem by further setting

$A_{n1} = A_{f1} = B_o = I_l$ and $A_o = A_1$. Then, the spectral factorization problem (3.54) and the pair of equations (3.56)-(3.57) become respectively

$$\begin{aligned} DD^* &= C_d \Phi_{\nu 1} C_d^* + A_1 \Phi_{\nu_2} A_1^* \\ D\bar{D} &= C_d \Phi_{\nu 1} \bar{C}_d + A_1 \Phi_{\nu 2} \bar{A}_1 \end{aligned} \tag{3.75}$$

and

$$\begin{aligned} \hat{Y}\bar{D} + A_1 \hat{Z} &= C_d \Phi_{\nu 1} \bar{C}_d \\ \hat{X}\bar{D} - \hat{Z} &= \Phi_{\nu_2} \bar{A}_1 . \end{aligned} \tag{3.76}$$

It is easy to verified that

$$\hat{X}_o = (A_1(0)\Phi_{\nu_2} D^{-1}(0))', \quad \hat{Z}_o = \hat{X}_o - \Phi_{\nu_2}\bar{A}_1, \quad \hat{Y}_o = D - A_1\hat{X}_o,$$

is the optimal solution, $A_1(0)$ and $D(0)$ being the 0-order matrix coefficients of the polynomial matrices A_1 and, respectively D. This can be easily verified by considering that, for (3.58), $\partial\bar{D} = \partial\bar{A}_1$ and, in turn, the only solution of (3.76) with $\partial\hat{Z} < \partial\bar{D}$ can be obtained for $\partial\hat{X}_o = 0$, by imposing that $\hat{X}_o\bar{D}_p = \Phi_{\nu_2}\bar{A}_{1p}$, with $\bar{D}_p = D'(0)$ and $\bar{A}_{1p} = A_1'(0)$ the higher-order matrix coefficient of the polynomial matrices $\bar{D}$ and $\bar{A}_1$ respectively. $\hat{Z}_o$ and $\hat{Y}_o$ can be verified by substitution in (3.76).

As a result, the following innovations model is derived

$$\begin{aligned} \hat{z}(t) &= A_1^{-1}(D - A_1\hat{X}_o)e(t) + A_1^{-1}B_1 u(t) \\ y(t) &= A_1^{-1}De(t) + A_1^{-1}B_1 u(t). \end{aligned} \tag{3.77}$$

In order to apply the polynomial procedure for the optimal controller design, (3.77) has to be reformulated in the form (3.7). The following r.c.m.f.d. is then determined

$$\begin{aligned} \Sigma_i &= \begin{bmatrix} I_p & (A_1 - I) \\ 0 & A_1 \end{bmatrix}^{-1} \begin{bmatrix} D - \hat{X}_o & B_1 \\ D & B_1 \end{bmatrix} \\ &= \begin{bmatrix} A_1^{-1}(D - A_1\hat{X}_o) & A_1^{-1}B_1 \\ A_1^{-1}D & A_1^{-1}B_1 \end{bmatrix} . \end{aligned} \tag{3.78}$$

Then, the optimal controller can be carried out, by solving the equation pair (3.25)-(3.35) for (X, Y, Z) with $\partial Z < \partial\bar{E}$. For the problem at hands, (3.25)-(3.35) reduce to

$$\begin{aligned} \bar{E}Y + ZA_3 &= \bar{B}_2\Psi_z(D_2 - \hat{X}_o A_3) \\ \bar{E}X - ZB_3 &= \bar{A}_2\Psi_u D_1 + \bar{B}_2\Psi_z\hat{X}_o B_3 . \end{aligned} \tag{3.79}$$

where E is the Schur factor solution of

$$E^*E = A_2^*\Psi_u A_2 + B_2^*\Psi_z B_2 . \tag{3.80}$$

Equations (3.79) can be easily verified by considering that $A_{1s} = A_{3s} = A_{4s} = I$, $A_{pl} = (A_1 - I)$, $[(D - \hat{X}_o)A_3 - A_{pl}D_2]_2 = ((D - \hat{X}_o)A_3 - A_{pl}D_2) = D_2 - \hat{X}_o A_3$ and $[B_1 D_1 - (D - \hat{X}_o)B_3]_1 = (B_1 D_1 - (D - \hat{X}_o)B_3) = \hat{X}_o B_3$. The above relations can be directly verified by taking into account (3.23) and (3.33). In order to compare the present procedure with the one derived in [1], one can verify that (3.79) can be reformulated as follows

$$\begin{aligned} \bar{E}Y + Z_1 A_3 &= \bar{B}_2 \Psi_z D_2 \\ \bar{E}X - Z_1 B_3 &= \bar{A}_2 \Psi_u D_1, \end{aligned} \tag{3.81}$$

with $Z_1 = Z + \bar{B}_2 \Psi_z \hat{X}_o$. Notice that (3.79) and (3.81) have the same optimal solution, because $\partial \bar{B}_2 \Psi_z \hat{X}_o < \partial \bar{E}$.

Last, the optimal control law can be carried out by

$$\mathcal{K} = D_1 X^{-1} Y D_2^{-1}. \tag{3.82}$$

Equations (3.75), (3.81), (3.80) and (3.82) are exactly the same equations derived in [1], 1979 - chapter 6, section 2 - equations (5), (16), (7) and (17) respectively) for the same control problem. □

Case 2: $A_n \neq 0$ and $C_n \neq 0$ - This is a more general class of regulation problems, allowing one to consider coloured measurement noise. A polynomial solution for this problem can be found in [8], to which we refer to in order to investigate the existing connections.

As above, this problem can be embedded in the MMSE-MD solution by setting $A_n = A_{n1}$ and $A_{f1} = I$, so that $A_n A_1^{-1} = A_o^{-1} B_o$ results from (3.55), with B_o square and nonsingular.

In order to obtain the solution, the innovations model of (3.48) has firstly to be determined. This can be done, by solving the spectral factorization problem (3.54) for D and the pair of Diophantine equations (3.56)-(3.57) for $(\hat{Y}, \hat{X}, \hat{Z})$, with $\partial \hat{Z} < \partial \bar{D}$. As a result, the innovations representation (3.63)-(3.64) is achieved, that can be expressed in the form (3.7) as follows

$$\Sigma = \begin{bmatrix} A_n & A_n(B_o A_1 - I) \\ 0 & B_o A_1 \end{bmatrix}^{-1} \begin{bmatrix} A_n D - \hat{X} & B_o B_1 \\ D & B_o B_1 \end{bmatrix} = \begin{bmatrix} A_1^{-1}\hat{Y} & A_1^{-1} B_1 \\ A_1^{-1} B_o^{-1} D & A_1^{-1} B_1 \end{bmatrix}.$$

Then, the design procedure depicted in theorem 3.3 can be directly applied to the above innovations representation in order to achieved the optimal control.

However, in order to compare our procedure with the one determined in [8], a more comfortable way is to use a different, but equivalent, innovations representation. To this end, consider

$$\begin{aligned} \hat{z}(t) &= A_1^{-1} \hat{Y}_n e(t) + A_1^{-1} B_1 u(t) \\ y(t) &= A_1^{-1} D e(t) + A_1^{-1} B_1 u(t), \end{aligned} \tag{3.83}$$

where $A_1 \leftarrow B_o A_1$, $B_1 \leftarrow B_o B_1$ and $\hat{Y}_n = B_o \hat{Y}$ for notational simplicity.

Then, (3.83) can be reformulated in the form (3.7) as follows

$$\Sigma = \begin{bmatrix} I_p & (A_1 - I) \\ 0 & A_1 \end{bmatrix}^{-1} \begin{bmatrix} \hat{D} & B_1 \\ D & B_1 \end{bmatrix} = \begin{bmatrix} A_1^{-1}\hat{Y}_n & A_1^{-1} B_1 \\ A_1^{-1} D & A_1^{-1} B_1 \end{bmatrix}, \tag{3.84}$$

provided that $\hat{D}$ is obtainable by solving the following matrix polynomial linear equation

$$A_1\hat{D} - (A_1 - I)D = \hat{Y}_n \tag{3.85}$$

Notice that (3.85) admits polynomial solutions for $\hat{D}$ only if A_1 is a left divisor of $(d\hat{Y}_n - D)$. However, (3.85) need not be really solved for the present discussion. Anyway, let us assume for a while that $\hat{D}$ is a polynomial solution of (3.85).

The controller design procedure can now be directly applied to the system (3.84). Then, taking into account (3.23), (3.33), (3.84) and (3.85), equations (3.25) and (3.35) become

$$\bar{E}Y + ZA_3 = \bar{B}_2\Psi_z D_2 - \bar{B}_2\Psi_z(D - \hat{D})A_3 \tag{3.86}$$
$$\bar{E}X - ZB_3 = \bar{A}_2\Psi_u D_1 + \bar{B}_2\Psi_z(D - \hat{D})B_3, \tag{3.87}$$

considering that $A_{1s} = A_{3s} = A_{4s} = I$, $A_{pl} = A_1 - I$, $[\hat{D}A_3 - A_{pl}D_2]_2 = (\hat{D}A_3 - A_{pl}D_2)$ and $[\hat{B}_1D_1 - \hat{D}B_3]_1 = (\hat{B}_1D_1 - \hat{D}B_3)$. As above, E is the Schur spectral factor solution of (3.80). The optimal control law is given by

$$\mathcal{K} = D_1X^{-1}YD_2^{-1} \tag{3.88}$$

whit X and Y solution of (3.86)-(3.87) with $\partial Z < \partial\bar{E}$.

An alternative, but equivalent, way to express (3.86)-(3.87) is considering the three following Diophantine equations

$$\bar{E}Y_1 + Z_1A3 = \bar{B}_2\Psi_z D_2 \tag{3.89}$$
$$\bar{E}X_1 - Z_1B_3 = \bar{A}_2\Psi_u D_1 \tag{3.90}$$
$$\bar{E}L + Z_2 = \bar{B}_2\Psi_z(D - \hat{D}) \tag{3.91}$$

The equivalence can easily be checked by verifying that (3.89)-(3.91) can be derived by (3.86)-(3.87) with the following setting

$$Y = Y_1 - LA_3, \quad X = X_1 + LB_3, \quad Z = Z_1 - Z_2$$

Equations (3.89)-(3.91) have to be solved with the usual degree conditions $\partial Z_1 < \partial\bar{E}$ and $\partial Z_2 < \partial\bar{E}$ in order to yield the correct solution.

It is interesting to notice that (3.89)-(3.91) generalize the solution of the previous example in that $L = 0$ and $Z_2 = \bar{B}_2\Psi_z[A_1(0)\Phi_{\nu_2}D^{-1}(0)]'$, $\partial Z_2 = 0$, is the optimal solution for (3.91) when $\hat{D} = D - [A_1(0)\Phi_{\nu_2}D^{-1}(0)]'$. This agrees with the previous results.

Then, by considering (3.85), it follows that

$$(D - \hat{D}) = A_1^{-1}(D - \hat{Y}_n) = A_1^{-1}(D\bar{D} - \hat{Y}_n\bar{D})\bar{D}^{-1}.$$

Further, by taking into account (3.57) and (3.58), (3.91) can be suitably reformulated as

$$\bar{E}L + Z_2 = \bar{B}_2\Psi_z A_1^{-1}(A_1\hat{Z} + C_n\Phi_{\nu_2}\bar{C}_n)\bar{D}^{-1}, \tag{3.92}$$

where $C_n \leftarrow A_o C_n$. Moreover, as done in [8], define the following polynomial matrices $\bar{D}_{fw}$, $\bar{L}_3$ and A_{wf} such that

$$\bar{D}_{fw}^{-1}\bar{L}_3 A_{wf}^{-1} = \bar{B}_2 \Psi_z A_1^{-1} C_n \Phi_{\nu_2} \bar{C}_n \bar{D}^{-1}, \tag{3.93}$$

and define the polynomial matrix S such that

$$\bar{D}_{fw}^{-1} S = \bar{B}_2 \Psi_z \hat{Z} \bar{D}^{-1}.$$

This is possible because, as pointed out in [8], in solving (3.93), $\bar{D}_{fw}$ is carried out in such a way that $\det \bar{D} \propto \det \bar{D}_{fw}$.

Finally, by simple substitutions, (3.92) becomes

$$\begin{aligned} \bar{E}L + \bar{D}_{fw}^{-1}(\bar{D}_{fw} Z_2 - S) &= \bar{D}_{fw}^{-1}\bar{L}_3 A_{wf}^{-1} \\ \bar{D}_{fw}\bar{E}L + P A_{wf} &= \bar{L}_3 \end{aligned} \tag{3.94}$$

with $P = \bar{D}_{fw} Z_2 - S$, to be determined, together L, from (3.94), with the condition $\partial P < \partial(\bar{D}_{fw}\bar{E})$. Notice that Z_2 and S have the correct degrees in order for P to satisfy the above degree constraint.

Then, the optimal control law turns out to be

$$\mathcal{K} = D_1(X_1 + L A_{wf}^{-1} B_3)^{-1}(Y_1 - L A_{wf}^{-1} A_3) D_2^{-1}. \tag{3.95}$$

Equations (3.54), (3.80), (3.89)-(3.90), (3.93), (3.94) and (3.95) equal respectively the following equations determined in [8]: (28), (29), (30)-(31), (37), (36) and (38) with obvious notational changes.

3.3 Tracking, servo and accessible disturbance problems

In this section it is shown how different kinds of accessible exogenous signals can be appropriately taken into account wherein the "standard" structure. These problems were recently addressed for a 2DOF control structure in an H_2 context in [18] and extended to a joint H_2/H_∞ context in [20]. Further, in [1] it was proved that the separation property between feedback and feedforward actions holds true for a large class of 2DOF H_2 LQG stochastic tracking and servo problems. This section extends the previous regulation solution to a general tracking and accesible disturbance control scheme.

In our formulation, the reference to be tracked and the accessible disturbance to be attenuated are assumed to be known up to the generic times $t + \tau$ and $t + \nu$ respectively, τ and ν being arbitrary integers, positive or negative. Thus, our formulation unifies the servo problem ($\tau > 0$) and the tracking one ($\tau \leq 0$) into a single framework. In particular, if τ (ν) is positive and large enough, the optimal reference feedforward (accessible disturbance) input can be tightly approximated without requiring any stochastic dynamic model for the reference (accessible disturbance).

Earlier contributions to the LQG accessible disturbance rejection problem may be found in [23] for the SISO case and in [13, 14] for the MIMO one. They solve the problem by means of an opportune Diophantine equation and their solutions require a stochastic modelling for the accessible disturbance and cover only the $\nu = 0$ case. More recently, in [19] was given a complete solution for the standard LQG control problem for any arbitrary integer ν.

3.3.1 Problem formulation

Consider the innovations representation of the discrete-time multivariable stochastic system of figure 3.5, described by the interconnection matrix Σ_i in (3.96). Σ_i embeds all the signal flows between the inputs and the outputs of the system.

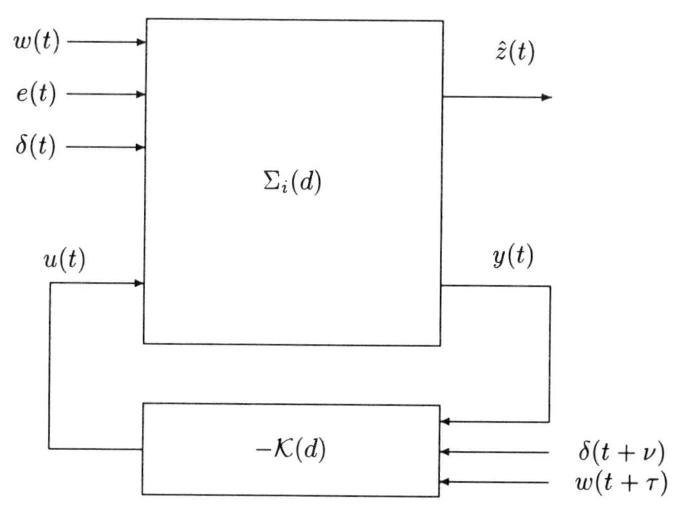

Figure 3.5: Innovations system structure for tracking problems.

The vector $\Upsilon(t) := [w'(t) \mid e'(t) \mid \delta'(t)]'$ denotes the exogenous inputs. In particular $w(t) \in \mathrm{Re}^{\,p}$ denotes the reference signal, $e(t) \in \mathrm{Re}^{\,l}$ the innovations process of $y(t)$ and represents disturbances and measurement noise, and $\delta(t) \in \mathrm{Re}^{\,q}$ the accessible disturbance. The vector $u(t) \in \mathrm{Re}^{\,m}$ is the control input; $\hat{z}(t) \in \mathrm{Re}^{\,p}$ is the regulated variable; and $Y(t) := [w'(t+\tau) \mid \delta'(t+\nu) \mid y'(t)]' \in \mathrm{Re}^{\,(p+q+l)}$ is the vector of the measured variables. The interconnection matrix $\Sigma_i(d)$ in the unit backward shift operator d, relates the variables $\hat{z}$, y, e , δ and u as follows

$$\begin{bmatrix} z(t) \\ y(t) \end{bmatrix} = \Sigma_i(d) \begin{bmatrix} e(t) \\ \delta(t) \\ u(t) \end{bmatrix} = \mathcal{Q}(d)e(t) + \mathcal{H}(d)\delta(t) + \mathcal{P}(d)u(t) \tag{3.96}$$

with

$$\Sigma_i = \begin{bmatrix} \hat{Q}(d) \; \hat{H}(d) \; \hat{P}(d) \\ Q(d) \; H(d) \; P(d) \end{bmatrix}, \; \mathcal{P} := \begin{bmatrix} \hat{P}(d) \\ P(d) \end{bmatrix}, \; \mathcal{Q} := \begin{bmatrix} \hat{Q}(d) \\ Q(d) \end{bmatrix}, \; \mathcal{H} := \begin{bmatrix} \hat{H}(d) \\ H(d) \end{bmatrix}$$

with $\hat{P}(d) \in R^{pm}(d)$, $P(d) \in R^{lm}(d)$, $\hat{Q}(d) \in R^{pl}(d)$, $Q(d) \in R^{ll}(d)$, $\hat{H}(d) \in R^{pq}(d)$ and $H(d) \in R^{lq}(d)$, are causal rational transfer matrices. Moreover, the following assumptions are adopted:

(A.3.5)
- $\{e(t)\}$ is a white sequence of random vectors $\in \mathrm{Re}^{\,l}$ with zero-mean and unitary covariance .
- $\{w(t)\}$ and $\{\delta(t)\}$ are second-order stationary processes.
- $\{w(t)\}$, $\{\delta(t)\}$ and $\{e(t)\}$ are mutually uncorrelated.
- $\hat{P}(d)$ and $P(d)$ are strictly causal transfer matrices.

Let $I^t := \{y^t, u^{t-1}, w^{t+\tau}, \delta^{t+\nu}\}$, $y^t := \{y(t), y(t-1), ..., \}$, and $\sigma(z)$ denote the set of random variables measurable with respect to the $\sigma-$algebra generated by z. The problem that is addressed hereafter is to find, amongst the *admissible (non anticipative) control laws*

$$u(t) \in \sigma(I^t), \tag{3.97}$$

the one that, in stochastic steady-state (s.s.s.), stabilizes the plant (3.96) and minimizes the quadratic cost

$$J = \varepsilon\{\|\hat{z}(t) - w(t)\|^2_{\Psi_z} + \|u(t)\|^2_{\Psi_u}\}. \tag{3.98}$$

In (3.98), $\Psi_z = \Psi'_z \geq 0$, $\Psi_u = \Psi'_u \geq 0$, $\|\nu\|^2_\Psi := \nu'\Psi\nu$; the prime denotes transpose and ε stands for expectation.

Let $(A_\Sigma \, , \, B_\Sigma)$ be a right coprime matrix fraction description (r.c.m.f.d.) of $\Sigma_i(d)$, $\Sigma_i(d) = B_\Sigma A_\Sigma^{-1}$, with $A_\Sigma \in R^{(l+q+m)(l+q+m)}[d]$ and $B_\Sigma \in R^{(p+l)(l+q+m)}[d]$ polynomial matrices with elements in $R[d]$. Further, let $(A_p \, , \, B_p)$ be a left coprime matrix fraction description (l.c.m.f.d.) of $P(d)$, $P = A_p^{-1}B_p$, with $A_p \in R^{ll}[d]$ and $B_p \in R^{lm}[d]$. Further, let assumption (A.3.2) be fulfilled for stabilizability.

The stabilizability conditions (A.3.2) have repercussions on the structures of A_Σ and B_Σ. The following result, which proof can be given as for lemma 3.1, establishes the permissible structures for all stabilizable Σ_i's.

Lemma 3.6. *Let the system (3.96) be stabilizable by a dynamic control law of the form $u = -\mathcal{K}y$, $\mathcal{K} \in R^{ml}(d)$, viz. (A.3.2) be fulfilled. Then, there exist a r.c.m.f.d. of Σ_i of the form*

$$\Sigma_i = \begin{bmatrix} \hat{C}_q \; \hat{C}_h \; \hat{B}_2 \\ C_q \; C_h \; B_2 \end{bmatrix} \begin{bmatrix} A_{2s} & 0 \\ A_{(l+q)m} & A_2 \end{bmatrix}^{-1} \tag{3.99}$$

with $A_{2s} \in \mathrm{Re}^{(l+q)(l+q)}[d]$ *strictly Schur, and a l.c.m.f.d. of* Σ_i *of the form*

$$\Sigma_i = \begin{bmatrix} A_{1s} & A_{pl} \\ 0 & A_1 \end{bmatrix}^{-1} \begin{bmatrix} \hat{D}_q & \hat{D}_h & \hat{B}_1 \\ D_q & D_h & B_1 \end{bmatrix} \tag{3.100}$$

with $A_{1s} \in \mathrm{Re}^{pp}[d]$ *strictly Schur.* (A_2, B_2) *and* (A_1, B_1) *are respectively a right and left m.f.d.'s of* P*, whose* only *possible right and, respectively, left divisors are* strictly Schur. □

From (3.99) derives that $\left(\begin{bmatrix} \hat{B}_2 \\ B_2 \end{bmatrix}, A_2\right)$ is a right coprime matrix fraction description of $\mathcal{P}$,

$$\mathcal{P} = \begin{bmatrix} \hat{B}_2 \\ B_2 \end{bmatrix} A_2^{-1}, \tag{3.101}$$

with $A_2 \in R^{mm}[d]$, $\hat{B}_2 \in R^{pm}[d]$ and $B_2 \in R^{lm}[d]$.

Define a right-spectral factor $E(d) \in R^{mm}[d]$ by the relation

$$\boxed{E^* E = A_2^* \Psi_u A_2 + \hat{B}_2^* \Psi_z \hat{B}_2} \tag{3.102}$$

with $A^*(d) := A'(d^{-1})$.

It is further assumed that

(A.3.6) $\left\{\right.$ • $E(d)$ is stable, i.e. $E^{-1}(d)$ is analytic in the closed unit disk.

One condition for (A.3.3) to be satisfied is given by (A.3.4).

Let

$$\mathcal{M}_1(d)u(t) = -\mathcal{N}_1(d)y(t)$$

be the solution of (3.98) for the pure regulation problem ($w \equiv \delta \equiv 0$), where $\mathcal{M}_1 \in R^{mm}(d)$ and $\mathcal{N}_1 \in R^{ml}(d)$ are stable causal transfer matrices that factorize $\mathcal{K}$ as $\mathcal{K} = \mathcal{M}_1^{-1}\mathcal{N}_1$ and satisfy the following identity

$$\mathcal{M}_1 A_2 + \mathcal{N}_1 B_2 = I_m \tag{3.103}$$

in order to stabilize the subsystem P.

Consider now the following equation

$$\mathcal{M}_1(d)u(t) = -\mathcal{N}_1(d)y(t) + v(t). \tag{3.104}$$

Since $\mathcal{M}_1$ and $\mathcal{N}_1$ are specific stable and causal transfer matrices, for each $v(t) \in \sigma(I^t)$, (3.104) yields an admissible control law. Thus, the stated LQG stochastic tracking and servo problems amount to finding a process $\{v(t)\} \in \sigma(I^t)$ such that the corresponding control law (3.104) stabilizes the plant and minimizes (3.98).

Under the closed-loop control law (3.104), $z(t)$ and $u(t)$ can be decomposed as follows:

$$\begin{aligned} u(t) &= \left(I_m + \mathcal{M}_1^{-1}\mathcal{N}_1 P\right)^{-1}\left[\mathcal{M}_1^{-1}v(t) - \mathcal{M}_1^{-1}\mathcal{N}_1(Qe(t) + H\delta(t))\right] \\ &:= u_v(t) + u_e(t) + u_\delta(t), \end{aligned} \tag{3.105}$$

$$\begin{aligned} z(t) &= \hat{P}\left(I_m + \mathcal{M}_1^{-1}\mathcal{N}_1 P\right)^{-1}\left[\mathcal{M}_1^{-1}v(t) - \mathcal{M}_1^{-1}\mathcal{N}_1(Qe(t) + H\delta(t))\right] \\ &\quad + \hat{Q}e(t) + \hat{H}\delta(t) \\ &:= z_v(t) + z_e(t) + z_\delta(t). \end{aligned} \tag{3.106}$$

By exploiting (3.101) and (3.103), one gets

$$\begin{aligned} u_v(t) &= \left(I_m + \mathcal{M}_1^{-1}\mathcal{N}_1 B_2 A_2^{-1}\right)^{-1}\mathcal{M}_1^{-1}v(t) \\ &= A_2\left(\mathcal{M}_1 A_2 + \mathcal{N}_1 B_2\right)^{-1}\mathcal{M}_1\mathcal{M}_1^{-1}v(t) \\ &= A_2 v(t), \end{aligned} \tag{3.107}$$

$$u_\delta(t) = -A_2\mathcal{N}_1 H\delta(t), \tag{3.108}$$

$$\begin{aligned} z_v(t) &= \hat{B}_2 A_2^{-1}\left(I_m + \mathcal{M}_1^{-1}\mathcal{N}_1 B_2 A_2^{-1}\right)^{-1}\mathcal{M}_1^{-1}v(t) \\ &= \hat{B}_2 A_2^{-1} A_2\left(\mathcal{M}_1 A_2 + \mathcal{N}_1 B_2\right)^{-1}\mathcal{M}_1\mathcal{M}_1^{-1}v(t) \\ &= \hat{B}_2 v(t). \end{aligned} \tag{3.109}$$

$$= -\hat{B}_2\mathcal{N}_1 H\delta(t) + \hat{H}\delta(t). \tag{3.110}$$

With reference to the decomposition (3.105)-(3.106) , the cost term (3.98) can be split as follows

$$J = J_{ee} + J_{ev} + J_{ew} + J_{ww} + J_{vw} + J_{vv} + J_{e\delta} + J_{\delta\delta} + J_{v\delta} + J_{\delta w} \tag{3.111}$$

where

$$\begin{aligned} J_{ee} &:= \varepsilon\{\|y_e(t)\|^2_{\Psi_z} + \|u_e(t)\|^2_{\Psi_u}\}; \\ J_{ev} &:= 2\varepsilon\{z_e'(t)\Psi_z z_v(t) + u_e'(t)\Psi_u u_v(t)\}; \\ J_{ww} &:= \varepsilon\{\|w(t)\|^2_{\Psi_z}\}; \\ J_{ew} &:= -2\varepsilon\{w'(t)\Psi_z z_e(t)\} = 0 \qquad [(A.3.1)]; \\ J_{vw} &:= -2\varepsilon\{v'(t)\hat{B}_2^*\Psi_z w(t)\}; \\ J_{vv} &:= \varepsilon\{\|z_v(t)\|^2_{\Psi_z} + \|u_v(t)\|^2_{\Psi_u}\}. \\ J_{e\delta} &:= 2\varepsilon\{z_e'(t)\Psi_z z_\delta(t) + u_e'(t)\Psi_u u_\delta(t)\} = 0 \qquad [(A.3.1)]; \\ J_{\delta\delta} &:= \varepsilon\{\|z_\delta(t)\|^2_{\Psi_z} + \|u_\delta(t)\|^2_{\Psi_u}\}; \\ J_{v\delta} &:= 2\varepsilon\{z_v'(t)\Psi_z z_\delta(t) + u_v'(t)\Psi_u u_\delta(t)\}; \\ J_{\delta w} &:= -2\varepsilon\{z_\delta'(t)\Psi_z w(t)\} = 0 \qquad [(A.3.1)]; \end{aligned} \tag{3.112}$$

A key property, whose proof was given in [18] is expressed by the following

Lemma 3.7. *Let (A.3.2), (A.3.5)-(A.3.6) be fulfilled. Then, with reference to the plant (3.96), controlled according to (3.104), $J_{ev} \equiv 0$.* □

The main result of this section can now be stated:

Theorem 3.8. *Let the assumptions (A.3.2), (A.3.4)-(A.3.6) be fulfilled. Then, the optimal control law for the LQG stochastic tracking, servo and accessible disturbance problem (3.96)-(3.98) is given by*

$$\boxed{\mathcal{M}_1 u(t) = -\mathcal{N}_1 y(t) + E^{-1}u_w^c(t) + E^{-1}u_\delta^c(t)} \tag{3.113}$$

where $\mathcal{M}_1$ and $\mathcal{N}_1$ are the optimal transfer matrices solving the underlying LQ pure regulation problem, viz. (3.96)-(3.98) with $w(t) \equiv \delta(t) \equiv 0$, and $u_w^c(t)$ and $u_\delta^c(t)$ are respectively the output reference and accessible disturbance feedforward inputs defined by

$$\boxed{u_w^c(t) = \varepsilon\{E^{-*}\hat{B}_2^*\Psi_z w(t) \mid w^{t+\tau}\}} \tag{3.114}$$

$$\boxed{u_\delta^c(t) = \varepsilon\{(E^{-*}\hat{B}_2^*\Psi_z \hat{H} - E\mathcal{N}_1 H)\delta(t) \mid \delta^{t+\nu}\}} \tag{3.115}$$

Proof. Because of (3.112), J_{ee}, $J_{\delta\delta}$ and J_{ww} are not affected by $\{v(t)\}$. Thus, the optimization problem (3.96)-(3.98) can be conveniently rewritten as

$$\min_{v(t)\in\sigma(I^t)} J = J_{ee} + J_{ww} + J_{\delta\delta} + \min_{v_w(t)\in\sigma(w^{t+\tau})}(J_{v_w v_w} + J_{v_w w}) + \min_{v_\delta(t)\in\sigma(\delta^{t+\nu})}(J_{v_\delta v_\delta} + J_{v_\delta w}).$$

where $v(t) = v_w(t) + v_\delta(t)$ is a convenient orthogonal decomposition such that $v_w(t) \in \sigma(w^{t+\tau})$ and $v_w(t) \in \sigma(\delta^{t+\nu})$.

The optimal control law is given by $\mathcal{M}_1 u(t) = -\mathcal{N}_1 y(t) + v(t)$, with $v_w(t)$ and $v_\delta(t)$ respectively minimizing

$$\begin{aligned}
J_{v_w v_w} + J_{v_w w} &= \varepsilon\left\{\|u_{v_w}(t)\|^2_{\Psi_u} + \|z_{v_w}(t)\|^2_{\Psi_z} - 2v_w'(t)\hat{B}_2^*\Psi_z w(t)\right\} \\
&= \varepsilon\left\{v_w'(t)\left(A_2^*\Psi_u A_2 + \hat{B}_2^*\Psi_z\hat{B}_2\right)v_w(t) - 2v_w'(t)\hat{B}_2^*\Psi_z w(t)\right\} \\
&= \varepsilon\left\{v_w'(t)E^*Ev_w(t) - 2v_w'(t)\hat{B}_2^*\Psi_z w(t)\right\} \\
&= \varepsilon\left\{\|Ev_w - E^{-*}\hat{B}_2^*\Psi_z w\|^2\right\} - \varepsilon\left\{\|E^{-*}\hat{B}_2^*\Psi_z w\|^2\right\}. \qquad (3.116) \\
J_{v_\delta v_\delta} + J_{v_\delta \delta} &= \varepsilon\left\{\|z_{v_\delta}(t)\|^2_{\Psi_z} + \|u_{v_\delta}(t)\|^2_{\Psi_u} + 2z_{v_\delta}'(t)\Psi_z z_\delta(t) + 2u_{v_\delta}'(t)\Psi_u u_\delta(t)\right\} \\
&= \varepsilon\left\{v_\delta'(t)E^*Ev_\delta(t) + 2v_\delta' E^*E\mathcal{N}_1 H\delta(t) - 2v_\delta'(t)\hat{B}_2^*\Psi_z\hat{H}\delta(t)\right\} \\
&= \varepsilon\left\{\|Ev_\delta - (E^{-*}\hat{B}_2^*\Psi_z\hat{H} - E\mathcal{N}_1 H)\delta\|^2\right\} \\
&\quad -\varepsilon\left\{\|(E^{-*}\hat{B}_2^*\Psi_z\hat{H} - E\mathcal{N}_1 H)\delta\|^2\right\}. \qquad (3.117)
\end{aligned}$$

Then, the minima are attained at $v_w(t) = E^{-1}u_w^c(t)$ and $v_\delta(t) = E^{-1}u_w^\delta(t)$, with $u_w^c(t)$ and $u_\delta^c(t)$ given as in (3.114) and (3.115) respectively. □

Remark 3.9. Due to assumption (A.3.5) on $\hat{P}(d)$ and (A.3.4) on $E(d)$, the transfer matrix $[\hat{B}_2(d^{-1})E^{-1}(d^{-1})]'$ is a strictly anticausal and stable transfer matrix. Hence, $u_w^c(t)$ is a mean-square (m.s) bounded process because it is obtained as a bounded sum of conditional expectation of future samples of $w(t)$, that is a bounded m.s. process, w.r.t. $w^{t+\tau}$.

Remark 3.10. The above property does not hold true for $u_\delta^c(t)$ because H and $\hat{H}$ are arbitrary causal transfer matrices. Therefore, $u_\delta(t)$ depends on the full samples sequence of $\{\delta(t)\}$. Further, in order to obtain an assessment of the m.s. boundedness of u_δ^c, it is convenient to express (3.115) in a different way. To this end, consider the general solution for the pure regulation problem ($w \equiv \delta \equiv 0$) as given in the previous sections.

The resulting optimal $\mathcal{N}_1$ is given by

$$\mathcal{N}_1 = E^{-1} Y A_{3s}^{-1} D_{2q}^{-1} \tag{3.118}$$

where the polynomial matrices triplet (Y, X, Z) is the minimum degree solution w.r.t. Z, that is $\partial Z < \partial \bar{E}$, of the following couple of Diophantine equations

$$\bar{E}Y + ZA_3A_{3s} = \bar{\hat{B}}_2 \Psi_u [\hat{D}_q A_3 - A_{pl} D_{2q}]_2 \tag{3.119}$$

$$\bar{E}X + ZB_3A_{4s} = \bar{\hat{A}}_2 \Psi_u D_{1q} A_{4s} + \bar{\hat{B}}_2 \Psi_u [\hat{B}_1 D_{1q} - \hat{D}_q B_3]_1 . \tag{3.120}$$

In the above equations the following coprime matrix fraction descriptions are assumed to hold:

$$\begin{aligned} D_{2q} A_3^{-1} &= A_1^{-1} D_q \\ [\hat{D}_q A_3 - A_{pl} D_{2q}]_2 A_{3s}^{-1} &= A_{1s}^{-1} [\hat{D}_q A_3 - A_{pl} D_{2q}] \\ B_3 D_{1q}^{-1} &= D_q^{-1} B_1 \\ [\hat{B}_1 D_{1q} - \hat{D}_q B_3]_1 A_{4s}^{-1} &= A_{1s}^{-1} [\hat{B}_1 D_{1q} - \hat{D}_q B_3] \end{aligned}$$

with A_{3s} and A_{4s} strictly Schur since A_{1s} is such. Further, $\bar{E} := d^p E^*$, $\bar{\hat{B}}_2 := d^p \hat{B}_2^*$, $\bar{\hat{A}}_2 := d^p \hat{A}_2^*$, with $p = \max\{\partial E, \partial \hat{B}_2, \partial \hat{A}_2\}$. Notice that $E^{-*} \hat{B}_2^* = \bar{E}^{-1} \bar{\hat{B}}_2$.

By using (3.118) and (3.119) in (3.115) and considering (3.96), one obtains after direct steps

$$\begin{aligned} \bar{E}^{-1} \bar{\hat{B}}_2 \Psi_z \hat{H} - E\mathcal{N}_1 H &= \begin{bmatrix} \bar{E}^{-1} \bar{\hat{B}}_2 \Psi_z & E\mathcal{N}_1 \end{bmatrix} \begin{bmatrix} \hat{D}_h A_{3h} - A_{pl} D_{2h} \\ D_{2h} \end{bmatrix} A_{3h}^{-1} \\ &= \bar{E}^{-1} \bar{\hat{B}}_2 \Psi_z (\hat{D}_h - \hat{D}_q D_q^{-1} D_h) \\ &\quad + \bar{E}^{-1} Z D_q^{-1} D_h \end{aligned} \tag{3.121}$$

where $A_1^{-1} D_h = D_{2h} A_{3h}^{-1}$ with (D_{2h}, A_{3h}) right coprime polynomial matrices. In (3.121), we have used $A_3 D_{2q}^{-1} D_{2h} A_{3h}^{-1} = D_q^{-1} D_h$.

From (3.121) it follows that u_δ^c is the conditional expectation w.r.t. $\delta^{t+\nu}$ of a process, said r, that is m.s. bounded. In fact, since $\bar{E}^{-1} \bar{\hat{B}}_2$ and $\bar{E}^{-1} Z$ are both strictly anticausal and stable transfer matrices and $D_q^{-1} D_h$ is a causal and stable transfer matrix, r is obtained by passing a m.s. bounded process δ through a causal plus anticausal stable block. In turn, it implies that also u_δ^c is a m.s. bounded process.

Equation (3.121) generalizes a similar equation obtained for the standard LQ problem solved in [19]. In fact, the more general system description here considered permits each pair ($\hat{D}_q$ and D_q) and ($\hat{D}_h$ and D_h) to be different. □

As shown in detail in [18, 19], once stochastic models for the reference and accessible disturbance are given, (3.20) and (3.21) can be evaluated by solving suitable bilateral Diophantine equations. However, as it was also shown in the above references, when τ and ν are positive and large enough, $u^c_w(t)$ and $u^c_\delta(t)$ can be tightly approximated without requiring any stochastic modelling.

Example 3.2. In this example we consider a 3DOF control problem with coloured measurement noise on the observation signal y. To this end, consider the following scalar system

$$\begin{aligned} z(t) &= \tfrac{d}{1-3d}u(t) + \zeta(t) + \delta(t) \\ y(t) &= \tfrac{d}{1-3d}u(t) + \tfrac{1}{1-3d}\psi(t) + \delta(t). \end{aligned} \tag{3.122}$$

The above system can be easily embedded in the interconnection matrix description (3.96) by making the following positions

$$\mathcal{P} = \begin{bmatrix} \frac{d}{1-3d} \\ \frac{d}{1-3d} \end{bmatrix}, \quad \mathcal{Q} = \begin{bmatrix} 1 & 0 \\ 0 & \frac{1}{1-3d} \end{bmatrix}, \quad \mathcal{H} = \begin{bmatrix} 1 \\ 1 \end{bmatrix}, \quad e = [\zeta \; \psi]' \tag{3.123}$$

where $\{\zeta(t)\}$ and $\{\psi(t)\}$ are uncorrelated white noise sources with unitary variance. In (3.122), $\{\zeta(t)\}$ acts as a disturbance on the plant output and $\{\frac{1}{1-3d}\psi(t)\}$ is a coloured measurement noise. $\delta(t)$ is an accessible disturbance that acts contemporary on z and y.

By choosing $\Psi_z = \Psi_u = 1$ in the quadratic cost (3.98), E in (3.102) results to equal $E = 0.943(3.370 - d)$.

The stabilizing control law minimizing (3.98) for $w \equiv \delta \equiv 0$ (regulation problem) are determined in

$$\mathcal{M}_1 = \frac{-65.107 - 0.3d}{0.943(3.370 - d)(d-3)^2}, \quad \mathcal{N}_1 = \frac{-30.138 + 2.693d - 0.297d^2}{0.943(3.370 - d)(d-3)^2} \tag{3.124}$$

In such a case, the solution of the servo problem (viz. minimizing (3.98) for a given reference and accessible disturbance sequences w and δ) is given by

$$(-65.107 - 0.3d)u(t) = (30.138 - 2.693d + 0.297d^2)y(t) + (d-3)^2(u^c_w(t) + u^c_\delta(t) \tag{3.125}$$

with

$$\begin{aligned} u^c_w(t) &= \varepsilon\left\{\frac{d^{-1}}{E(d^{-1})}w(t) \mid w^{t+\tau}\right\} \\ &= (3.18)^{-1}\varepsilon\left\{\sum_{j=1}^{\infty}(3.370)^{-j+1}w(t+j) \mid w^{t+\tau}\right\} \end{aligned} \tag{3.126}$$

$$\begin{aligned} u^c_\delta(t) &= \varepsilon\left\{\left(\frac{d^{-1}}{E(d^{-1})} + \frac{-30.138 + 2.693d - 0.297d^2}{(d-3)^2}\right)\delta(t) \mid \delta^{t+\nu}\right\} \\ &= (3.18)^{-1}\varepsilon\left\{\sum_{j=1}^{\infty}(3.370)^{-j+1}\delta(t+j) \mid \delta^{t+\nu}\right\} \end{aligned} \tag{3.127}$$

$$+ \varepsilon \left\{ \sum_{i=0}^{\infty} \alpha_i \delta(t-i) \mid \delta^{t+\nu} \right\} \tag{3.128}$$

where α_i are the series coefficients of the z-transform of $E\mathcal{N}_1 = \sum_{i=0}^{\infty} \alpha_i d^i$.

Notice that if $\tau > 0$ and/or $\nu > 0$, $u_w^c(t)$ and/or $u_\delta^c(t)$ can be tightly approximated without using any stochastic model for $\{w(t)\}$ and/or $\{\delta(t)\}$, by merely retaining the first few samples of the sum in (3.126) and/or (3.127)-(3.128). This amounts in assuming that the controller knows the reference and the accessible disturbance few steps in advance. Further, if $\nu > 0$, (3.128) does not depend on the statistical properties of the disturbance.

3.4 Conclusions

LQG Regulation and Tracking problems have been solved via polynomial equation approach for a general system configuration.

The polynomial solution allows one to obtain an irreducible fractional factorization description of the LQG regulator, in contrast with the Wiener-Hopf solution.

This work does not confirm the common belief that the polynomial equation approach allows one to obtain the LQG regulator equations without explicitly solving a Kalman filtering problem. On the contrary, the procedure that appears suitable for general system configurations consists of a two stage procedure which is reminiscent of the Certainty-Equivalence property of Stochastic Dynamic Programming for the LQG.

3.5 Appendix

Proof of lemma 3.1. Σ_i can be factorized w.l.o.g. [12] as follows

$$\Sigma_i = A_\Sigma^{-1} B_\Sigma = \begin{bmatrix} A_{1s} & A_{pl} \\ 0 & A_1 \end{bmatrix}^{-1} \begin{bmatrix} \hat{D} & \hat{B}_1 \\ D & B_1 \end{bmatrix}, \tag{3.129}$$

with (A_Σ, B_Σ) a l.c.m.f.d. of Σ_i and $A_1^{-1} B_1$ a l.m.f.d. of P *not necessarily coprime*. Because of (A.3.2), Σ_i can be stabilized by feedback control laws of the form $\begin{bmatrix} e \\ u \end{bmatrix} = -\mathcal{C} \begin{bmatrix} z \\ y \end{bmatrix}$, $\mathcal{C} \in \text{Re}^{(l+m)(p+l)}[d]$,

$$\mathcal{C} = \mathcal{N}_c \mathcal{M}_c^{-1} = \begin{bmatrix} 0 & 0 \\ 0 & \mathcal{N}_2 \end{bmatrix} \begin{bmatrix} \mathcal{M}_o & 0 \\ 0 & \mathcal{M}_2 \end{bmatrix}^{-1}, \tag{3.130}$$

with $(\mathcal{N}_2, \mathcal{M}_2)$ such that $\mathcal{K} = \mathcal{N}_2 \mathcal{M}_2^{-1}$, $\mathcal{N}_2 \in \text{Re}^{ml}(d)$, $\mathcal{M}_2 \in \text{Re}^{ll}(d)$ and $\mathcal{M}_o \in \text{Re}^{pp}(d)$ stable transfer-matrices such that [1]

$$I_{p+l} = A_\Sigma \mathcal{M}_c + B_\Sigma \mathcal{N}_c = \begin{bmatrix} A_{1s}\mathcal{M}_o & A_{pl}\mathcal{M}_2 \\ 0 & A_1 \mathcal{M}_2 \end{bmatrix} + \begin{bmatrix} 0 & \hat{B}_1 \mathcal{N}_2 \\ 0 & B_1 \mathcal{N}_2 \end{bmatrix}. \tag{3.131}$$

By taking into account the determinant on both sides of (3.131) one has

$$\begin{aligned}\det I_{p+l} &= \det(A_{1s}\mathcal{M}_o)\det(A_1\mathcal{M}_2 + B_1\mathcal{N}_2)\\ &= \det A_{1s}\det\mathcal{M}_o\det L\det(A_{1o}\mathcal{M}_2 + B_{1o}\mathcal{N}_2)\end{aligned} \tag{3.132}$$

where the polynomial matrix L is a greatest left common divisor of ($A_1 = LA_{1o}, B_1 = LB_{1o}$). From (3.132), it follows that the system Σ_i is stabilizable by $\mathcal{K}$, provided that A_{1s} and L are strictly Schur polynomial matrices. Since (A.3.2) ensures that P has no unstable hidden modes and, hence, L is strictly Schur, one has to impose that A_{1s} is strictly Schur in order to guarantee the closed-loop stabilizability.

Equivalent conditions can be similarly established for (3.6) by taking into account a r.c.m.f. description of Σ_i and a l.c.f. representation of $\mathcal{C}$. □

Proof of theorem 3.5. By referring to the system description (3.48) with the additional assumption $u(t) \equiv 0$, the error $\tilde{z}(t)$ can be formulated as follows

$$\tilde{z}(t) = \mathcal{H}_{\tilde{z}\zeta}\zeta(t) + \mathcal{H}_{\tilde{z}\xi}\xi(t), \tag{3.133}$$

where $\mathcal{H}_{\tilde{z}\zeta}$ and $\mathcal{H}_{\tilde{z}\xi}$ are the following transfer-matrices

$$\mathcal{H}_{\tilde{z}\xi} = (I - \mathcal{H}_{\hat{z}y}A_f^{-1}B_f)A_1^{-1} \tag{3.134}$$

$$\mathcal{H}_{\tilde{z}\zeta} = -\mathcal{H}_{\hat{z}y}A_f^{-1}A_{n1}^{-1}. \tag{3.135}$$

Thus, the covariance error (3.51) can be conveniently re-expressed as

$$J = tr \prec \Phi_{\tilde{z}} \succ = tr\{\prec \mathcal{H}_{\tilde{z}\xi}\Phi_\xi\mathcal{H}_{\tilde{z}\xi}^* + \prec \mathcal{H}_{\tilde{z}\zeta}\Phi_\zeta\mathcal{H}_{\tilde{z}\zeta}^* \succ\}. \tag{3.136}$$

After a direct application of the usual "completing the square" argument and taking into account (3.134) and (3.135), (3.136) can be split in two terms

$$J = J_1 + J_2 \tag{3.137}$$

with

$$J_1 := tr \prec LL^* \succ \tag{3.138}$$

$$J_2 := tr \prec A_1^{-1}(\Phi_\xi - \Phi_\xi B_o^* D^{-*}D^{-1}B_o\Phi_\xi)A_1^{-*} \tag{3.139}$$

$$L := \mathcal{H}_{\hat{z}y}A_f^{-1}A_{n1}^{-1}A_o^{-1}D - A_1C_d\Phi_{\nu_1}C_d^*B_o^*D^{-*}, \tag{3.140}$$

where D is the Schur polynomial matrix solving the following left spectral factorization problem

$$DD^* = B_oC_d\Phi_{\nu_1}C_d^*B_o^* + A_oA_{f1}C_n\Phi_{\nu_2}C_n^*A_{f1}^*A_o^*, \tag{3.141}$$

The matrix D exists if and only if [1]

$$\text{rank}[B_oC_d \quad A_oA_{f1}C_n](e^{j\omega}) = \dim y \qquad \forall\omega \in [0, 2\pi].$$

Consequently, since J_2 does not depend on $\mathcal{H}_{\hat{z}y}$, only J_1 needs to be minimized. As just done in previous sections, the optimal solution can be computed by splitting L into its causal and strictly anticausal components. To this end, let p be the integer such that

$$p = \max\{\partial D, \partial(B_o C_d), \partial(A_o A_{f1} C_n)\} \tag{3.142}$$

and define

$$\bar{D} := d^p D^*, \quad \overline{C_d B_o} := d^p C_d^* B_o^*, \quad \overline{C_n A_{f1} A_o} := d^p C_n^* A_{f1}^* A_o^* \tag{3.143}$$

Consequently $C_d^* B_o^* D^{-*} = \overline{C_d B_o} \bar{D}^{-1}$. Now, the first term of L in (3.140) can be uniquely decomposed in two terms, by introducing the following Diophantine equation

$$\hat{Y}\bar{D} + A_1 \hat{Z} = C_d \Phi_{\nu_1} \overline{C_d B_o}, \tag{3.144}$$

that has to be solved for a polynomial matrix pair $(\hat{Y}, \hat{Z})$ with $\partial\hat{Z} < \partial\bar{D}$. In fact, according to (3.144), L can be rewritten as $L = (\mathcal{H}_{\hat{z}y} A_f^{-1} A_{n1}^{-1} A_o^{-1} D - A_1^{-1}\hat{Y}) - \hat{Z}\bar{D}^{-1}$, that is in a suitable form, being $\hat{Z}\bar{D}^{-1}$ a stable and strictly anticausal transfer-matrix. Then J_1 can be decomposed in

$$J_1 = J_3 + J_4 + J_5$$

$$J_3 := tr \prec \hat{Z}\bar{D}^{-1}\bar{D}^{-*}\hat{Z}^* \succ \tag{3.145}$$

$$J_4 := tr\{\prec \hat{Z}\bar{D}^{-1}L_+^* + L_+\bar{D}^{-*}\hat{Z}^* \succ\} = 0 \tag{3.146}$$

$$J_5 := tr \prec L_+ L_+^* \succ \tag{3.147}$$

$$L_+ := \mathcal{H}_{\hat{z}y} A_f^{-1} A_{n1}^{-1} A_o^{-1} D - A_1^{-1}\hat{Y} \tag{3.148}$$

Notice that $J_4 = 0$ since $\hat{Z}\bar{D}^{-1}L_+^*$ and $L_+\bar{D}^{-*}\hat{Z}^*$ are both stable and respectively strictly anticausal and strictly causal transfer matrices. Thus, the minimum is attained at $L_+ = 0$, and the optimal deconvolution filter results

$$\mathcal{H}_{\hat{z}y} = A_1^{-1}\hat{Y}D^{-1}A_o A_{n1} A_f. \tag{3.149}$$

The corresponding optimal cost turns out to be

$$J_{min} = tr \prec A_1^{-1}(C_d \Phi_{\nu_1} C_d^* - \hat{Y}\hat{Y}^*)A_1^{-*} \succ \tag{3.150}$$

Provided that (3.144) has an unique solution with $\partial\hat{Y} < \partial\bar{D}$, (3.149) uniquely specifies the minimum mean-square error deconvolution filter $\mathcal{H}_{\hat{z}y}$. In general, when A_1 is non-Schur, a second Diophantine equation needs in order to guarantee the mean-square boundedness of the deconvolution filter.

In order to derive the second Diophantine equation, one can proceed as follows. Define the signals

$$s(t) := A_{n1} A_f y(t) \tag{3.151}$$

$$\tilde{s}(t) := s(t) - A_{n1} B_f \hat{z} = (I - A_{n1} B_f \mathcal{H}_{\hat{z}y} A_f^{-1} A_{n1}^{-1}) A_{n1} A_f y(t) \tag{3.152}$$

Then, taking into account (3.152) and (3.48), $\mathcal{H}_{\hat{z}y}$ can be rewritten as

$$\mathcal{H}_{\hat{z}y} = \mathcal{H}_{\hat{z}\tilde{s}}(I + A_{n1}B_f\mathcal{H}_{\hat{z}\tilde{s}})^{-1}A_{n1}A_f \tag{3.153}$$

with $\mathcal{H}_{\hat{z}y}$ implicitly defined by $\hat{z}(t) = \mathcal{H}_{\hat{z}\tilde{s}}\tilde{s}(t)$. A convenient parameterization for $\mathcal{H}_{\hat{z}\tilde{s}}$ is the following

$$\mathcal{H}_{\hat{z}\tilde{s}} = A_1^{-1}\mathcal{N}_o\mathcal{M}_o^{-1}, \tag{3.154}$$

where $\mathcal{N}_o$ and $\mathcal{M}_o$ are stable and causal transfer-matrices. Then, by explicating (3.154) in (3.153), one gets

$$A_1^{-1}\hat{Y}D^{-1}A_oA_{n1}A_f = A_1^{-1}\mathcal{N}_o\mathcal{M}_o^{-1}(I + A_{n1}B_fA_1^{-1}\mathcal{N}_o\mathcal{M}_o^{-1})A_{n1}A_f,$$
$$\hat{Y}D^{-1} = \mathcal{N}_o(A_o\mathcal{M}_o + B_o\mathcal{N}_o)^{-1} \tag{3.155}$$

from which it can be argued that the mean-square boundedness of J_{min} is satisfied, provided that

$$A_o\mathcal{M}_o + B_o\mathcal{N}_o = I. \tag{3.156}$$

Consequently, the optimal $\mathcal{N}_o$ results given by

$$\mathcal{N}_o = \hat{Y}D^{-1}, \tag{3.157}$$

that is stable and causal as assumed.

Thus, $\mathcal{M}_o$ can be obtained from (3.156), taking into account (3.144), as follows

$$\begin{aligned} A_o\mathcal{M}_o &= I - B_o\hat{Y}D^{-1} = (D\bar{D} - B_o\hat{Y}\bar{D})\bar{D}^{-1}D^{-1} \\ &= (D\bar{D} - B_oC_d\Phi_{\nu_1}\overline{C_dB_o} + B_oA_1\hat{Z})\bar{D}^{-1}D^{-1} \\ &= A_o(A_{f1}C_n\Phi_{\nu_2}\overline{C_nA_{f1}A_o} + A_{n1}B_f\hat{Z})\bar{D}^{-1}D^{-1}. \end{aligned} \tag{3.158}$$

Further, for ensuring the stability of $\mathcal{M}_o$, the following Diophantine equation

$$\hat{X}\bar{D} - A_{n1}B_f\hat{Z} = A_{f1}C_n\Phi_{\nu_2}\overline{C_nA_{f1}A_o}, \tag{3.159}$$

has to be solved together (3.144), for some matrix polynomial $\hat{X}$. As consequence,

$$\mathcal{M}_o = \hat{X}D^{-1}. \tag{3.160}$$

Finally, summing (3.144) to (3.159) and considering (3.157) and (3.160), the following unilateral Diophantine equation is also derived

$$A_o\hat{X} + B_o\hat{Y} = D \tag{3.161}$$

Equation (3.161) is a necessary requirement for mean-square boundedness, that, in turn, is fulfilled only if D is strictly Schur. Notice that mean-square boundedness implies stability of $\mathcal{H}_{\hat{z}\tilde{s}}$, which, in turn, implies the stability of the convolution filter $\mathcal{H}_{\hat{z}y}$.

Then, one finds for $\hat{z}$ and y the following expression

$$y(t) = \mathcal{H}_{\hat{z}e}e(t) = A_f^{-1}A_{n1}^{-1}A_o^{-1}De(t)$$
$$\hat{z}(t) = \mathcal{H}_{\hat{z}y}\mathcal{H}_{\hat{z}e}e(t) = A_1^{-1}\hat{Y}e(t)$$

where $e(t)$ is a zero-mean, unitary variance innovations process of $y(t)$.

It remains to be proved that (3.144) and (3.159) are uniquely solvable for $(\hat{Y}, \hat{Z}, \hat{X})$ with $\partial\hat{Z} < \partial\bar{D}$. A condition under which this can be proved is that the greatest-common-right-divisors (GCRD) of A_1 and $A_{n1}B_f$ is strictly Schur. This is the case in the present discussion since A_n is strictly Schur and B_f and A_1 have no unstable common factors for stabilizability. The proof can be obtained, *mutatis mutandis*, following the one reported in [1] for the LQG problem.

3.6 References

[1] Casavola A. and E. Mosca. "Joint H_2/H_∞ LQ Stochastic Tracking and Servo Problems", Proceedings of *NTST Conference, Genova July 9-11, 1990*, in Progress in System and Control Theory Series, Birkhauser Boston Inc, 1990.

[2] Casavola A., M.J. Grimble, E. Mosca and P. Nistri. "Continuous-Time LQ Regulator Design by Polynomial Equations", *Automatica*, 27, No. 3, pp. 555-557, 1989.

[3] Chisci L. and E.Mosca. "Polynomial Solution to MMSE Multichannel Deconvolution", Proceedings of *ECC '91 Control Conference*, 1991.

[4] Chisci L. and E.Mosca. "Polynomial Equations for the Linear MMSE State Estimation", *IEEE Trans. Automat. Contr.*, AC-37, N. 5, pp. 623-626, 1992.

[5] Doyle J.C., K. Glover,P.P. Khargonekar and B.A. Francis. "State-space Solutions to Standard H_2 and H_∞ Control Problems", *IEEE Trans. Automat. Contr.*, AC. 34, pp. 831-846, 1990.

[6] Francis B.A. and J.C. Doyle. "Linear Control Theory with an H_∞ Optimality Criterion", *SIAM J. Contr. and Opt.*, 25, pp. 815-844, 1987.

[7] Francis B.A.. *A Course in H_∞ Control Theory*, In lecture Notes in Control and Information Science, Vol. 88, Springer-Verlang, Berlin, 1987.

[8] Grimble M.J.. "Multivariable Controllers for LQG Self-tuning Applications with Coloured Measurement Noise and Dynamic Cost Weighting", *Int. J. Contr.*, Vol. 17, N. 4, pp. 543-557, 1986.

[9] Grimble M.J.. "Two-degrees of Freedom Feedback and Feedforward Optimal Control of Multivariable Stochastic Systems", *Automatica*, Vol. 24, N. 6, pp. 809-817, 1988.

[10] Grimble M.J.. "Polynomial Matrix Solution to the Standard H_2 Optimal Control Problem", *Int. J. Systems Sci.*, Vol. 22, pp. 793-806, 1991.

[11] Grimble M.J. and M.A. Johnson. *Optimal Control and Stochastic Estimation Theory: Theory and Applications*, Part I and II, John Wiley, London, 1988.

[12] Gundes A.N. and C.A. Desoer. *Algebraic Theory of Linear Feedback Systems with Full and Decentralized Compensators*, Lecture Notes in Control and Information Sciences, N. 142, Springer-Verlang, New-york, 1990.

[13] Hunt K.J. and M. Šebek. "Optimal Multivariable Regulation with Disturbance Measurement Feedforward", *Int. Journ. of Cont.*, 49, No. 1, pp. 373-378, 1989.

[14] Hunt K.J. and M. Šebek. "Multivariable LQG Self-tuning Control with Disturbance Measurement Feedforward", Proceedings of *ICASSP Conference*, pp. 359-364, Glasgow, 1989.

[15] Hunt K.J., M. Šebek and V. Kučera. "Polynomial Approach to H_2-Optimal Control: The Multivariable Standard Problem", Proceedings of *30th CDC*, pp. 1261-1266, Brighton, England, 1991.

[16] Kučera V.. *Discrete Linear Control : the Polynomial Equation Approach*, (Chichester: Wiley), 1979.

[17] Mosca E., L.Giarre' and A.Casavola. "On the Polynomial Equations For the MIMO LQ Stochastic Regulator", *IEEE Trans. Automat. Contr.*, Vol. AC-35, N. 3, pp. 320-322, 1990.

[18] Mosca E. and G. Zappa. "Matrix Fraction Solution to the Discrete-time LQ Stochastic Tracking and Servo Problem", *IEEE Trans. Automat. Contr.*, AC-34, No. 2, pp. 240-242, 1989.

[19] Mosca E. and L. Giarré. "A Polynomial Approach to the MIMO LQ Servo and Disturbances Rejection Problems", *Automatica*, 8, No. 1, pp. 209-213, 1989.

[20] Mosca E., A. Casavola and L. Giarré. "Minimax LQ Stochastic Tracking and Servo Problems", *IEEE Trans. Automat. Contr.*, AC-35, No. 1, pp. 95-97, 1990.

[21] Park K. and J.J. Bongiorno, Jr.. "A General Theory for the Wiener-Hopf Design of Multivariable Control Systems", *IEEE Trans. Automat. Contr.*, Vol. AC-34, N. 6., pp.619-626, 1989.

[22] Shaked, U.. "A General Transfer Function Approach to Linear Stationary Filtering and Steady-state Optimal Control Problem", *Int. J. Contr.*, Vol. 24, pp. 741-770, 1976.

[23] Sternard M. and T. Söderstrom. "LQG-optimal Feedforward Regulators", *Automatica*, Vol. 24, No. 4, pp. 557-561, 1988.

4

A Game Theory Polynomial Solution to the H_∞ Control Problem

D. Fragopoulos and M. J. Grimble

4.1 Abstract

A new solution is given to the H_∞ multivariable control synthesis problem. The system is represented in polynomial matrix form and the derivation follows a game theory approach. This provides physical intuition regarding the form of the solution obtained. Moreover, the polynomial equations can be solved by a straightforward numerical algorithm even in the multivariable case. The links to LQG optimal control are also apparent by comparison with the LQG polynomial equations.

4.2 Introduction

A differential game problem is formulated and solved using polynomial methods. A linear plant is used which is excited by a stochastic disturbance. The treatment is carried out in the frequency domain. The solution uses a similar type of analysis to Kučera's (1979) solution of the LQG problem with the optimal control obtained over all linear, stabilizing output feedback controls. Unlike the LQG problem, both the optimal controller, minimizing the cost, as well as the worst case disturbance controller, maximizing the cost, are obtained by the game problem. The treatment here extends the results of Grimble (1994) that were obtained for the state-feedback case, exploiting the link with the Riccati equation solutions (Glover and Doyle, 1988), to higher order (non-Markov) plant descriptions. The work on this paper is based on Fragopoulos (1994).

A related problem is the Linear/Exponential/Quadratic/Gaussian (LEQG) control problem (Whittle, 1990). This problem is solved by converting it to a deterministic differential game. In a polynomial context, the game problem was first considered by Whittle and Kuhn (1986) and was further developed in Whittle (1990). The treatment there is in the time domain using a path integral method which is equivalent to the Pontryagin maximum principle for higher order systems. Although the method is quite powerful, to obtain the controller in closed-loop form, the plant model has to be restricted (B_1 assumed constant in (4.3)) resulting in loss of generality relative to the formulation used here.

Another related problem, equivalent to the LEQG problem in its infinite horizon form (Glover and Doyle, 1988), is the minimum entropy problem (Mustafa and Glover, 1990). A parameterization of all solutions satisfying an H_∞ bound, using the polynomial approach, has been obtained by Kwakernaak (1990, 1994) and Meinsma (1993). The solution is based on the J-factorization approach (see Green (1993) for a treatment based on rational fractional representations). These solutions however do not pick out the unique game/LEQG/minimum entropy solution.

Next it is required to show that the solution to the game problem obtained here satisfies an H_∞ bound. This is normally easy to show when solving over all worst case disturbance signals (Yaesh, 1992). Here however the worst case disturbance signal is a priori assumed to be of a linear output feedback form. This assumption is a plausible one given the linear, quadratic and Gaussian nature of the problem but it is nevertheless a conjecture for the indefinite case considered here. (In the definite case this result is a consequence of the Certainty Equivalence Principle (Whittle, 1990).) An alternative way for proving the H_∞ boundedness, and one which is used here, is by showing that the game solution is one of the H_∞ solutions obtained by the J-factorization approach. This way has the added advantage of obtaining conditions for closed loop stability that already exist with the J-factorization approach. Next the minimum entropy controller is obtained for the polynomial model using some general results from Mustafa and Glover (1990). The resulting controller is shown to be the same as the game controller (Fragopoulos, 1994).

The paper structure is as follows. In section 2 the problem is defined. In 3 the game problem is solved. The relation to the J-factorization problem is demonstrated in section 4 where the stability conditions are given and the parameterization of all H_∞ bounding controllers obtained. In section 5 the minimum entropy solution is given and it is shown to be the same as the game solution. The solution is applied to the mixed sensitivity design problem and a numerical example is solved in section 6. The paper concludes with section 7. The appendix includes the LQG formulas which are themselves original. These are an improvement to Kučera's (1979) solution as they dispense with the cancellations introduced in the controller formula by the Kučera solution.

4.3 Problem definition

Consider the following linear plant:

$$Ay = B_1 w = B_2 u \tag{4.1}$$

with signals y, u, w corresponding to measurements, control and disturbance respectively and where A, B_1, B_2 are polynomial matrices in the Laplace variable, s, with A and B_1 square and non-singular. Replacing s by $\mathcal{D} = \frac{d}{dt}$ in (4.1) results in a set of differential equations which must be well-posed. For this to hold the matrix A must be row reduced. Moreover without loss of generality it may be assumed that the pair $(A, [B_1\ B_2])$ is as in Rosenbrock (1970).

Define the error signal, z:

$$z = C_1 y + D_{12} u. \tag{4.2}$$

where C_1, D_{12} are polynomial matrices in s. The system may be cast in 'standard form', with the following system realization left coprime.

$$\Sigma_P := \left[\begin{array}{c|cc} -A & B_1 & B_2 \\ \hline C_1 & 0 & D_{12} \\ I & 0 & 0 \end{array}\right], \quad \begin{bmatrix} 0 \\ z \\ y \end{bmatrix} = \Sigma_P \begin{bmatrix} y \\ w \\ u \end{bmatrix} \tag{4.3}$$

In the above it has been assumed that $D_{11} = 0$. This is taken w.l.o.g. for constant D_{11}. The plant of (4.3) has the transfer P:

$$\begin{bmatrix} z \\ y \end{bmatrix} = P \begin{bmatrix} w \\ u \end{bmatrix} \quad \text{with} \quad P := \begin{array}{c} \\ p_1 \\ p_2 \end{array} \begin{array}{c} \begin{array}{cc} m_1 & m_2 \end{array} \\ \begin{bmatrix} P_{11} & P_{12} \\ P_{21} & P_{22} \end{bmatrix} \end{array}$$

The dimensions of the various signals are as above with $m_1 = p_2$. The following assumptions will be required for the above system.

Assumptions 1

(i) $\Sigma_{P_{12}} = \begin{bmatrix} -A & B_2 \\ C_1 & D_{12} \end{bmatrix}$ must have full column rank on the imaginary axis.

(ii) B_1 must be non-singular on the imaginary axis. Moreover without loss of generality B_1 will be assumed strictly Hurwitz.

The above assumptions ensure that P_{12}, P_{21} are free from invariant zeros on the imaginary axis and hence they ensure the existence of an L_∞ bound for the $w \to z$ transfer. Assumption 1 (ii) in particular is strengthened for the two-block case.

Assumptions 2

(i) $[A\ B_2]$ has full row rank on the closed RHP.

The above assumption is equivalent to Σ_{p22} being stabilizable.
Now define right coprime fraction

$$\bar{B}\bar{A}^{-1} := A^{-1}B \quad \text{with} \quad B := [B_1 \; B_2] \quad \text{and} \quad \bar{A} := \begin{bmatrix} \bar{A}_1 \\ \bar{A}_2 \end{bmatrix} \tag{4.4}$$

Under the assumption that $A^{-1}B$ is coprime, the system of (4.3) is *strictly equivalent* to

$$\Sigma_P := \left[\begin{array}{c|c} -\bar{A}_1 & I \;\; 0 \\ -\bar{A}_2 & 0 \;\; I \\ \hline C_1\bar{B} + D_{12}\bar{A}_2 & 0 \;\; 0 \\ \bar{B} & 0 \;\; 0 \end{array}\right]$$

(Kailath, 1980). The above assumptions may now be expressed equivalently as follows.

Assumptions 1′

(i) $\begin{bmatrix} \bar{A}_1 \\ C_1\bar{B} + D_{12}\bar{A}_2 \end{bmatrix}$ must have full column rank on the imaginary axis.

(ii) $\begin{bmatrix} \bar{B} \\ \bar{A}_2 \end{bmatrix}$ must be non-singular on the imaginary axis. Moreover without loss of generality $\begin{bmatrix} \bar{B} \\ \bar{A}_2 \end{bmatrix}$ will be assumed strictly Hurwitz.

Assumptions 2′

(i) $\begin{bmatrix} \bar{A}_1 \\ \bar{B} \end{bmatrix}$ has full row rank on the closed RHP.

Assumption 2 is adapted to cope with the zeros in the RHP.
Now split the disturbance signal w into two components

$$w = w_0 + w_1 \tag{4.5}$$

with w_0 representing a white noise signal of unit power spectral density matrix. The game cost-function may now be written as

$$I(K_2, K_1) = \frac{1}{2\pi j}\text{tr}\oint_D \left(\Phi_z(s) - \gamma^2\Phi_{w_1}(s)\right)ds \tag{4.6}$$

where the contour integral is evaluated around the D contour in the right-half plane, for some constant γ, $\gamma > 0$. Here Φ_z, Φ_{w_1} are the power spectral densities of z and w_1 respectively and

$$\begin{bmatrix} w_1 \\ u \end{bmatrix} = -\begin{bmatrix} K_1 \\ K_2 \end{bmatrix} y = -Ky \tag{4.7}$$

with K some stabilizing transfer function matrix. The game problem to be solved therefore becomes:

$$\min_{K_2} \max_{K_1} I\ (K_2, K_1) \tag{4.8}$$

In the game problem above 'nature' is trying to maximize the cost with K_1, while the controller, K_2, is trying to minimize the cost under the worst case (closed loop) disturbance.

In the above the disturbance w_0 is used to provide appropriate excitation to the system. This compares with the role the initial condition vector plays in the state-space methods.

4.4 The game problem

4.4.1 Main result

Recall the definitions

$$\bar{B}\bar{A}^{-1} := A^{-1}B \quad \text{with} \quad B := [B_1\ B_2] \quad \text{and} \quad \bar{A} := \begin{bmatrix} \bar{A}_1 \\ \bar{A}_2 \end{bmatrix} \tag{(4.4)}$$

and define

$$V := \begin{bmatrix} V_1 \\ V_2 \end{bmatrix} := \begin{bmatrix} C_1 & 0 & D_{12} \\ 0 & I_{m_1} & 0 \end{bmatrix} \tag{4.9}$$

so that the freedom in (4.4) is used to make $V \begin{bmatrix} \bar{B} \\ \bar{A} \end{bmatrix}$ column reduced. Next define the indefinite spectral factor D_c using:

$$D_c^* J D_c = \begin{bmatrix} \bar{B} \\ \bar{A} \end{bmatrix}^* \begin{bmatrix} V_1 \\ V_2 \end{bmatrix}^* \begin{bmatrix} I_{p_1} & 0 \\ 0 & -\gamma^2 I_{m_1} \end{bmatrix} \begin{bmatrix} V_1 \\ V_2 \end{bmatrix} \begin{bmatrix} \bar{B} \\ \bar{A} \end{bmatrix} \tag{4.10}$$

and let there exist $J = \begin{bmatrix} -I_{m_1} & 0 \\ 0 & I_{m_2} \end{bmatrix}$ and D_c strictly Hurwitz and column reduced, with the same column degrees as in $V \begin{bmatrix} \bar{B} \\ \bar{A} \end{bmatrix}$. Such a D_c is called a *canonical factor* (Kwakernaak and Sebek, 1994).

Define the right coprime matrix fraction

$$N_3 D_3^{-1} := B_1^{-1}[A\ -B_1\ -B_2] \tag{4.11}$$

Also introduce the bilateral diophantine equation:

$$D_c^*[G_0\ H_0] + F_0 N_3 = \begin{bmatrix} \bar{B} \\ \bar{A} \end{bmatrix}^* \begin{bmatrix} V_1 \\ V_2 \end{bmatrix}^* \begin{bmatrix} I_{p_1} & 0 \\ 0 & -\gamma^2 I_{m_1} \end{bmatrix} \begin{bmatrix} V_1 \\ V_2 \end{bmatrix} D_3 \tag{4.12}$$

in the unknowns F_0, G_0, H_0 for minimum degree with respect to F_0. Then define the left coprime matrix fraction:

$$D_4^{-1}[G\ H] := [G_0\ H_0] D_3^{-1} \tag{4.13}$$

The optimal controller may be obtained as follows:

$$K_{opt} = H^{-1}G \tag{4.14}$$

The stability of the optimal solution is established from the following so called implied equation

$$[G\, H]\begin{bmatrix}\bar{B}\\ \bar{A}\end{bmatrix} = D_4 J D_c \tag{4.15}$$

which is obtained from (4.12) after the right multiplication by $D_3^{-1}\begin{bmatrix}\bar{B}\\ \bar{A}\end{bmatrix}$ and simplification. The implied equation (4.15) defines the characteristic polynomial of the system when both the minimizing (actual) controller K_2 and the maximizing (fictitious) controller K_1 are used. The stability of the actual closed-loop system needs to be established separately. To this end define right coprime matrix fraction:

$$\bar{G}\bar{H}^{-1} := H^{-1}G \tag{4.16}$$

with $\begin{bmatrix}\bar{G}_1\\ \bar{G}_2\end{bmatrix} := \bar{G}$. Then the stability of the closed-loop system with $K_1 = 0$ may be determined using:

$$\Xi := A\bar{H} + B_2\bar{G}_2 \tag{4.17}$$

which is the closed-loop polynomial matrix of the actual closed-loop system (without the presence of the worst case disturbance).

Remark 4.1. The formulas above compare with the results for the LQG problem, theorem 4.12.

The following theorem summarizes the main results.

Theorem 4.2. *Given the plant of (4.3) satisfying the assumptions 1–2, the problem of (4.8) is solved for a given value of γ, $\gamma \neq 0$, if*

(i) There exists a canonical J-spectral factor D_c defined by (4.9)

(ii) T_0 defined by (4.25),is in L_2

(iii) The matrix H, defined in (4.14), is non-singular ($\det(H) \not\equiv 0$).

The optimal min-max controller, K_{opt}, is obtained from equations (4.4)–(4.14) and it is unique.

Proof. The proof of the theorem requires proof of: (i) stationarity of the cost w.r.t the controller at the optimal solution, (ii) convexity of the cost w.r.t. K_1 and concavity of the maximized cost w.r.t. K_2 and (iii) solvability of the bilateral diophantine equation (4.12). Proceed next by proving stationarity.

Stationarity conditions

First define the *sensitivity function*, S, and the *control sensitivity function*, M, as follows

$$S := (I + P_{2\bullet}K)^{-1} \quad \text{and} \quad M := KS. \tag{4.18}$$

M may be partitioned conformably with (w, u) as

$$\begin{bmatrix} M_1 \\ M_2 \end{bmatrix} := M.$$

After closing the loop with a controller K the signals z and w_1 are obtained

$$z = (P_{11} - P_{1\bullet}MP_{21})w_0 \tag{4.19}$$

$$w_1 = -M_1 P_{21} w_0 \tag{4.20}$$

The cost integrand in (4.6) now takes the form

$$\begin{aligned} I_c &:= \text{tr}(\Phi_z - \gamma^2 \Phi_{w_1}) \\ &= \text{tr}\left((P_{11} - P_{1\bullet}MP_{21})(P_{11} - P_{1\bullet}MP_{21})^* - \gamma^2 (M_1 P_{21})(M_1 P_{21})^*\right) \\ &= \text{tr}\big(P_{11}^* P_{11} - P_{21}^* M^* P_{1\bullet}^* P_{11} - P_{11}^* P_{1\bullet} M P_{21} \\ &\quad + P_{21}^* (M^* P_{1\bullet}^* P_{1\bullet} M - \gamma^2 M_1^* M_1) P_{21}\big) \end{aligned} \tag{4.21}$$

The last term in the expression above may be written as

$$\begin{aligned} M^* P_{1\bullet}^* P_{1\bullet} M - \gamma^2 M_1^* M_1 &= M^* (P_{1\bullet}^* P_{1\bullet} - \Gamma) M \\ &= M^* Y_c^* J Y_c M \end{aligned}$$

where

$$\Gamma := \begin{bmatrix} \gamma^2 I_{m_1} & 0 \\ 0 & 0 \end{bmatrix} \quad \text{and} \quad Y_c^* J Y_c = P_{1\bullet}^* P_{1\bullet} - \Gamma.$$

The Y_c is a J-spectral factor having the polynomial representation:

$$Y_c := D_c \bar{A}^{-1}. \tag{4.22}$$

Thus I_c may be written as

$$I_c = \text{tr}\left(P_{11}^* P_{11} - P_{21}^* M^* P_{1\bullet}^* P_{11} - P_{11}^* P_{1\bullet} M P_{21} + P_{21}^* M^* Y_c^* J Y_c M P_{21}\right). \tag{4.23}$$

Completing the squares in I_c with respect to M obtain

$$I_c = \text{tr}\left((Y_c M P_{21} - J^{-1} Y_c^{-*} P_{1\bullet}^* P_{11})^* J (Y_c M P_{21} - J^{-1} Y_c^{-*} P_{1\bullet}^* P_{11}) + T_0\right) \tag{4.24}$$

where T_0 has the definition

$$T_0 := P_{11}^* (I - P_{1\bullet} Y_c^{-1} J^{-1} Y_c^{-*} P_{1\bullet}^*) P_{11} \tag{4.25}$$

At this point it is appropriate to give some intuitive explanation of the solution procedure and explain what is to follow. In the expression for I_c above the only quantity depending on the controller is M. Thus T_0 is independent of the controller. In expression

$$X := Y_c M P_{21} - J^{-1} Y_c^{-*} P_{1\bullet}^* P_{11} \tag{4.26}$$

the first term is causal while the second term contains both causal and anti-causal factors. A standard Wiener argument states that a necessary condition for stationarity is the following

$$Y_c M P_{21} = \left[J^{-1} Y_c^{-*} P_{1\bullet}^* P_{11} \right]_+ \tag{4.27}$$

leading to the optimal control sensitivity function:

$$M = Y_c^{-1} \left[J^{-1} Y_c^{-*} P_{1\bullet}^* P_{11} \right]_+ P_{21}^{-1} \tag{4.28}$$

In the above note that both Y_c and P_{21} are minimum phase and thus M is stable. However it is well known (Grimble and Johnson, 1988; or Youla et al., 1976) that stability of M is not enough to guarantee closed loop stability (e.g. with unstable open loop plants). The polynomial equations (4.4)–(4.14) that have been introduced have the dual purpose of satisfying (4.28) while at the same time guaranteeing closed-loop stability (Kučera, 1979).

So much for the intuition. Proceed now with obtaining polynomial expressions for the term X appearing in the integrand, I_c. Substitution for the polynomial definitions of the terms in (4.26) gives:

$$X = D_c \bar{A}^{-1} K (A + BK)^{-1} B_1 - J^{-1} D_c^{-*} \begin{bmatrix} \bar{B} \\ \bar{A} \end{bmatrix}^* V_1^* C_1 A^{-1} B_1$$

Using equations (4.4), (4.11)–(4.13) and after some algebraic manipulations obtain

$$X = D_c \bar{A}^{-1} K (A + BK)^{-1} B_1 - (J^{-1} D_4^{-1} G A^{-1} B_1 + J^{-1} D_c^{-*} F_0).$$

Further with the help of (4.15) and after some more algebra obtain

$$\begin{aligned} X &= J^{-1} D_4^{-1} (H K_n - G K_d)(A K_d + B K_n)^{-1} B_1 - J^{-1} D_c^{-*} F_0 \\ &= J^{-1} D_4^{-1} H (K - H^{-1} G)(A + BK)^{-1} B_1 - J^{-1} D_c^{-*} F_0 \end{aligned} \tag{4.29}$$

with $K = K_n K_d^{-1}$. The RHS of (4.29) in the above consists of a causal term, provided K is stabilizing, and an anti-causal (strictly proper and anti-stable) term which is independent of K. Thus

$$X = T_+ + T_-$$

with the causal term

$$T_+ := J^{-1} D_4^{-1} H (K - H^{-1} G)(A + BK)^{-1} B_1 \tag{4.30}$$

and the anti-causal term

$$T_- := -J^{-1} D_c^{-*} F_0. \tag{4.31}$$

If T_+ is proper as well as stable and T_- is strictly proper and anti-stable, as a consequence of Cauchy's Theorem T_+ and T_- are orthogonal in the following sense:

$$\frac{1}{2\pi j}\text{tr}\oint_D T_-^* J T_+ ds = 0 \tag{4.32}$$

From the above results it follows that

$$\frac{1}{2\pi j}\text{tr}\oint_D X^* J X ds = \frac{1}{2\pi j}\text{tr}\oint_D T_+^* J T_+ ds + \frac{1}{2\pi j}\text{tr}\oint_D T_-^* J T_- ds \tag{4.33}$$

Now a variational argument will be used to complete the stationarity proof. Let

$$K = K_0 + \varepsilon\Delta \tag{4.34}$$

where K_0 is the stationary value of the controller and $\varepsilon\Delta$ is a variation such that both K and K_0 are stabilizing in the sense that $(A + BK)$ is strictly Hurwitz, with ε a real. Clearly for stationarity it is required that

$$\frac{\partial}{\partial\varepsilon}\left(\text{tr}\oint_D T_+^* J T_+ ds\right)\bigg|_{\varepsilon=0} = 0 \tag{4.35}$$

Now take

$$K_0 = H^{-1}G \tag{4.36}$$

and substitute in the RHS of (4.30) to obtain an expression for T_+ as a function of ε:

$$\begin{aligned} T_+ &= J^{-1}D_4^{-1}H\varepsilon\Delta(I + A^{-1}B(K_0 + \varepsilon\Delta))^{-1}A^{-1}B_1 \\ &= J^{-1}D_4^{-1}H\varepsilon\Delta(I + P_{2\bullet}(K_0 + \varepsilon\Delta))^{-1}P_{21} \\ &= J^{-1}D_4^{-1}H\varepsilon\Delta(I + S_0 P_{2\bullet}\varepsilon\Delta)^{-1}S_0 P_{21} \end{aligned} \tag{4.37}$$

where S_0 is the sensitivity function evaluated for K_0, i.e. $S_0 = (I + P_{2\bullet}K_0)^{-1}$. Employing a Taylor series expansion of the inverse term in (4.37) obtain

$$T_+ = J^{-1}D_4^{-1}H\Delta\varepsilon(I - \varepsilon S_0 P_{2\bullet}\Delta + 0(\varepsilon))S_0 P_{21} \tag{4.38}$$

where $0(\varepsilon)$ stands for zero of order ε as $\varepsilon \to 0$. Substitution for the above expression in $\text{tr}\oint_D T_+^* J T_+ ds$ results in

$$\text{tr}\oint_D T_+^* J T_+ ds = \varepsilon^2 \text{tr}\oint_D \left(D_4^{-1}H\Delta S_0 P_{21}\right)^* J^{-1}\left(D_4^{-1}H\Delta S_0 P_{21}\right) ds + 0(\varepsilon^2) \tag{4.39}$$

Observe that the above is a polynomial in ε, involving only second or higher order terms. Thus, since the first order terms in ε are all zero, it becomes apparent that the stationarity condition of (4.35) is satisfied.

Nature of the saddle point

Now it is necessary to show that the stationary point obtained in the previous section corresponds to the solution of the min-max problem of (4.8), taking first the max over K_1 and then the min over K_2 in this order.

Partition the variation Δ conformably with K_1, K_2 so that:

$$\begin{bmatrix}\Delta_1\\ \Delta_2\end{bmatrix} := \Delta \tag{4.40}$$

Next consider the minimax problem

$$\min_{\Delta_2}\max_{\Delta_1}\frac{1}{2\pi j}\oint_D I_c ds = \min_{\Delta_2}\max_{\Delta_1}\frac{1}{2\pi j}\mathrm{tr}\oint_D (T_+^* J T_+ + T_-^* J T_- + T_0)ds \tag{4.41}$$

In (4.41) above T_- and T_0 are independent of Δ_1, Δ_2 and thus it is enough to consider

$$\min_{\Delta_2}\max_{\Delta_1}\frac{1}{2\pi j}\mathrm{tr}\oint_D (T_+^* J T_+)ds \tag{4.42}$$

Define

$$Q := H^* D_4^{-*} J^{-1} D_4^{-1} H \tag{4.43}$$

and partition Q conformably to (w_1, u) in Q_{ij}, $i,j = 1,2$. The integrand in (4.42) after substitution from (4.43) becomes

$$T_+^* J T_+ = \varepsilon^2 (S_0 P_{21})^* \begin{bmatrix}\Delta_1\\ \Delta_2\end{bmatrix}^* \begin{bmatrix}Q_{11} & Q_{12}\\ Q_{12}^* & Q_{22}\end{bmatrix}\begin{bmatrix}\Delta_1\\ \Delta_2\end{bmatrix}(S_0 P_{21}) \tag{4.44}$$

A completion of the squares argument employing Schur's formula results in

$$\begin{aligned} T_+^* J T_+ &= \varepsilon^2 (S_0 P_{21})^* \\ &\times((\Delta_1 + Q_{11}^{-1}Q_{12}\Delta_2)^* Q_{11}(\Delta_1 + Q_{11}^{-1}Q_{12}\Delta_2) + \Delta_2{}^*(Q_{22} - Q_{12}^* Q_{11}^{-1} Q_{12})\Delta_2) \\ &\times(S_0 P_{21}) \end{aligned}$$

The maximization of (4.42) with respect to Δ_1 is possible only if $Q_{11} \le 0$ over the imaginary axis. Here

$$Q_{11} = (D_4^{-1} H_1)^* J^{-1} (D_4^{-1} H_1) \tag{4.45}$$

where H_1 is defined from

$$[H_1\ H_2] := H \tag{4.46}$$

with H partitioned conformably to (w_1, u) . Indeed by lemma 4.9 (see appendix) we have that

$$Q_{11} < 0 \tag{4.47}$$

with Q_{11} a constant matrix. The strict negativity of Q_{11} implies that the maximizing Δ_1, Δ_{1*}, is unique. Δ_{1*} is given by

$$\Delta_{1*} = -Q_{11}^{-1} Q_{12} \Delta_2 \tag{4.48}$$

and the maximum value of the cost, (4.42), is given by

$$\max_{\Delta_1}\mathrm{tr}\oint_D T_+^* J T_+ ds = \mathrm{tr}\oint_D \varepsilon^2 (S_0 P_{21})^* \Delta_2{}^*(Q_{22} - Q_{12}^* Q_{11}^{-1} Q_{12})\Delta_2 (S_0 P_{21}) ds \tag{4.49}$$

Next it is required to show that (4.49) above may be minimized w.r.t. Δ_2. From (4.49) it is apparent that the minimization is possible only if $Q_{22} - Q_{12}^* Q_{11}^{-1} Q_{12} \geq 0$ on the imaginary axis. This condition may be shown to hold because of the following reasoning:

(i) $Q \sim J$ a. e. on the imaginary axis since D_4 is strictly Hurwitz, non-singular and H must also be non-singular if the stationary controller K_0 is to exist.

(ii) $Q \sim \text{diag}\{Q_{11}, Q_{22} - Q_{12}^* Q_{11}^{-1} Q_{12}\}$

From (i) and (ii) it is implied that $J \sim \text{diag}\{Q_{11}, Q_{22} - Q_{12}^* Q_{11}^{-1} Q_{12}\}$ a. e. on the imaginary axis, and given that:

$$\dim Q_{11} = \text{neg} J$$

it follows that

$$Q_{22} - Q_{12}^* Q_{11}^{-1} Q_{12} > 0 \quad \text{a. e. on the } j\omega\text{-axis} \tag{4.50}$$

with the above being ≥ 0 everywhere on the imaginary axis, even if H has zeros on the axis. (In the case when H has no zeros on the imaginary axis the strict inequality in (4.50) holds everywhere on the axis.) Assuming that $\Delta_2(s)$ is analytic in s, as it is normal to do and is certainly the case for rational functions in s, then the minimizing Δ_2, Δ_{2*}, is *uniquely* determined as

$$\Delta_{2*} = 0 \tag{4.51}$$

Thus the minimax problem has the unique solution $\begin{bmatrix} \Delta_{1*} \\ \Delta_{2*} \end{bmatrix} = 0$ for ε: small. Thus the unique optimal controller to the minimax problem is K_0 as defined in (4.36).

Solvability of the diophantine equation and simplifications

In order to prove the solvability of (4.10) it will be necessary to exploit the structure of the right to left fraction conversions (4.4) and (4.11). In particular from (4.11) write

$$\begin{bmatrix} B_1 \vdots -A \; B \end{bmatrix} \begin{bmatrix} N_3 \\ \cdots \\ D_{31_\bullet} \\ D_{32_\bullet} \end{bmatrix} = 0 \tag{4.52}$$

where D_3 is partitioned in two block rows $D_{31_\bullet}$, $D_{32_\bullet}$. Suppose for a while that (A, B) were left coprime. Then for the left to right coprime fraction conversion (4.11) the following is true (Kailath, 1980)

$$\text{Smith}(N_3) = \text{Smith}([-A \; B]) = [I \; 0] \tag{4.53}$$

It is possible to make $N_3 = [I\,0]$ by multiplying (4.52) from the right by an appropriate unimodular matrix. Thus (4.52) takes the form:

$$\begin{bmatrix} B_1 \vdots -A \; B_1 \; B_2 \end{bmatrix} \begin{bmatrix} I & 0 \\ \cdots & \cdots \\ 0 & \bar{B} \\ -I & \bar{A}_1 \\ 0 & \bar{A}_2 \end{bmatrix} = 0 \tag{4.54}$$

Now the above relation holds even if (A, B) are not coprime as it is valid after left multiplying by the left g.c.d. of (A, B).

Comparing (4.52) with the above obtain D_3. Substitution of N_3 and D_3 in the diophantine equation (4.12) results in the following two block column partitions:

$$D_C^* G_0 + F_0 = \gamma^2 \begin{bmatrix} \bar{B} \\ \bar{A} \end{bmatrix}^* \begin{bmatrix} V_1 \\ V_2 \end{bmatrix}^* \begin{bmatrix} 0 \\ I_{m_1} \end{bmatrix} = \gamma^2 \bar{A}_1^*. \tag{4.55}$$

and

$$D_c^* H_0 + 0 = D_c^* J D_c. \tag{4.56}$$

The last equation of the above does always have a solution

$$H_0 = JD_c \tag{4.57}$$

while the first one is just the left division of $\gamma^2 A_1^*$ by D_c^*, and as such it is solvable. Now from a comparison with (4.82) in the appendix obtain

$$G_0 = -D_4^{-1} H_1 \tag{4.58}$$

which is a constant by lemma 4.9. Next left multiply the implied equation (4.15) by D_4^{-1}, and after substitution by the above obtain

$$D_4^{-1}[G\; H_2] = (JD_c + G_0 \bar{A}_1) \begin{bmatrix} \bar{B} \\ \bar{A}_2 \end{bmatrix}^{-1} \tag{4.59}$$

from which the LHS may be evaluated as a right to left fraction conversion.

Remark 4.3. The simplifications obtained here are a result of the special structure of the problem. The noise signal w_0 and the worst case disturbance signal w_1 which is treated as part of the control signal enter the autoregressive equation of (4.3) with the same matrix, B_1. The special structure of the game problem becomes apparent by comparing the game formulation (4.3) with that of the H_2 problem in the appendix (4.97).

Remark 4.4. In theorem 4.2 condition (ii) is necessary for the cost to be finite while condition (iii) is required for the controller to exist. A necessary condition for H to be non-singular is that $(D_c^{-1}\bar{A}_1)(\jmath_\infty)$ has full column rank (see (4.83)). Cases when the optimal control problem is well defined with H singular are however possible. An example of such a case is the sensitivity minimization of stable plants: this results in zero cost for an infinite gain controller. This implies that the cost may be arbitrarily reduced for some appropriate controller. In this sense the condition on the non-singularity of H may be removed when one is interested on the inf-sup problem.

Condition (i) is used to ensure that $V \begin{bmatrix} \bar{B} \\ \bar{A} \end{bmatrix} D_c^{-1}$ lives in H_∞.The behaviour of the spectral factor, D_c, is thought to be the following. Canonical factors D_c exist in (γ_0, ∞) for some positive real γ_0 with the exception of the specific values of γ (a countable set that corresponds to an optimal cost in the AAK sense) where D_c loses its column reducedness. (γ_0 is the lowest value of γ for which the signature of the RHS of (4.12) is constant over the frequency range.) For details on properties of the spectral factor see Kwakernaak and Sebek (1994).

Remark 4.5. In addition to internal stability, a continuous time plant is normally desirable to have a closed-loop system that is internally proper, so that the response is free of impulsive modes. In theorem 4.2 the issue of internal properness has not been considered. Nevertheless the solution to the optimal problem in theorem 4.2 is uniquely determined and it is obtained irrespective of internal properness. Thus internal properness may be checked retrospectively, a sufficient condition for which is that the plant, P_{22}, and controller, K, be proper and one of the two sensitivity functions ($(I + KP_{22})^{-1}$ or $(I + P_{22}K)^{-1}$) exist and be proper.

4.4.2 Summary of the simplified solution procedure

Let assumptions 1–2 hold. The solution procedure may now be summarized as follows:

(i) Evaluate the left to right coprime matrix fraction conversion

$$\bar{B}\bar{A}^{-1} := A^{-1}B \quad \text{with} \quad B := [B_1\ B_2] \quad \text{and} \quad \begin{bmatrix} \bar{A}_1 \\ \bar{A}_2 \end{bmatrix} := \bar{A} \qquad ((4.4))$$

so that $V \begin{bmatrix} \bar{B} \\ \bar{A} \end{bmatrix}$ is column reduced, where V is defined as

$$V := \begin{bmatrix} V_1 \\ V_2 \end{bmatrix} := \begin{bmatrix} C_1 & 0 & D_{12} \\ 0 & I_{m_1} & 0 \end{bmatrix}. \qquad ((4.9))$$

(ii) Solve J-spectral factorization

$$D_c^* J D_c = \begin{bmatrix} \bar{B} \\ \bar{A} \end{bmatrix}^* \begin{bmatrix} V_1 \\ V_2 \end{bmatrix}^* \begin{bmatrix} I_{p_1} & 0 \\ 0 & -\gamma^2 I_{m_1} \end{bmatrix} \begin{bmatrix} V_1 \\ V_2 \end{bmatrix} \begin{bmatrix} \bar{B} \\ \bar{A} \end{bmatrix} \qquad ((4.10))$$

with D_c Hurwitz.

(iii) Perform polynomial left division by D_c^*

$$D_C^* G_0 + F_0 = \gamma^2 \bar{A}_1^* \qquad ((4.55))$$

resulting in F_0 such that $D_C^{-*} F_0$ is strictly proper and in G_0: constant.

(iv) Obtain G, H_2, D_4 from the right to left fraction conversion

$$D_4^{-1}[G\ H_2] = (JD_c + G_0\bar{A}_1)\begin{bmatrix}\bar{B}\\ \bar{A}_2\end{bmatrix}^{-1} \tag{(4.59)}$$

and H_1 by substitution

$$H_1 = -D_4G_0.$$

(v) The controller is given by

$$K_{opt} = [H_1\ H_2]^{-1}G.$$

4.4.3 Comments

On the numerical solution

Of the above steps (i) and (iv) involve standard right to left fraction conversions of which the first one is independent of γ. In step (iii) G_0 may be obtained by dividing the row coefficient matrix of $\bar{A}_1^*$ having the row degrees of D_c^* by the leading row degree coefficient matrix of D_c^*. Here recall that D_c^* is row reduced. The polynomial matrix F_0 may be obtained by subtraction. Thus the solution of the diophantine equation is greatly simplified. Last but not least, step (ii) requires the solution of a J-spectral factorization. This amounts to the most demanding computation in the solution of the game problem. The J-factorization is analogous to the Riccati equation in state-space formulations. For a review on the numerical solution of the J-factorization see Kwakernaak and Sebek (1994). A more recent and promising method for solving the J-spectral factorization is described in Kwakernaak (1994).

On the min-max problem

It becomes apparent from the solution to the game problem considered in theorem 4.2 that in general the max and min operation of the game problem may not be interchanged, with the maximization over w_1 having to take place first. A case where the max and min operations may be interchanged is that of the first order (Markov) problem. Thus the corollary below is obtained.

Corollary 4.6. *Consider the Markov case where $A = sI - A_0$ and B, C_1, D_{12} are all constant matrices. Then in the min-max problem of (4.8) the min and max operations may be interchanged without affecting the solution.*

Proof. See appendix.

4.5 Relations to the J-factorization H_∞ problem

4.5.1 Introduction

It is now interesting to establish the relationship between the game theoretic solution developed here and the J-factorization solution (Green et al., 1990;

Green, 1993). For a polynomial version of the latter that is more appropriate for such a comparison, see Meinsma (1993), Kwakernaak (1994). There are good reasons for seeking to establish such a relationship. It is well known that there is a whole class of solutions satisfying a given H_∞ bound (see the references above). Although it is argued in the state-space literature that the unique game solution, which is also equivalent to the minimum entropy (Mustafa and Glover, 1990) and the LEQG solution (Whittle, 1990), is one of particular interest, it is useful to know what all the solutions satisfying an H_∞ bound are. Another question of some interest is to determine conditions for closed-loop stability directly from the J-factorization without explicitly having to check the closed loop polynomial. Both these questions are answered in the J-factorization approach.

4.5.2 The J-factorization solution

The system of (4.3) has the following right fractional representation:

$$P = \begin{bmatrix} C_1\bar{B} + D_{12}\bar{A}_2 \\ \bar{B} \end{bmatrix} \begin{bmatrix} \bar{A}_1 \\ \bar{A}_2 \end{bmatrix}^{-1} \tag{4.60}$$

Applying the formulas in Kwakernaak (1994), noting that $\begin{bmatrix} \bar{B} \\ \bar{A}_2 \end{bmatrix}$ is square and strictly Hurwitz as $\det \begin{bmatrix} \bar{B} \\ \bar{A}_2 \end{bmatrix} = \det B_1$, all H_∞ strictly bounding controllers $K_2 = K_n K_d^{-1}$ are given by

$$\begin{bmatrix} -K_d \\ K_n \end{bmatrix} = \begin{bmatrix} \bar{B} \\ \bar{A}_2 \end{bmatrix} D_c^{-1} \begin{bmatrix} U_d \\ U_n \end{bmatrix} \tag{4.61}$$

where U_d is non-singular, square, stable rational and U_n is stable rational satisfying

$$\left\| U_n U_d^{-1} \right\|_\infty < 1. \tag{4.62}$$

Define (lower) linear fractional representation $F(\cdot,\cdot)$: $F(P,K) := P_{11} + P_{12}K(I - P_{22}K)^{-1}P_{21}$. The resulting closed-loop transfer function, $F(P, -K_2)$, satisfies

$$\|F(P, -K_2)\|_\infty < \gamma.$$

Moreover there exists a stabilizing controller satisfying the above inequality iff the matrix

$$\begin{bmatrix} \bar{A}_1 \\ D_{c,2} \end{bmatrix} \quad \text{is strictly Hurwitz} \tag{4.63}$$

with $D_{c,2}$ being the lower partition of D_c and having the same dimensions as $\bar{A}_2$.

The stability condition is of particular importance as it is expressed only in terms of the spectral factor D_c. This result stems from the solution of a two-block distance problem associated to the J-factorization (4.10) (see Green et al., 1990; Meinsma, 1993).

4.5.3 Connection with the game solution

Define a right fractional representation of the controller

$$\bar{G}\bar{H}^{-1} := H^{-1}G = K \tag{4.64}$$

or equivalently

$$[G\ H_1\ H_2]\begin{bmatrix}-\bar{H}\\ \bar{G}_1\\ \bar{G}_2\end{bmatrix} = 0 \tag{4.65}$$

where $\bar{G}$ is partitioned conformably with H.

In the sequel the simplified formulas will be used. Right multiply (4.59) by $\begin{bmatrix}-\bar{H}\\ \bar{G}_2\end{bmatrix}$ and use equations (4.65) and (4.58) to obtain

$$D_4^{-1}H_1\bar{G}_1 = (JD_c + G_0\bar{A}_1)\begin{bmatrix}\bar{B}\\ \bar{A}_2\end{bmatrix}^{-1}\begin{bmatrix}-\bar{H}\\ \bar{G}_2\end{bmatrix}.$$

Substitution from (4.58) in the above results in

$$G_0\bar{G}_1 = (JD_c + G_0\bar{A}_1)\begin{bmatrix}\bar{B}\\ \bar{A}_2\end{bmatrix}^{-1}\begin{bmatrix}-\bar{H}\\ \bar{G}_2\end{bmatrix} \tag{4.66}$$

and after rearranging obtain

$$J^{-1}G_0\left(\bar{G}_1 - \bar{A}_1\begin{bmatrix}\bar{B}\\ \bar{A}_2\end{bmatrix}^{-1}\begin{bmatrix}-\bar{H}\\ \bar{G}_2\end{bmatrix}\right) = D_c\begin{bmatrix}\bar{B}\\ \bar{A}_2\end{bmatrix}^{-1}\begin{bmatrix}-\bar{H}\\ \bar{G}_2\end{bmatrix} \tag{4.67}$$

or equivalently

$$\begin{bmatrix}-\bar{H}\\ \bar{G}_2\end{bmatrix} = \begin{bmatrix}\bar{B}\\ \bar{A}_2\end{bmatrix}D_c^{-1}J^{-1}G_0\left(\bar{G}_1 - \bar{A}_1\begin{bmatrix}\bar{B}\\ \bar{A}_2\end{bmatrix}^{-1}\begin{bmatrix}-\bar{H}\\ \bar{G}_2\end{bmatrix}\right) \tag{4.68}$$

Note that $\bar{G}_1 - \bar{A}_1\begin{bmatrix}\bar{B}\\ \bar{A}_2\end{bmatrix}^{-1}\begin{bmatrix}-\bar{H}\\ \bar{G}_2\end{bmatrix}$ is stable and has the Rosenbrock system matrix

$$\begin{bmatrix}\bar{B} & -\bar{H}\\ \bar{A}_2 & \bar{G}_2\\ \bar{A}_1 & \bar{G}_1\end{bmatrix} \tag{4.69}$$

which determines the closed-loop stability and will thus be strictly Hurwitz, non-singular by the implied equation (4.15). Thus the transfer

$$\bar{G}_1 - \bar{A}_1\begin{bmatrix}\bar{B}\\ \bar{A}_2\end{bmatrix}^{-1}\begin{bmatrix}-\bar{H}\\ \bar{G}_2\end{bmatrix}$$

is square, non-singular, stable and minimum phase and may thus be ignored in the evaluation of the controller in (4.68). Now by lemma 4.9 and (4.58) obtain

$$G_0^*J^{-1}G_0 < 0. \tag{4.70}$$

The left hand side of (4.67) may thus be identified with a $\begin{bmatrix} U_d \\ U_n \end{bmatrix}$ in (4.61). Then the controller $K_2 = K_n K_d^{-1}$ takes the form $K_2 = \bar{G}_2 \bar{H}^{-1}$.

The above results are summarized as the following theorem.

Theorem 4.7. *The game solution satisfies an H_∞ bound $\|F(P, -K_2)\| < \gamma$. The optimal controller is obtained by the substitution:*

$$\begin{bmatrix} U_d \\ U_n \end{bmatrix} := J^{-1} G_0$$

in (4.61).

The controller K_2 obtained by solving the game problem stabilises the (actual) closed-loop iff the matrix $\begin{bmatrix} \bar{A}_1 \\ D_{c,2} \end{bmatrix}$ is strictly Hurwitz.

4.6 Relations to the minimum entropy control problem

The minimum entropy problem has been studied extensively (see Mustafa and Glover (1990) and the references therein). A particularly interesting result that motivates investigation of minimum entropy problem is the following: the solution minimizing the entropy (at infinity) has the desirable property of minimizing the H_2 norm of the error transfer under the constraint that the H_∞ norm of the (same) error transfer is bound by some constant, γ (Mustafa and Glover, 1990). Moreover for the relation between the entropy problem (at infinity) and the Linear Exponential of Quadratic/Gaussian (LEQG) problem in its time average form see Glover and Doyle (1988), and Whittle (1990). In this section a theorem is quoted stating that the controller that minimizes an entropy integral is the same as the controller obtained from solving the game problem.

Define the *entropy at* s_0

$$I(H;\gamma;s_0) := -\frac{\gamma^2}{2\pi} \int_{-\infty}^{\infty} \ln \left| \det \left(I - \gamma^{-2} H^*(j\omega) H(j\omega) \right) \right| \frac{s_0^2}{s_0^2 + \omega^2} d\omega \tag{4.71}$$

with $s_0 > 0$ and a transfer function, H, so that $H \in L_\infty$, $\|H\|_\infty < \gamma$.

Of particular interest is the *entropy at infinity*, defined by

$$I(H;\gamma;\infty) := \lim_{s_0 \to \infty} \{ I(H;\gamma;s_0) \}. \tag{4.72}$$

From the definitions above it is clear that for the entropy integral to exist, H must be contractive: $\|H\|_\infty < \gamma$. Moreover for the entropy at infinity to be finite, a rational H must be strictly proper.

It is the minimization of the entropy at *infinity* that relates to the game and the LEQG problem.

The following provides the main result.

Theorem 4.8 (theorem 5.4 in Fragopoulos, 1994). *The controller solving the game problem also minimizes the entropy at infinity* $s_0 \to \infty$ *with*

$$U_n U_d^{-1} = -G_{0,2} G_{0,1}^{-1}$$

The proof employs the idea of all-pass embedding implemented in Meinsma (1993) by the use of J-factorizations, and results from Mustafa and Glover (1990) for minimizing entropy of an error transfer: $H = F(Q,U)$ where $Q^*Q = I$ with $\|Q_{22}(\infty)\| < 1$ where $U \in H_\infty$, $\|U\|_\infty < 1$. For a complete proof see Fragopoulos (1994).

4.7 A design example: mixed sensitivity

4.7.1 Mixed sensitivity problem formulation

The solution to the perfect observation game problem developed is readily applicable to the multivariable mixed sensitivity problem (Kwakernaak, 1986). Consider the standard plant, P,

$$P : \begin{bmatrix} z_1 \\ z_2 \\ \cdots \\ y \end{bmatrix} = \begin{bmatrix} W_q V & W_q G \\ 0 & W_r \\ V & G \end{bmatrix} \begin{bmatrix} w \\ \cdots \\ u \end{bmatrix} \tag{4.73}$$

where

$$[V \; G] = D^{-1}[E \; N] \tag{4.74}$$

is the combined plant/disturbance model with D, E, N polynomial matrices and E is strictly Hurwitz. The error and control weightings W_q, W_r respectively have the following polynomial fractional representations

$$W_q = B_q A_q^{-1} \quad \text{and} \quad W_r = B_r A_r^{-1}. \tag{4.75}$$

A loop transformation is required to put the system P in the system form of (4.3). Post multiply the last column of P by A_r and premultiply the last row of P by A_q^{-1} to obtain P_l

$$P_l := \begin{bmatrix} B_q A_q^{-1} D^{-1} E & B_q A_q^{-1} D^{-1} N A_r \\ 0 & B_r \\ A_q^{-1} D^{-1} E & A_q^{-1} D^{-1} N A_r \end{bmatrix}. \tag{4.76}$$

P_l has the system representation

$$\Sigma_{P_l} := \left[\begin{array}{c|cc} -DA_q & E & NA_r \\ \hline B_q & 0 & 0 \\ 0 & 0 & B_r \\ I & 0 & 0 \end{array}\right] \tag{4.77}$$

which is in the form of (4.3). Let the controller for P_l be K_l and the controller for P be K. The controller K may be recovered by

$$K = A_r K_l A_q^{-1}. \tag{4.78}$$

The above reveals a structural property of the controller. Using this structure it is easy to see for example that any integrators may be introduced in the controller through the zeros of A_q. Moreover it is often convenient to take the control weighting W_r to be improper in order to force the controller to roll-off at high frequencies. (The controller, $K(j\omega)$, is asymptotically bounded from above by $\bar{\sigma}\big(\gamma W_r^{-1} V^{-1}(j\omega)\big)$, as $\omega \to \infty$, with V usually taken to be biproper.) Thus it is demonstrated that improper weightings are both meaningful and easily accommodated in a polynomial formulation. The improper system may be loop transformed (by multiplying the control or output by an appropriate polynomial matrix) to be made proper. This however incurs cancellations in the controller (Meinsma, 1993) and is thus undesirable.

4.7.2 Numerical example

Consider the plant/disturbance model with

$$D = \begin{bmatrix} s^2 & -s^2 \\ 0 & s(s+2) \end{bmatrix}, \quad E = \begin{bmatrix} s+1 & -s \\ 0 & s+2 \end{bmatrix}, \quad N = \begin{bmatrix} 1 & 0 \\ 0 & 1 \end{bmatrix}$$

and weights $W_q = I_2$ and $W_r = 0.1 I_2$. The optimal value of γ is obtained as $\gamma_{opt} = 0.44762$. As the spectral factor D_c is not canonical at the optimal value of γ, take $\gamma = 0.5$.

Evaluation of the right fraction conversion (4.4), yields

$$\bar{A} = \begin{bmatrix} -1 & 0 & -1 & -1 \\ 0 & 0 & 0 & -1 \\ 1 & 0 & 1 & 1 \\ 0 & 0 & 0 & 2 \end{bmatrix} + \begin{bmatrix} 1 & 0 & 0 & 0 \\ 0 & 1 & 0 & 0 \\ 0 & 0 & 1 & 0 \\ 0 & 0 & 0 & 1 \end{bmatrix} s$$

$$\bar{B} = \begin{bmatrix} 1 & 0 & 0 & 0 \\ 0 & 1 & 0 & 0 \end{bmatrix}$$

The spectral factor has the evaluation

$$D_c = \begin{bmatrix} -1.7170 & -0.6881 & -0.6614 & -1.0623 \\ 4.3726 & -3.9571 & 1.4349 & -0.4383 \\ 2.9340 & -0.1800 & 0.8690 & 0.6916 \\ -3.7708 & 4.1352 & -1.2252 & 0.6263 \end{bmatrix} + \begin{bmatrix} 0.3905 & 0.3166 & 0 & 0.0104 \\ -0.3195 & 0.9825 & 0 & 0.0085 \\ -0.0546 & 0 & 0.0585 & 0.0818 \\ -0.0394 & 0 & -0.0811 & 0.0590 \end{bmatrix} s$$

The stability of the resulting closed loop may be checked at this point by finding the zeros of $\begin{bmatrix} \bar{A}_1 \\ D_{c_2} \end{bmatrix}$. These are: -11.5277, -2.7476, $-4.1339 \pm 2.8628j$, all stable. Proceeding with the solution of the division (4.55) gives F_0, G_0

$$F_0 = \begin{bmatrix} 0.9028 & -0.5743 \\ -0.8121 & 1.2369 \\ 0.2699 & -0.1997 \\ -0.0133 & 0.0905 \end{bmatrix}, \quad G_0 = \begin{bmatrix} 0.3905 & 0.3166 \\ -0.3195 & 0.3870 \\ 0.0546 & 0 \\ 0.0394 & 0 \end{bmatrix}.$$

The negativity condition of lemma 4.9 may be verified

$$G_0^T J^{-1} G_0 = \begin{bmatrix} -0.25 & 0 \\ 0 & -0.25 \end{bmatrix} = -\gamma^2 I < 0$$

The shortest way for obtaining the actual controller, K_2, is form (4.68) so that

$$\bar{H} = \begin{bmatrix} 12.7989 & -3.5796 \\ -3.1797 & 10.8449 \end{bmatrix} + \begin{bmatrix} 1 & 0 \\ 0 & 1 \end{bmatrix} s,$$

$$\bar{G}_2 = \begin{bmatrix} 31.1813 & -5.4185 \\ -7.5326 & 51.1863 \end{bmatrix} + \begin{bmatrix} 46.7465 & -34.5913 \\ -3.7663 & 25.5932 \end{bmatrix} s$$

The above $\bar{H}$, $\bar{G}_2$ are determined within right multiplication by a stable, minimum phase, square, non-singular matrix. The corresponding closed-loop zeros are:

$$-11.9483,\ -2.7380,\ -3.9787 \pm 2.8178j,\ -2.0000,\ -1.0000.$$

The singular value plot of the closed loop transfer, $w \to z$, appears in figure 4.1.

4.8 Conclusions

In this paper a polynomial, perfect observation, output feedback, differential game problem is solved in the frequency domain. The worst case disturbance controller as well as the actual optimal controller are obtained. The optimal solution is unique. It becomes clear from the solution that the order of the minimization and the maximization may not be interchanged in general (unlike the state-feedback problem, corollary 4.6) with the maximization over the worst case disturbance taking place first. The solution was originally obtained by using equations that are adapted from the LQG formulas. (The LQG formulas used here are themselves an improvement on Kučera's (1979) solution.) These equations are then simplified by taking advantage of the structure of the particular game problem considered. The closed-loop is shown to satisfy an H_∞ bound by establishing a link with the J-factorization approach. Also from this link the stability conditions for the actual closed loop (where the worst case controller is omitted) are obtained.

The game controller is the same as the minimum entropy controller. An outline of the proof for this result is given.

Finally a mixed sensitivity design is formulated and a numerical example solved.

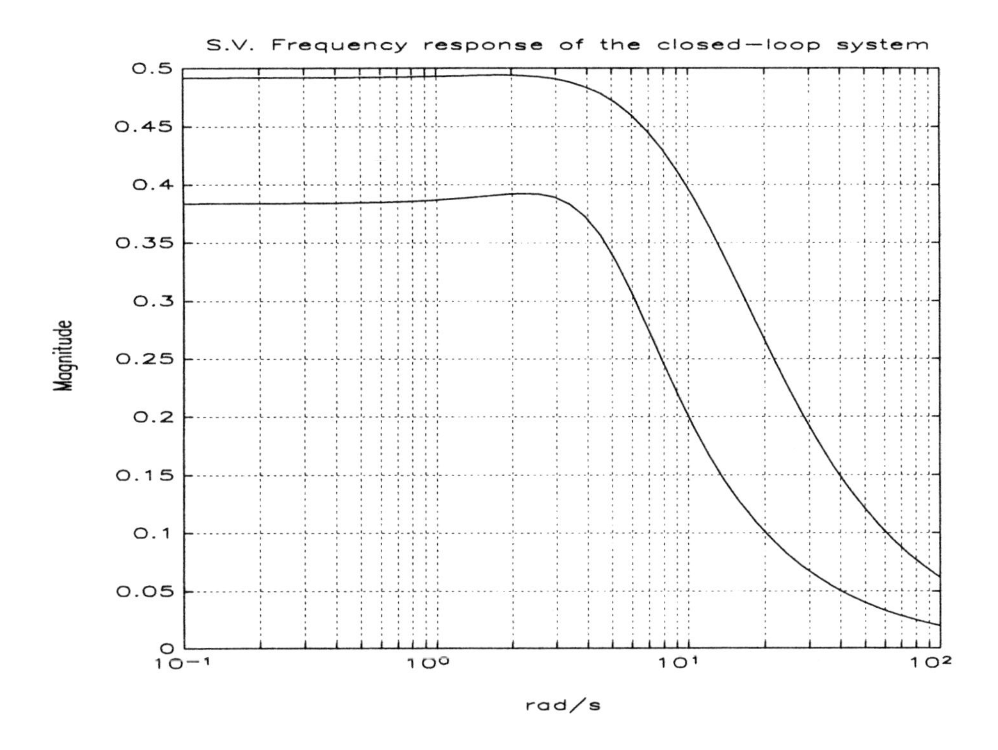

Figure 4.1: Singular value plots of the closed-loop transfer $w \to z$.

4.9 Appendix

Lemma 4.9. *The following matrix relation holds*

$$(D_4^{-1}H_1)^* J^{-1}(D_4^{-1}H_1) \leq -\gamma^2 \tag{4.79}$$

with $D_4^{-1}H_1$ *a constant matrix.*

Proof. Multiply (4.12) on the right by D_c^{-*} and on the left by D_3^{-1} to obtain

$$[G_0\ H_0]D_3^{-1} + D_c^{-*}F_0N_3D_3^{-1} = D_c^{-*}\begin{bmatrix}\bar{B}\\ \bar{A}\end{bmatrix}^*\begin{bmatrix}V_1\\ V_2\end{bmatrix}^*\begin{bmatrix}I_{p_1} & 0\\ 0 & -\gamma^2 I_{m_1}\end{bmatrix}\begin{bmatrix}V_1\\ V_2\end{bmatrix} \tag{4.80}$$

Substitute from (4.11) and (4.13) so that

$$\begin{aligned}&D_4^{-1}[G\ H_1\ H_2] + D_c^{-*}F_0B_1^{-1}[A\ -B_1\ -B_2]\\ &= D_c^{-*}\begin{bmatrix}C_1\bar{B} + D_{12}\bar{A}_2\\ \bar{A}_1\end{bmatrix}^*\begin{bmatrix}I_{p_1} & 0\\ 0 & -\gamma^2 I_{m_1}\end{bmatrix}\begin{bmatrix}C_1 & 0 & D_{12}\\ 0 & I_{m_1} & 0\end{bmatrix}\end{aligned} \tag{4.81}$$

Consider the second block column of the above which simplifies to the following

$$\begin{aligned}D_4^{-1}H_1 &= D_c^{-*}F_0 - \gamma^2 D_c^{-*}\begin{bmatrix}C_1\bar{B} + D_{12}\bar{A}_2\\ \bar{A}_1\end{bmatrix}^*\begin{bmatrix}0\\ I_{m_1}\end{bmatrix}\\ &= D_c^{-*}F_0 - \gamma^2 D_c^{-*}\bar{A}_1^*\end{aligned} \tag{4.82}$$

Note that in (4.82) the RHS consists of a strictly proper term and a proper term, thus being proper overall. Moreover the LHS is stable while the RHS is anti-stable. Thus they both have to be constant so that they have no (finite) poles while being proper at the same time. Hence

$$D_4^{-1}H_1 = -\gamma^2 D_c^{-*}\bar{A}_1^*|_{s=\infty} \tag{4.83}$$

Note that for a canonical factor D_c of (4.10), the expression $\gamma D_c^{-*}\bar{A}_1^*$ is part of a J-unitary matrix. In particular define U_1, $U_1 \in L_\infty$

$$\begin{aligned}U_1 &:= D_c^{-*}\begin{bmatrix}\gamma\bar{A}_1\\ C_1\bar{B} + D_{12}\bar{A}_2\end{bmatrix}^*\\ &= D_c^{-*}\begin{bmatrix}\bar{B}\\ \bar{A}\end{bmatrix}^*\begin{bmatrix}\gamma V_2\\ V_1\end{bmatrix}^*\end{aligned} \tag{4.84}$$

which by (4.10) satisfies

$$U_1\begin{bmatrix}J & 0\\ 0 & I_{p_1-m_2}\end{bmatrix}U_1^* = J. \tag{4.85}$$

Consider the following lemma.

Lemma 4.10 (lemma A4.4 in Fragopoulos, 1994). *Given the* L_∞ *matrix* U_1*:* $U_1JU_1^* = J_1$ *with* $J = \text{diag}\{J_1, -I_3\}$ *there exists* U_3 *such that*

$$U := \begin{bmatrix}U_1\\ U_3\end{bmatrix},\quad UJU^* = J.$$

By lemma 4.10 there exists J-unitary complement of U_1, U_3, such that: $U := \begin{bmatrix} U_1 \\ U_3 \end{bmatrix} \in L_\infty$ satisfies $UJ_pU^* = J_p$ with

$$J_p := \begin{bmatrix} J & 0 \\ 0 & I_{p_1-m_2} \end{bmatrix} = \left[\begin{array}{cc:c} -I_{m_1} & 0 & 0 \\ 0 & I_{m_2} & 0 \\ \hdashline 0 & 0 & I_{p_1-m_2} \end{array}\right].$$

But

$$\begin{aligned} UJ_pU^* &= J_p \Leftrightarrow \\ U^*J_p^{-1}UJ_pU^* &= U^* \Leftrightarrow \\ U^*J_p^{-1}U &= J_p^{-1} \end{aligned} \tag{4.86}$$

Partition U in a block 3×3 matrix conformably with J_p

$$\left[\begin{array}{ccc} U_{11} & U_{12} & U_{13} \\ U_{21} & U_{22} & U_{23} \\ \hdashline U_{31} & U_{32} & U_{33} \end{array}\right] = \left[\begin{array}{c} U_1 \\ \hdashline U_3 \end{array}\right] = U$$

and consider the $(1,1)$ partition of (4.86) yielding

$$\begin{bmatrix} U_{11} \\ U_{21} \end{bmatrix}^* J \begin{bmatrix} U_{11} \\ U_{21} \end{bmatrix} = -I_{m_1} - U_{31}^*U_{31} \leq -1 < 0 \tag{4.87}$$

The inequality above holds for all $s = j\omega$ including $s = j\infty$, since U lives in L_∞. Now identify

$$\begin{bmatrix} U_{11} \\ U_{21} \end{bmatrix} = U_1 \begin{bmatrix} I_{m_1} \\ 0 \\ 0 \end{bmatrix}$$

evaluated at $s = j\infty$ as the matrix part of the RHS of (4.83) to prove the statement of the lemma.

Lemma 4.11. *Consider the J-spectral factor, D_c, defined in (4.10), having the partitioned form* $\begin{bmatrix} D_{c,1} \\ D_{c,2} \end{bmatrix} := D_c$ *and define the $L\infty$ matrix*

$$Q_{22} = D_{c,1} \begin{bmatrix} \gamma \bar{A}_1 \\ D_{c,2} \end{bmatrix}^{-1} \begin{bmatrix} 0 \\ I \end{bmatrix}.$$

Then

$$\|Q_{22}\|_\infty < 1.$$

Proof. Define

$$X := \begin{bmatrix} Q_{22} \\ I \end{bmatrix} = \begin{bmatrix} D_{c,1} \\ D_{c,2} \end{bmatrix} \begin{bmatrix} \gamma \bar{A}_1 \\ D_{c,2} \end{bmatrix}^{-1} \begin{bmatrix} 0 \\ I \end{bmatrix}.$$

Thus it is enough to prove that $X^*JX > 0$. Proceeding write

$$\begin{aligned} X^*JX &= \begin{bmatrix} 0 \\ I \end{bmatrix}^* \begin{bmatrix} \gamma\bar{A}_1 \\ D_{c,2} \end{bmatrix}^{-*} D_c^* J D_c \begin{bmatrix} \gamma\bar{A}_1 \\ D_{c,2} \end{bmatrix}^{-1} \begin{bmatrix} 0 \\ I \end{bmatrix} \\ &= \left(\left(\begin{bmatrix} \gamma\bar{A}_1 \\ D_{c,2} \end{bmatrix} (D_c^* J D_c)^{-1} \begin{bmatrix} \gamma\bar{A}_1 \\ D_{c,2} \end{bmatrix}^* \right)^{-1} \right) \\ &= \left(I - D_{c,2} (D_c^* J D_c)^{-1} \gamma \bar{A}_1^* \left(\gamma^2 \bar{A}_1^* (D_c^* J D_c)^{-1} \bar{A}_1 \right)^{-1} \gamma \bar{A}_1 (D_c^* J D_c)^{-1} D_{c,2}^* \right) \end{aligned}$$

Note that from the proof of lemma 4.9 $\gamma^2 \bar{A}_1^* (D_c^* J D_c)^{-1} \bar{A}_1 < -I$. Thus the expression above is positive definite on the imaginary axis.

Theorem 4.12 (LQG Solution; Fragopoulos, 1994). *Consider the following plant defined by*

$$\begin{bmatrix} 0 \\ z \\ y \end{bmatrix} = \begin{bmatrix} -A & B_0 & B \\ C & 0 & D \\ I & 0 & 0 \end{bmatrix} \begin{bmatrix} y \\ w_0 \\ u \end{bmatrix} \tag{4.88}$$

where w_0 is a white noise signal of unity power spectral density, u is the control and y is the measurement. The plant has transfer function P

$$\begin{bmatrix} z \\ y \end{bmatrix} = P \begin{bmatrix} w_0 \\ u \end{bmatrix} \quad \text{with} \quad P := \begin{bmatrix} P_{11} & P_{12} \\ P_{21} & P_{22} \end{bmatrix}$$

Assume that:

(i) $\begin{bmatrix} -A & B \\ C & D \end{bmatrix}$ *is full column rank on the imaginary axis.*

(ii) B_0 *is non-singular and strictly Hurwitz.*

(iii) $[A\ B]$ *is full row rank on the closed RHP.*

Consider the following minimization problem

$$\min_{K_2:\,\text{stabilizing}} \operatorname{tr} \oint_D \Phi_z(s) ds \tag{4.89}$$

where Φ_z is the power spectral density of z.

Define the right coprime matrix fraction

$$\bar{B}\bar{A}^{-1} := A^{-1}B \tag{4.90}$$

so that $[C\ D] \begin{bmatrix} \bar{B} \\ \bar{A} \end{bmatrix}$ *is column reduced. Moreover define the right coprime matrix fraction*

$$N_3 D_3^{-1} := B_0^{-1}[A\ -B] \tag{4.91}$$

and Hurwitz spectral factor D_c *satisfying:*

$$D_c^* D_c = \begin{bmatrix} \bar{B} \\ \bar{A} \end{bmatrix}^* [C \; D]^* [C \; D] \begin{bmatrix} \bar{B} \\ \bar{A} \end{bmatrix} \tag{4.92}$$

and note that D_c *is column reduced. Solve the bilateral diophantine equation*

$$D_c^* [G_0 \; H_0] + F_0 N_3 = \begin{bmatrix} \bar{B} \\ \bar{A} \end{bmatrix}^* [C \; D]^* [C \; D] D_3 \tag{4.93}$$

in the unknowns F_0, G_0, H_0 *for minimum degree with respect to* F_0. *The optimal controller is*

$$K_{opt} = H^{-1} G \tag{4.94}$$

with G, H *defined by the left fraction*

$$D_4^{-1} [G \; H] := [G_0 \; H_0] D_3^{-1}. \tag{4.95}$$

The closed-loop stability is guaranteed by the polynomial equation

$$[G \; H] \begin{bmatrix} \bar{B} \\ \bar{A} \end{bmatrix} = D_4 D_c \tag{4.96}$$

which is implied from the equations above and which determines the closed-loop polynomial.

The optimal cost, if it exists, is given by

$$\frac{1}{2\pi j} \mathrm{tr} \oint_D \left(\left(D_c^{-*} F_0\right)^* \left(D_c^{-*} F_0\right) + P_{11}^* \left(I - P_{12} Y_c^{-1} Y_c^{-*} P_{12}^*\right) P_{11} \right) ds \tag{4.97}$$

where $Y_c = D_c \bar{A}^{-1}$.

Proof. This result may be obtained in a similar way to the game problem solution with the only difference that the convexity condition (4.47) is irrelevant here.

The LQG equations presented above take the form of Kučera's (1979) equations if D_3 is written in a block diagonal form

$$D_3 = \begin{bmatrix} D_1 & 0 \\ 0 & D_2 \end{bmatrix}$$

where the partitions are square and conformable with $[A \; - \; B]$. As a result of (4.91), generically, $\det(D_1) \propto \det(D_2) \propto \det(B_0)$ resulting in $\det(D_3) \propto \det^2(B_0)$ compared with $\det(D_3) \propto \det(B_0)$ if the structure of D_3 is not restricted to being block diagonal. In (4.95) however $\det(D_4) \propto \det(B_0)$. Thus in Kučera's case a cancellation of a common factor with determinant $\det(B_0)$ is necessary while in the solution presented here no cancellation is needed. Thus it is believed that the equations above should be numerically easier to solve as well as conceptually more attractive. The formulas for the H_2 solution are original.

Proof of corollary 4.6 *Consider the Markov case where* $A = sI - A_0$ *and* B, C_1, D_{12} *are all constant matrices. Then in the min-max problem of (4.8) the min and max operations may be interchanged without affecting the solution.*

Proof. From equation (4.81) obtain

$$D_4^{-1}H = D_c^{-*}F_0B_1^{-1}B + D_c^{-*}\left[(C_1\bar{B} + D_{12}\bar{A}_2)^*\bar{A}_1^*\right]\begin{bmatrix} 0 & D_{12} \\ -\gamma^2 I_{m_1} & 0 \end{bmatrix}$$

In the above the RHS is antistable and proper while the LHS is stable. Thus they are both constant. Using the strict properness of $D_C^{-*}F_0$ and of $A^{-1}B = \bar{B}\bar{A}^{-1}$ and thus of $\bar{B}D_c^{-1}$, obtain:

$$D_4^{-1}H = D_c^{-*}\bar{A}^*R\big|_{s=\infty}$$

where

$$R := \begin{bmatrix} D_{12}^*D_{12} & 0 \\ 0 & -\gamma^2 I_{m_1} \end{bmatrix}.$$

Now from (4.43) we have

$$\begin{aligned} Q &= H^*D_4^{-*}J^{-1}D_4^{-1}H \\ &= R^*\bar{A}D_c^{-1}JD_c^{-*}\bar{A}^*R\big|_{s=\infty} \end{aligned} \tag{4.98}$$

Note however that from the definition of D_c follows

$$\bar{A}^{-*}D_c^*JD_c\bar{A}^{-1}\big|_{s=\infty} = R$$

which if substituted in the expression (4.98) results in

$$Q = R = \begin{bmatrix} D_{12}^*D_{12} & 0 \\ 0 & -\gamma^2 I_{m_1} \end{bmatrix}. \tag{4.99}$$

Thus the part of the proof of theorem 4.2 concerning the second order variations may be modified to prove the statement of the lemma.

4.10 References

Fragopoulos D. (1994) *H_∞ synthesis theory using polynomial system representations*, submitted for the degree of Ph.D., University of Strathclyde, Scotland, August.

Glover, K. and J. C. Doyle (1988). 'State-space formulae for all stabilizing controllers that satisfy an H_∞-norm bound and relations to risk sensitivity', System and Control Letters, Vol. 11, pp. 167–172.

Green, M., K. Glover, D. Limebeer and J. Doyle (1990). 'A J-Factorization Approach to H_∞ Control', SIAM J. Contr. Optimization, Vol. 28, N.6, pp 1350–1371, November.

Grimble, M. J. (1994). *Robust Industrial Control: Optimal Design Approach for Polynomial Systems*, Prentice Hall.

Grimble, M. J. and M. A. Johnson (1988). *Optimal Control and Stochastic Estimation: Theory and Applications*, John Wiley & Sons, Chichester, UK.

Kailath, T. (1980). *Linear Systems*, Prentice-Hall, Englewood Cliffs, N. J.

Kučera, V. (1979). *Discrete Linear Control: The Polynomial Equation Approach*, Wiley.

Kwakernaak, H. (1986). 'A polynomial approach to minimax frequency domain optimization of multivariable feedback systems', Int. J. Contr., Vol. 44, pp. 117–156.

Kwakernaak, H. (1990). 'The Polynomial Approach to H_∞-Optimal Regulation', to appear Lecture notes 1990 CIME Course on Recent Developments in H_∞ Control Theory, Springer Verlag.

Kwakernaak, H. (1994). 'Frequency domain solution of the standard H_∞ problem', IFAC Symposium on Robust Design, Rio de Janeiro (to appear).

Kwakernaak, H. and M. Sebek (1994). 'Polynomial J-spectral factorization', IEEE Trans. Aut. Control, Vol. AC-39, No. 2, pp. 315–328, February.

Meinsma, G. (1993). *Frequency Domain Methods in H_∞ Control*, PhD Thesis, University of Twente, The Netherlands.

Redheffer, R. M., (1960). 'On a certain linear fractional transformation', J. Math. Physics, Vol. 39, pp. 269–286.

Rosenbrock, H. H. (1970). *State-space and Multivariable Theory*, Nelson, London.

Whittle, P. and J. Khun (1986). 'A Hamiltonian formulation of the risk-sensitive Linear/ Quadratic/Gausian control', Int. J. Contr., Vol. 43, pp. 1–12.

Whittle, P. (1990). *Risk Sensitive Optimal Control*, John Wiley & Sons, Chichester, UK.

Yaesh, I. (1992) *Linear Optimal Control and Estimation in the Minimum H_∞-Norm Sense*, PhD Thesis, Tel Aviv University, Israel.

Youla, D. C., H. A. Jabr and J. J. Bongiorno (1976). 'Modern Wiener-Hopf design of optimal controllers-Part II: The multivariable case', IEEE Trans. Aut. Control, Vol. AC-21, No. 3, pp. 319–338, June.

Young, N. (1988). *An Introduction to Hilbert spaces*, Cambridge University Press.

4.11 Acknowledgements

The second author is grateful for the support of the Engineering Research & Physical Sciences Council on the hot rolling mill robust control applications project, No. GR/J/54864.

5

H_2 Design of Nominal and Robust Discrete Time Filters

Mikael Sternad and A. Ahlén

5.1 Abstract

Polynomial methods were originally developed with control applications in mind [1, 2], but have turned out to be very useful within digital signal processing and communications. The present chapter [1] will outline a polynomial equations framework for nominal amnd robust multivariable linear filtering and, at the same time, illustrate its utility for signal processing problems in digital communications.

5.2 Introduction

Wiener and Kalman techniques for model-based filter design have been used extensively by electrical engineers for decades. Today, these methods are well known as tools for the design of $\mathcal{H}_2$-optimal estimators. The use of Kalman filters has been the common choice within the control community. A primary reason is the development of the state feedback control theory [3], where Kalman observers

[1]Parts of the chapter include material from [4], [5] and [6]. It is reprinted with permission by IEEE and by Academic Press. Pages 20–25 contain material from the conference paper [5], *Narrowband and broadband multiuser detection using a multivariable DFE*, from IEEE PIMRC'95, Toronto, 1995, 732–736. Section 5.3 is modified from reference [4], from IEEE Transactions on Automatic Control, 1995; 40:405–418. The chapter describes work supported by the Swedish National Board for Technical Development, NUTEK, under contracts 87-01583 and 9303294-2, as well as by the Swedish Research Council for Engineering Sciences (TFR) under grant 92-775.

constitute essential elements. Other reasons are flexibility and the ability to cope with time-varying systems, as well as the availability of Riccati equation solvers with good numerical behaviour.

Engineers working in the signal processing and communication fields have instead tended to prefer Wiener filters. One reason is their lower computational complexity; when only a few signals are to be estimated, the use of a time-varying estimator for the whole state vector is deemed unnecessary. The use of filters in input-output form will furthermore provide immediate engineering insight: A quick inspection of poles and zeros roughly indicate what filter properties can be expected.

There are additional differences between control and statistical signal processing. While rational transfer functions (IIR-models) are suitable as models of the often slow dynamics of industrial plants, complexity requirements and high speed applications have frequently forced designers in the signal processing and communication fields to restrict their attention to FIR-models and filters [7, 8, 9]. FIR-filters have the advantage of always being stable and they can readily be optimized, on line as well as off line [10, 11].

FIR filters can approximate any time-invariant impulse response, but this may require many filter coefficients. Long filters might be unacceptable from a complexity point of view. Furthermore, the use of many parameters may lead to overfitting [12]; the amount of information per parameter simply becomes too small. Superior results could therefore in many applications be achieved by the use of structures with fewer parameters, such as IIR-models and filters.

Realizable IIR Wiener filters, based on IIR-models, are conceptually easy to derive [13, 14, 15], but explicit solutions have been difficult to obtain. The polynomial systems framework has been of help here. The rather intractable causal bracket operation $\{\cdot\}_+$, which is central in the classical expression of realizable Wiener filters, can now be readily evaluated by means of a Diophantine equation [6, 17, 18]. The polynomial approach to the design of Wiener estimators for a wide class of filtering problems will be outlined in section 5.3. The polynomial solution provides not only the optimal adjustment, but also the optimal structure and degree of the estimator. The resulting filters have a structure in which numerator and denominator polynomial matrices of the signal models appear directly. Such expressions provide immediate engineering insight into the properties of the solution.

Wiener filters are obtained by minimizing mean square error criteria. Although this is highly relevant in numerous applications, other types of criteria have been suggested as well. In particular, the use of robust filtering is of interest, since signal models are rarely exactly known.

In order to attain robustness, criteria of minimax type have frequently been used in the past. See, for example, [19, 20, 21]. A motivation comes from situations where, perhaps for safety reasons, the effect of a worst case scenario must be minimized, possibly in the presence of model errors. Robust filters obtained in this way tend to be rather conservative.

Another concept which has been considered in the context of robust design is $\mathcal{H}_\infty$-optimization. The design is then conducted with respect to signals of bounded power and unknown spectral density. See chapters 2 and 4. In the control literature, a main motivation for the development of $\mathcal{H}_\infty$-optimization during the 1980's [22, 23] was to achieve robust stability for feedback systems. This motivation is absent in open-loop filtering problems. Applications of the $\mathcal{H}_\infty$ concept also to estimation problems have, however, been proposed. See, for instance, [24, 25] and [26]. The use of $\mathcal{H}_\infty$-optimization for filtering simply implies the minimization of the largest principal gain of the transfer function between unknown norm-bounded signals and the estimation error. The resulting filters will be inherently conservative, since they are designed to guard against extreme situations, where all disturbance energy is concentrated at the worst frequency. When such a design has to be robustified against modelling errors [25, 26], the conservatism will become even more pronounced. In view of this, it seems more promising to robustify a traditional $\mathcal{H}_2$ design, which focuses on the minimization of mean square estimation errors.

Robust $\mathcal{H}_2$-estimation can be formulated in minimax-$\mathcal{H}_2$ terms, see [19, 20, 21] and [27, 28, 29, 30], but we believe that in signal processing and communication applications, the *average* performance will be a more adequate measure of robust performance.

Robust filtering in an average $\mathcal{H}_2$ sense can be attained by parametrizing model uncertainties by sets of random variables. The average, with respect to these variables, of the mean square estimation error is then minimized. The result will be a single robust filter, designed with respect to the specified set of possible systems. The use of averaged $\mathcal{H}_2$ filtering criteria has been suggested previously in the literature in [31, 32, 33] and more recently in [34, 35] and [4]. The design of robust (cautious) Wiener filters, as presented in the last three references, will be summarized in section 5.4. A corresponding design of robustified Kalman filters, based on a combined use of state-space and polynomial methods, was presented in [36]. This method is outlined in Sect.=C45.5. A comprehensive treatment can be found in the thesis [37] by Öhrn.

The robust filters derived in section 5.4 and 5.5 are suitable for model-based design problems with moderate spectral uncertainties. They are also capable of accommodating slow time variations. An illustrative example can be found in [38]. If the uncertainties become large, or if the time variations are rapid, then the use of a single robust filter will no longer be appropriate. The use of filter banks or adaptive methods is then required.

Robust filtering can be used to obtain acceptable performance for a set of systems. It can also be used when uncertain models are obtained by system identification. The parameter covariance matrix will, for validated models, constitute a useful indication of the actual amount of model error [39].

When the amount of data available for model adjustment is limited, it is important to improve the model quality in frequency regions which matter most for the subsequent filtering or control. In the control community, this is known as *identification for control* [40]. The subject has received increased interest during recent years primarily for control problems [41], but the issue is equally relevant for estimation problems.

The development of improved methods for model identification can be seen as a complement to robust filter design. In digital communications, for example, equalization of channel dynamics is essential. Based on a channel model, the input signal to the channel is to be estimated. It would be desirable to use an identification algorithm which concentrates its accuracy in the frequency regions of most importance for the equalization. Improved results could then be obtained in the subsequent filtering step, since the range of dynamics over which a robust filter needs to operate will thus be smaller.[2] Unfortunately, little effort has been spent on methods of identification for filtering, based on small data sets. For a preliminary investigation of such problems, see the thesis [42] by Bigi.

The area of digital communications poses many other new challenges for the model-based design of robust, adaptive and multivariable filters. We shall briefly outline some aspects which have recently received attention.

5.2.1 Digital communications: a challenging application area

Digital mobile radio communications [43, 44] is one of the most rapidly expanding areas within the growing field of digital communications [45, 46]. A major reason is, of course, the introduction of cellular telephone systems, such as GSM[3] and D-AMPS[4]. These systems are now capable of providing both voice, fax and data services. However, various categories of users impose considerable pressure on the development of the systems, by continuously requiring higher capacity, improved quality and more advanced services, some of which involve Internet access.

To meet these demands, operators and manufacturers are already planning for the third generation of systems. Today, two leading technologies for third generation systems can be discerned, namely Time Division Multiple Access (TDMA), such as GSM, which is narrowband, and Code Division Multiple Access (CDMA), which is broadband. (For some details, refer to section 5.3.4.) With either technology, third generation systems will be designed to operate at significantly higher carrier frequencies than the systems of today.

Currently, TDMA systems are most widely spread. Due to the often rather severe conditions and the limited time for calculations, algorithms are required to be highly efficient and of low complexity. Estimation problems occurring in TDMA and CDMA systems are challenging, for several reasons:

- *Multipath propagation.* In general, a signal travels to the receiver antenna along multiple paths with differing transmission delays. This is known as

[2] Connected to this problem is the question if filters should be based on estimated models (indirect tuning/adaptation) or if filter coefficients should be adjusted to the data directly. One aspect is that model estimation is performed by minimizing the ***output*** prediction error, while direct adjustment of an equalizer is performed by minimizing the ***input*** smoothing estimation error. The latter criterion is directly related to the purpose of an equalizer, namely input estimation. These questions are discussed further in section 5.3.4.

[3] Global System for Mobile communications. The standard is used in Europe, as well as in many other parts of the world, including North American PCS networks.

[4] Digital Advanced Mobile Phone System. The standard is used in North America and a similar standard is used in Japan.

multipath propagation. Received symbols may thus be smeared out over several symbol intervals, causing intersymbol interference. To retrieve the transmitted symbol sequence, channel estimation and symbol estimation, equalization, will then be required. The design of equalizers is closely connected to the deconvolution problem, formulated in section 5.3.3.

- *Short data records.* In TDMA systems, data are transmitted in bursts, where each burst is allocated to a specific user. A small fraction of the data, the training sequence, is known to the receiver. It is used for identification of the transmission channel. Since the training sequence is short, channel estimation errors are inevitable. The use of novel ways to perform identification for filtering, combined with a subsequent robust filter design, constitutes a promising path for improving the detection.

- *High disturbance levels.* The signal received from a particular mobile transmitter is contaminated by noise and interference, caused by other users in nearby cells and also on adjacent channels. The systems in use today often suffer from an inadequate transmission quality, with frequent interruptions of the radio connections. In some geographical areas, the capacity is also inadequate. A conceivable way to alleviate these problems, in CDMA as well as TDMA systems, is to use *many sensors* (antennas) on base stations and possibly also on mobile units. Multivariable methods for signal processing can then be utilized, to improve reception by nulling out interferers while increasing sensitivity in the direction of the transmitter. See, for example, [47, 48, 49] and [5]. In Sect. 5.3.4, we will discuss this problem further.

- *Rapidly time varying systems.* Mobiles which are travelling in urban areas will move through a standing wave pattern, due to radio waves reflected from surrounding nearby scatterers. The received signals will therefore have a time-varying amplitude, a phenomenon known as short-time *fading.* Depending on the speed of the mobile, the carrier frequency and the symbol rate, the fading will give rise to different degrees of time variability of the channel. In some systems, such as GSM, the time variations are slow or moderate. A detector which is robust to uncertain channel models may therefore be adequate. In other systems, such as D-AMPS, the time variations will be much faster, requiring detectors to be adaptive. As indicated in section 5.6, this motivates the study of improved adaptation algorithms, as in [50, 51, 52, 53] and [54]. For a detailed study of the channel tracking problem, see the thesis [54] by Lindblom, where also a novel systematic methodology for the design of adaptation algorithms, based on polynomial Wiener filtering concepts, is presented.

Summing up, estimation in mobile radio communications requires efficient algorithms, which not only provide insight, but also robustness and adaptivity as well as low complexity. It is our experience that the polynomial systems framework is an excellent tool for solving problems in this challenging area.

5.2.2 Remarks on the notation

Signals, matrices and polynomial coefficients may, in the following, be complex-valued. This is, for example, required in communication applications. Let p^* denote the complex conjugate of a scalar p and $\mathbf{P}^*$ the complex conjugate transpose of a matrix $\mathbf{P}$. Let tr$\mathbf{P}$ denote the trace of $\mathbf{P}$, while $\mathbf{P}^T$ is the transpose of $\mathbf{P}$.

For any complex-valued polynomial

$$P(q^{-1}) \;=\; p_0 + p_1 q^{-1} + \ldots + p_{np} q^{-np}$$

in the backward shift operator q^{-1}, where $q^{-1}y(k) = y(k-1)$, define the *conjugate polynomial*

$$P_*(q) \;\stackrel{\Delta}{=}\; p_0^* + p_1^* q + \ldots + p_{np}^* q^{np}$$

where q is the forward shift operator. A polynomial $P(q, q^{-1})$ in both positive and negative powers of q will be called *double-sided*. Rational matrices, or transfer function matrices, are denoted by boldface calligraphic symbols, for example as $\boldsymbol{\mathcal{R}}(q^{-1})$. Polynomial matrices are denoted by boldface symbols, for example $\boldsymbol{P}(q^{-1})$.

For polynomial matrices, $\boldsymbol{P}_*(q)$ denotes complex conjugate, transpose and substitution of q for q^{-1}. When appropriate, the complex variable z is substituted for the forward shift operator q. Arguments of polynomial and rational matrices are often omitted, when there is no risk for misunderstanding. The *degree* of a polynomial matrix $\boldsymbol{P}$, $\deg \boldsymbol{P}$ or np, is the highest degree of any of its polynomial elements. A polynomial (matrix) is called *monic* if it has a unit leading coefficient (matrix).

A rational matrix, $\boldsymbol{\mathcal{R}}(z^{-1})$, is defined as *stable* if all of its elements have poles within $|z| < 1$. A rational matrix is *causal* if all of its elements are causal transfer functions. Square *polynomial* matrices $\boldsymbol{P}(q^{-1})$ are called *stable* if all zeros of $\det \boldsymbol{P}(z^{-1})$ are located in $|z| < 1$. If $\boldsymbol{P}(z^{-1})$ is stable, then all poles of the elements of $\boldsymbol{P}^{-1}(z^{-1})$ will be located in $|z| < 1$, while all elements of $\boldsymbol{P}_*^{-1}(z)$ have poles in $|z| > 1$. For *marginally stable* square polynomial matrices, some zeros of $\det \boldsymbol{P}(z^{-1})$ are located on $|z| = 1$.

A rational matrix$\boldsymbol{\mathcal{G}}(z^{-1})$ may be represented by polynomial matrices as a *matrix fraction description* (MFD), either left or right:

$$\boldsymbol{\mathcal{G}}(q^{-1}) = \boldsymbol{A}_1^{-1}(q^{-1})\boldsymbol{B}_1(q^{-1}) = \boldsymbol{B}_2(q^{-1})\boldsymbol{A}_2^{-1}(q^{-1}) \; . \tag{5.1}$$

See [55]. It can also be converted to *common denominator form*

$$\boldsymbol{\mathcal{G}}(q^{-1}) = \frac{1}{A(q^{-1})}\boldsymbol{B}(q^{-1}) \tag{5.2}$$

where $\boldsymbol{B}(q^{-1})$ is a polynomial matrix. The scalar and monic polynomial $A(q^{-1})$ is then a common multiple of the denominators of all rational elements in $\boldsymbol{\mathcal{G}}(q^{-1})$.

A filter$\boldsymbol{\mathcal{V}}(q^{-1})$ which whitens a stochastic process $y(k)$, in the sense that

$$\varepsilon(k) = \boldsymbol{\mathcal{V}}(q^{-1})y(k) \quad , \quad E\varepsilon(i)\varepsilon(j)^T = \mathbf{0} \; , \; i \neq j$$

is called a *whitening filter*. The whitening filters considered in the discus[illegible] below are stably and causally invertible square rational matrices. The inverse [illegible] the above relation,

$$y(k) = \mathcal{V}^{-1}(q^{-1})\varepsilon(k) \tag{5.3}$$

represents an *innovations model* of the signal $y(k)$ [56].

5.3 Wiener filter design based on polynomial equations

The use of polynomial methods for the (nominal) design of Wiener filters has been discussed during the last decade by several authors [17, 18, 57, 58, 59, 60, 61, 62]. We shall begin this section by presenting a fairly general problem formulation in section 5.3.1 and section 5.3.2, which includes many of the previously considered filtering problems as special cases. Then, as an example of the general setup, a deconvolution problem (section 5.3.3) and a equalization problem (section 5.3.4) will be discussed in some detail. The resulting estimators constitute multivariable model-based filters, predictors or fixed-lag smoothers for the nominal case, without model errors.

5.3.1 A general $\mathcal{H}_2$ filtering problem

Based on measurements $d(k)$ up to time $k+m$, a vector

$$z(k) = (z_1(k) \dots z_\ell(k))^T$$

of ℓ signals is to be estimated. The signals are modelled as the outputs of the linear time-invariant discrete-time stochastic system

$$\begin{pmatrix} d(k) \\ z(k) \end{pmatrix} = \begin{pmatrix} \mathcal{G}_g(q^{-1}) \\ \mathcal{D}_g(q^{-1}) \end{pmatrix} u_g(k) \tag{5.4}$$

and the estimator is represented as a transfer function matrix, operating on measurement data $d(k+m)$

$$\hat{z}(k|k+m) = \mathcal{R}_d(q^{-1})d(k+m) \quad . \tag{5.5}$$

Here, $\mathcal{G}_g$, $\mathcal{D}_g$, and $\mathcal{R}_d$ are rational matrices of appropriate dimensions and $\{u_g(k)\}$ is a stochastic process, not necessarily white. Depending on the *smoothing lag* m, the estimator would constitute a predictor ($m < 0$), a filter ($m = 0$) or a fixed lag smoother ($m > 0$).

When the model (5.4) is assumed exactly known, we will consider the minimization of the estimation error covariance matrix

$$\mathbf{P} \triangleq E\varepsilon(k)\varepsilon^*(k) \tag{5.6}$$

where

$$\varepsilon(k) = (\varepsilon_1(k) \dots \varepsilon_\ell(k))^T \triangleq \mathcal{W}(q^{-1})(z(k) - \hat{z}(k|k+m)) \quad .$$

$^{-1}$) is a stable and causal transfer function weighting matrix, ·d to emphasize filtering performance in important frequency iance matrix (5.6) is to be minimized, in the sense that any ·r provides a covariance matrix $\bar{\mathbf{P}}$, for which $\bar{\mathbf{P}} - \mathbf{P}$, is nonneg- ..e minimization is performed under the constraint of realizability ..al stability and causality) of the filter $\mathcal{R}_d(q^{-1})$.

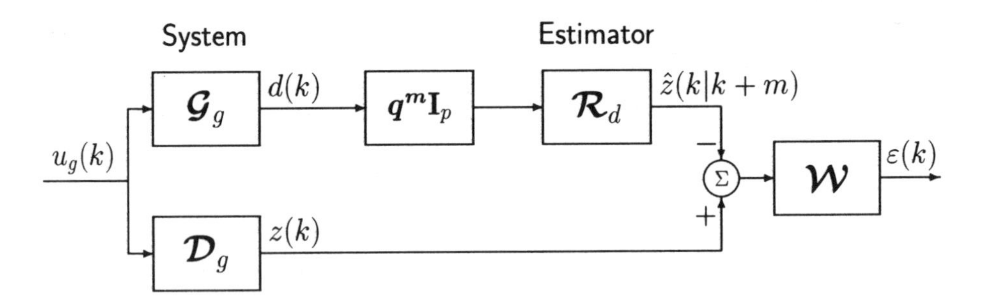

Figure 5.1: A general Wiener filtering problem.

Minimization of the covariance matrix (5.6) also implies the minimization of the sum of the elementwise mean square errors (MSE)'s:

$$J = \mathrm{tr}E(\varepsilon(k)\varepsilon^*(k)) = E(\varepsilon^*(k)\varepsilon(k)) = \sum_{i=1}^{\ell} E|\varepsilon_i(k)|^2 . \tag{5.7}$$

In section 5.4 and 5.5, where the model (5.4) is assumed to be uncertain, we will instead consider the minimization of an *averaged* MSE criterion

$$\bar{J} = \mathrm{tr}\bar{E}E(\varepsilon(k)\varepsilon^*(k)) \tag{5.8}$$

where $\bar{E}(\cdot)$ denotes an expectation over stochastic variables, which are used for parametrizing the set of admissible models.

5.3.2 A structured problem formulation

While the model (5.4) is general, it is frequently of advantage to introduce additional structure, to obtain solutions which provide useful engineering insight. For the purpose of this chapter, we will therefore introduce a more detailed structure, which encompasses a number of special cases, = some of which will be studied in more detail.

Let us partition the vector $u_g(k)$ in (5.4) into two parts

$$u_g(k) = \begin{pmatrix} u(k) \\ w(k) \end{pmatrix}$$

where $w(k)$ represents additive measurement noise, uncorrelated to the desired signal $z(k)$. The desired signal is assumed to be a filtered version of $u(k)$,

$$z(k) = \mathcal{D}_g(q^{-1})u(k) \ .$$

We also introduce explicit stochastic models for the vector $u(k)$ and the noise

$$u(k) = \mathcal{F}(q^{-1})e(k) \quad , \quad w(k) = \mathcal{H}(q^{-1})v(k),$$

with $\mathcal{F}$ and $\mathcal{H}$ being stable or marginally stable. The noise sequences $\{e(k)\}$ and $\{v(k)\}$ are assumed to be mutually uncorrelated, white and stationary. They have zero means and covariance matrices $\phi \geq 0$ and $\psi \geq 0$.[5]

If noise-free measurements are available, it can be of interest to handle them separately. We therefore partition the measurement vector as

$$d(k) \triangleq \begin{pmatrix} y(k) \\ a(k) \end{pmatrix} \tag{5.9}$$

where the noise $w(k)$ affects $y(k) = (y_1(k) \ldots y_p(k))^T$ additively, while $a(k) = (a_1(k) \ldots a_h(k))^T$ is uncorrupted by $w(k)$.

The model structure (5.4) is thus converted to the form

$$\begin{pmatrix} y(k) \\ a(k) \\ z(k) \end{pmatrix} = \begin{pmatrix} \mathcal{G}(q^{-1}) & \mathbf{I} \\ \mathcal{G}_a(q^{-1}) & \mathbf{0} \\ \mathcal{D}(q^{-1}) & \mathbf{0} \end{pmatrix} \begin{pmatrix} u(k) \\ w(k) \end{pmatrix}$$

$$\begin{pmatrix} u(k) \\ w(k) \end{pmatrix} = \begin{pmatrix} \mathcal{F}(q^{-1}) & \mathbf{0} \\ \mathbf{0} & \mathcal{H}(q^{-1}) \end{pmatrix} \begin{pmatrix} e(k) \\ v(k) \end{pmatrix} \tag{5.10}$$

See figure 5.2. Above, $\mathcal{G}$, $\mathcal{G}_a$, $\mathcal{F}$, $\mathcal{H}$, and $\mathcal{D}$ are transfer function matrices of appropriate dimensions. The transfer functions will, in the following, be parametrized either by state-space models or by polynomial matrices in q^{-1} as MFD's (5.1) or common denominator forms (5.2).

All of the subsystems will in the present chapter be assumed stable. Corresponding problems with marginally stable blocks are discussed in [37]. Structured or unstructured model uncertainty may be present in any subsystem.

Based on the measurements $d(k)$, up to time $k+m$, our aim is thus to optimize the linear estimator (5.5)

$$\hat{z}(k|k+m) = \mathcal{R}_d(q^{-1})z(k+m) \ = \ (\mathcal{R}(q^{-1}) \ \mathcal{R}_a(q^{-1})) \begin{pmatrix} y(k+m) \\ a(k+m) \end{pmatrix} \tag{5.11}$$

in which both $\mathcal{R}$ and $\mathcal{R}_a$ are required to be stable and causal transfer function matrices.

The structure depicted in figure 5.2 covers a large set of different problems. We shall in the present chapter discuss the following special cases:

[5]Frequently, it is convenient to normalize ϕ and ψ to unit matrices and include variance scaling in $\mathcal{F}$ and $\mathcal{H}$ respectively. This will be the case, for example, in section 5.3.3 in the robust estimation problems discussed in section 5.4

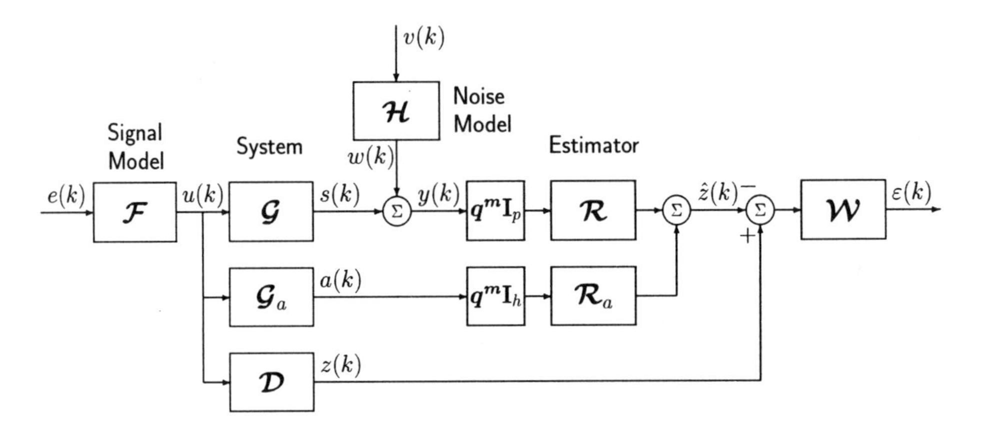

Figure 5.2: Unifying structure for a collection of $\mathcal{H}_2$ filtering problems. The signal $z(k)$ is to be estimated based on data $y(k)$ up to time $k+m$.

- Multisignal deconvolution and linear equalization (section 5.3.3);
- Decision feedback equalization of a white vector sequence of digital symbols: $\mathcal{W} = \mathbf{I}$, $\mathcal{F} = \mathbf{I}$, $\mathcal{D} = \mathbf{I}$, $\mathcal{G}_a = q^{-m-1}\,\mathbf{I}$ (section 5.3.4);
- Estimation based on uncertain input-output models (section 5.4);
- Estimation based on uncertain state-space models (section 5.5).

5.3.3 Multisignal deconvolution

The estimation of input signals to dynamic systems is known as *deconvolution*, or input estimation. In such problems, $\mathcal{G}$ in (5.10) constitutes a dynamic system. An application, discussed in more detail in section 5.3.4, is the equalization of communication channels [45, 46, 63, 64]. Another recent interesting application of multivariable deconvolution is the reconstruction of stereophonic sound, as described by Nelson *et. al.* in [65]. Numerical differentiation may be formulated as the problem of estimating the input to a discrete-time approximation of an integrator [57, 58]. Applications to seismic signal processing are described in [66], and the references therein.

We will consider the problem of deconvolution with multiple inputs and multiple outputs, assuming the involved dynamic systems to be exactly known. All problems described by figure 5.1 and figure 5.2 are included.

The deconvolution problem could be set up and solved using general MFD's, see [16, 17] or [24]. Here, we will approach the problem by representing some transfer functions by MFD's having *diagonal denominator matrices*, while others are represented in *common denominator* form.

Parametrizing the problem in this way has several advantages. First, no coprime factorizations will be required, which results in a transparent solution.

Thus, engineering insight is more easily obtained. Second, the solution involves a *unilateral* Diophantine equation instead of a bilateral one: The polynomial matrices to be determined appear on the same sides of different terms of the equation, instead of on opposite sides. This will make the solution attractive, both from a numerical and from a pedagogical point of view: Solving a unilateral Diophantine equation corresponds to solving a block-Toeplitz system of linear equations with multiple right-hand sides. For an example, see [4].

Let the measurement vector $y(k)$ and the input $u(k)$ be described by

$$\begin{aligned} y(k) &= \boldsymbol{A}^{-1}(q^{-1})\boldsymbol{B}(q^{-1})u(k) + \boldsymbol{N}^{-1}(q^{-1})\boldsymbol{M}(q^{-1})v(k) \\ u(k) &= \frac{1}{D(q^{-1})}\boldsymbol{C}(q^{-1})e(k) \ . \end{aligned} \tag{5.12}$$

Here, $\{\boldsymbol{A},\boldsymbol{B},\boldsymbol{N},\boldsymbol{M},\boldsymbol{C}\}$ are polynomial matrices of dimensions $p|p$, $p|s$, $p|p$, $p|r$, and $s|n$, respectively, while D is a scalar polynomial. The matrices $\boldsymbol{A}$ and $\boldsymbol{N}$ are assumed *diagonal.* As indicated in the previous section, $\{e(k)\}$ and $\{v(k)\}$ are mutually uncorrelated zero mean stochastic processes. Here, they are normalized to have *unit* covariance matrices of dimensions $n|n$ and $r|r$, respectively. Since rows of $\boldsymbol{M}$ are allowed to be zero, noise-free measurements can be included. The polynomial matrix $\boldsymbol{B}$ need not be stably invertible. It may not even be square.

From data $y(k)$ up to time $k+m$, an estimator

$$\hat{z}(k|k+m) = \boldsymbol{\mathcal{R}}(q^{-1})y(k+m) \tag{5.13}$$

of a filtered version $z(k)$ of the input $u(k)$

$$z(k) = \frac{1}{T(q^{-1})}\boldsymbol{S}(q^{-1})u(k)$$

is sought. The filter $\boldsymbol{S}/T$, with T scalar and $\boldsymbol{S}$ of dimension $\ell|s$, may represent additional dynamics in the problem description, cf. [58, 59], a frequency shaping weighting filter cf. [57], or the selection of particular states.

The covariance matrix (5.6), or the sum of MSE's (5.7), is to be minimized with dynamic weighting

$$\boldsymbol{\mathcal{W}}(q^{-1}) = \frac{1}{U(q^{-1})}\boldsymbol{V}(q^{-1}) \ .$$

This problem formulation corresponds to the choice

$$\boldsymbol{\mathcal{G}} = \boldsymbol{A}^{-1}\boldsymbol{B} \ , \ \boldsymbol{\mathcal{G}}_a = \boldsymbol{0} \ , \ \boldsymbol{\mathcal{F}} = \boldsymbol{C}/D \ , \ \boldsymbol{\mathcal{H}} = \boldsymbol{N}^{-1}\boldsymbol{M} \ , \ \boldsymbol{\mathcal{D}} = \boldsymbol{S}/T \ , \ \boldsymbol{\mathcal{W}} = \boldsymbol{V}/U$$

and $\boldsymbol{\mathcal{R}}_z = [\boldsymbol{\mathcal{R}} \ \boldsymbol{0}]$ in (5.6),(5.10) and (5.11). See figure 5.3.

When $\{u(k)\}$ is a sequence of digital symbols in a communication network, the estimator (5.13) will constitute a multivariable *linear equalizer.* In order to retrieve the transmitted symbols, the estimate $\hat{z}(k) = \hat{u}(k)$ is then fed into a decision device.

Introduce the following assumptions.

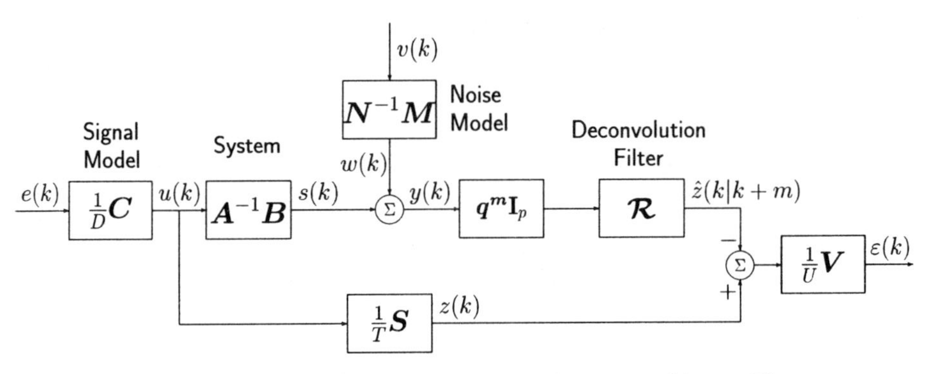

Figure 5.3: A generalized multi-signal deconvolution problem. The vector sequence $\{z(k)\}$ is to be estimated from the measurements $\{y(k)\}$, up to time $k+m$.

Assumption 5.1. *The polynomials $D(q^{-1})$, $T(q^{-1})$, and $U(q^{-1})$ are all stable and monic, while the polynomial matrices $\boldsymbol{A}(q^{-1}), \boldsymbol{N}(q^{-1})$ and $\boldsymbol{V}(q^{-1})$ have stable determinants and unit leading coefficient matrices. (Thus, they have stable and causal inverses.)*

Assumption 5.2. *The spectral density of $y(k)$, $\boldsymbol{\Phi}_y(e^{j\omega})$, is nonsingular for all ω.*

From (5.12), we now obtain the spectral density matrix $\boldsymbol{\Phi}_y$ as

$$\boldsymbol{\Phi}_y = \frac{1}{DD_*}\boldsymbol{A}^{-1}\boldsymbol{B}\boldsymbol{C}\boldsymbol{C}_*\boldsymbol{B}_*\boldsymbol{A}_*^{-1} + \boldsymbol{N}^{-1}\boldsymbol{M}\boldsymbol{M}_*\boldsymbol{N}_*^{-1} = \boldsymbol{\alpha}^{-1}\boldsymbol{\beta}\boldsymbol{\beta}_*\boldsymbol{\alpha}_*^{-1} \quad (5.14)$$

where

$$\boldsymbol{\beta}\boldsymbol{\beta}_* = \boldsymbol{N}\boldsymbol{B}\boldsymbol{C}\boldsymbol{C}_*\boldsymbol{B}_*\boldsymbol{N}_* + DD_*\boldsymbol{A}\boldsymbol{M}\boldsymbol{M}_*\boldsymbol{A}_* \quad (5.15)$$

and

$$\boldsymbol{\alpha} \triangleq D\boldsymbol{N}\boldsymbol{A} \ .$$

Note that $\boldsymbol{A}$ and $\boldsymbol{N}$ commute, since they are assumed diagonal.

Under assumption 5.2, a stable $p|p$ spectral factor $\boldsymbol{\beta}$, with $\det \boldsymbol{\beta}(z^{-1}) \neq 0$ in $|z| \geq 1$ and with nonsingular leading matrix $\boldsymbol{\beta}_0 = \boldsymbol{\beta}(0)$, can always be found. Thus, $\boldsymbol{\alpha}^{-1}\boldsymbol{\beta}$ constitutes an innovations model of the measurement vector, while $\boldsymbol{\beta}^{-1}\boldsymbol{\alpha}$ is a stable and causal whitening filter, cf. (5.3).

The optimal estimator can be derived in many ways, of which completing the squares, the variational approach, the inner-outer factorization approach [67] and the classical Wiener solution are the most well known. See [16] or [6] for a comparision. We shall here use the variational approach and a brief outline of this methodology is presented next.

Optimization by variational arguments[6, 16, 17]

Consider the estimator (5.5) and the criterion (5.6). Introduce an *alternative weighted estimate*

$$\hat{\delta}(k|k+m) = \mathcal{W}(q^{-1})\hat{z}(k|k+m)+\nu(k) = \mathcal{W}(q^{-1})\mathcal{R}_d(q^{-1})d(k+m)+\nu(k) \quad (5.16)$$

where the column vector $\nu(k)$ of stationary signals represents a modification of the (weighted) estimate. See figure 5.4. The estimate $\hat{z}(k)$ is optimal if and only if no admissible variation can improve upon the criterion value.

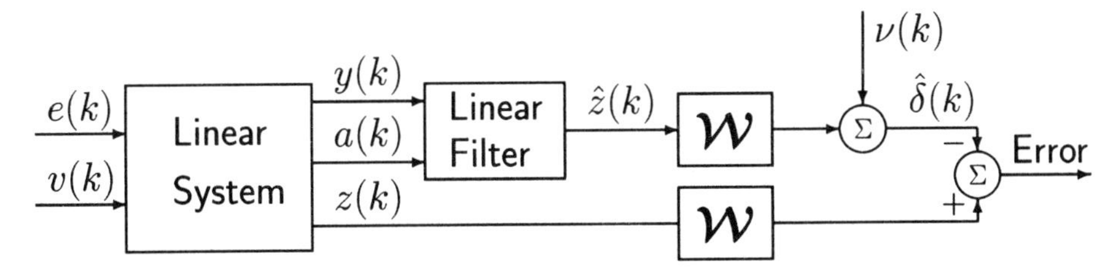

Figure 5.4: Setup for filter optimization via the variational approach in the general estimation problem defined by (5.10) and (5.11). The weighted estimate is perturbed by a variation $\nu(k)$.

All admissible variations can be represented by

$$\nu(k) = \mathcal{T}(q^{-1})d(k+m)$$

where $\mathcal{T}(q^{-1})$ is some stable and causal rational matrix. In problems with unstable models, any nonstationary mode of $d(k)$ must be cancelled by zeros of $\mathcal{T}(q^{-1})$. Except for these requirements, $\mathcal{T}(q^{-1})$ is an arbitrary.

The use of the modified estimator (5.16) results in the covariance matrix

$$\bar{\mathbf{P}} = E\{\mathcal{W}(q^{-1})z(k) - \hat{\delta}(k|k+m)\}\{\mathcal{W}(q^{-1})z(k) - \hat{\delta}(k|k+m)\}^*$$

$$= E\varepsilon(k)\varepsilon(k)^* - E\varepsilon(k)\nu(k)^* - E\nu(k)\varepsilon(k)^* + E\nu(k)\nu(k)^* \ . \quad (5.17)$$

If the *cross-terms* in (5.17) are zero, then $\nu(k) \equiv \mathbf{0}$ will evidently minimize $\bar{\mathbf{P}}$, since $E\nu(k)\nu(k)^*$ is positive semidefinite if any component of $\nu(k)$ has nonzero variance. Then, the estimator (5.5) is optimal.[6]

Of the two cross terms, it is sufficient to consider only $E\varepsilon(k)\nu(k)^*$, for symmetry reasons. The estimation error $\varepsilon(k)$ in (5.6) is required to be *stationary*. This will be the case if $z(k)$ and $d(k)$ are stationary, since $\mathcal{W}(q^{-1})$ and $\mathcal{R}_d(q^{-1})$ are required to be stable[7].

[6]By taking the trace of (5.17), it is evident that the scalar MSE criterion (5.7) is also minimized by $\nu(k) = \mathbf{0}$.

[7]If $z(k)$ or $d(k)$ were generated by marginally stable models, stationarity of the estimation error would have to be verified separately, *after* the derivation. See [6, 16, 37].

With $\{\varepsilon(k)\}$ and $\{\nu(k)\}$ being stationary sequences, Parsevals formula can now be used to convert the requirement $E\varepsilon(k)\nu(k)^* = \mathbf{0}$ into the frequency-domain relation

$$E\varepsilon(k)\nu(k)^* = \frac{1}{2\pi j}\oint_{|z|=1} \boldsymbol{\phi}_{\varepsilon\nu^*}\frac{dz}{z} = \mathbf{0} \quad . \tag{5.18}$$

The rational $\ell|\ell$-matrix $\boldsymbol{\phi}_{\varepsilon\nu^*}$ is the cross spectral density. The expression (5.18) corresponds to the elementwise orthogonality conditions[8]

$$E\varepsilon_v(k)\nu_n(k)^* = \frac{1}{2\pi j}\oint_{|z|=1} \boldsymbol{\phi}^{vn}_{\varepsilon\nu^*}\frac{dz}{z} = 0 \qquad v = 1\ldots\ell\ ,\ n = 1\ldots\ell\ . \tag{5.19}$$

These ℓ^2 conditions determine the estimator $\mathcal{R}_d(q^{-1})$. They are fulfilled if the integrands are made analytic inside the integration path $|z| = 1$. All poles of the integrands inside the unit circle should thus be cancelled by zeros.

Using the common denominator form or the left polynomial matrix fraction description, the relations (5.19) can be evaluated collectively, rather than individually, when $\ell > 1$. They then reduce to a linear polynomial (matrix) equation, a bilateral or unilateral *Diophantine equation.*

In robust design problems with uncertain models, the operator $\bar{E}E(\cdot)$ is substituted for $E(\cdot)$ in the reasoning above [4].

Derivation of the deconvolution estimator

The methodology outlined above will now be exemplified on the problem specified by (5.12)-(5.15) and by the assumptions 5.1 and 5.2. Let

$$\varepsilon(k) = \frac{1}{U}\boldsymbol{V}(z(k) - \hat{z}(k|k+m))$$

be the filtered error and $\nu(k) = \mathcal{T}(q^{-1})y(k+m)$ the variation. Since the noises $e(k)$ and $v(k)$ are assumed uncorrelated, and since all the involved systems are assumed stable, we obtain the cross covariance matrix

$$E\varepsilon(k)\nu^*(k) = E\frac{1}{U}\boldsymbol{V}\left[\left(\frac{1}{T}\boldsymbol{S} - q^m\mathcal{R}\boldsymbol{A}^{-1}\boldsymbol{B}\right)\frac{1}{D}\boldsymbol{C}e(k) - q^m\mathcal{R}\boldsymbol{N}^{-1}\boldsymbol{M}v(k)\right]$$

$$\left[\mathcal{T}q^m\left(\frac{1}{D}\boldsymbol{A}^{-1}\boldsymbol{B}\boldsymbol{C}e(k) + \boldsymbol{N}^{-1}\boldsymbol{M}v(k)\right)\right]^*$$

$$= \frac{1}{2\pi j}\oint_{|z|=1}\frac{1}{U}\boldsymbol{V}\left\{z^{-m}\frac{1}{TDD_*}\boldsymbol{S}\boldsymbol{C}\boldsymbol{C}_*\boldsymbol{B}_*\boldsymbol{A}_*^{-1}\right.$$

[8] When $\ell > 0$, these conditions imply (but are stronger than) orthogonality between the estimation error and any admissible perturbation of the estimate, which corresponds to $\mathrm{tr}E[\varepsilon(k)\nu(k)^*] = \mathrm{tr}E\nu(k)^*\varepsilon(k) = E\nu(k)^*\varepsilon(k) == 0$.

$$- \mathcal{R}\left[\frac{1}{DD_*}A^{-1}BCC_*B_*A_*^{-1} + N^{-1}MM_*N_*^{-1}\right]\Big\}\mathcal{T}_*\frac{dz}{z} \quad . \tag{5.20}$$

The use of the expression (5.14) in (5.20) gives, with $\alpha_*^{-1} = D_*^{-1}N_*^{-1}A_*^{-1}$,

$$E\varepsilon(k)\nu^*(k) = \frac{1}{2\pi j}\oint \frac{1}{U}\left\{\frac{z^{-m}}{TD}VSCC_*B_*N_* - V\mathcal{R}\alpha^{-1}\beta\beta_*\right\}\alpha_*^{-1}\mathcal{T}_*\frac{dz}{z} \quad . \tag{5.21}$$

The integral vanishes if the estimator $\mathcal{R}$ is adjusted so that no element of the integrand in (5.21) has poles inside the integration path $|z| = 1$. Since A, N and D are stable, $\alpha^{-1} = A^{-1}N^{-1}D^{-1}$ has poles only in $|z| < 1$. Elements of $\beta(z^{-1})$ may contribute poles at the origin.[9] These factors can be cancelled directly by $\mathcal{R}$. Moreover, if $\mathcal{R}$ contains V^{-1}/T as a left factor, the matrix V is cancelled and $1/TD$ can be factored out from the two terms of the integrand, to be cancelled later. Thus, we select

$$\mathcal{R} = \frac{1}{T}V^{-1}Q\beta^{-1}NA \tag{5.22}$$

where $Q(q^{-1})$, of dimension $\ell|p$, is undetermined. With the filter (5.22) inserted, the cross covariance matrix (5.21) becomes

$$E\varepsilon(k)\nu^*(k) = \frac{1}{2\pi j}\oint_{|z|=1}\frac{1}{UTD}\{z^{-m}VSCC_*B_*N_* - Q\beta_*\}\alpha_*^{-1}\mathcal{T}_*\frac{dz}{z} \quad .$$

All poles of every element of $\alpha_*^{-1}\mathcal{T}_*$ are located outside $|z| = 1$, since α is a stable polynomial matrix and the rational matrix $\mathcal{T}$ is causal and stable. The remaining factor of the integrand may contribute poles in $|z| < 1$. In order to attain orthogonality, we therefore require that

$$z^{-m}VSCC_*B_*N_* \;=\; Q\beta_* + zL_*UTD\,\mathbf{I}_p \tag{5.23}$$

for some polynomial matrix $L_*(z)$. We then obtain an integrand with only strictly unstable rational functions in z as elements, so the integral

$$E\varepsilon(k)\nu^*(k) = \frac{1}{2\pi j}\oint_{|z|=1} L_*(z)\alpha_*^{-1}(z)\mathcal{T}_*(z)dz \tag{5.24}$$

will vanish. Equation (5.23) is a linear polynomial matrix equation, a *unilateral* Diophantine equation. Here, $Q(q^{-1})$ and $L_*(q)$ are polynomial matrices, of dimension $\ell|p$, with generic degrees[10]

$$nQ = \max(nc + ns + nv + m, nt + nd + nu - 1)$$

$$nL = \max(nc + nb + nn - m, n\beta) - 1 \tag{5.25}$$

[9] A polynomial $\beta(z^{-1})$ can, alternatively, be represented as a rational function in z, by multiplying and dividing with $z^{n\beta}$. The elements of this rational function have poles at the origin $z = 0$. Thus, we have to eliminate all numerator polynomials having z^{-1} as argument.

[10] In special cases, the degrees may be lower.

where $nc = \deg \boldsymbol{C}, ns = \deg \boldsymbol{S}$ etc. Unique solvability of (5.23) with respect to $\boldsymbol{Q}$ and $\boldsymbol{L}_*$ can be demonstrated, see, for example, [4] or [6].

The design equations thus consist of the left spectral factorization (5.15), the Diophantine equation (5.23) and the filter (5.22).[11] For scalar systems, the solution reduces to the one presented in [57]. See also [59]. A numerical illustration can be found in [6].

A Wiener filter works by first whitening the measurement. By introducing D as a stable common factor, it is evident that the estimator (5.22) contains a whitening filter $\boldsymbol{\beta}^{-1}\boldsymbol{\alpha} = \boldsymbol{\beta}^{-1}\boldsymbol{NAD}$ as a right factor.

The spectral factor $\boldsymbol{\beta}$ is unique up to a right orthogonal matrix. (If $\boldsymbol{FF}_* = \mathbf{I}$ then $\boldsymbol{\beta\beta}_* = (\boldsymbol{\beta F})(\boldsymbol{F}_*\boldsymbol{\beta}_*)$.) There exist several efficient algorithms for polynomial matrix spectral factorization, some of which are based on state space methods. A survey of algorithms is presented in [68]. See also [69, 70].

The attainable MSE

The minimal (scalar) MSE criterion value is obtained by inserting (5.22), (5.15), and (5.23), in this order, into (5.7). We thus obtain, with $\boldsymbol{H} \triangleq \boldsymbol{VSC}$ and $P \triangleq UTD$,

$$\mathrm{tr}E(\varepsilon(k)\varepsilon^*(k))_{\mathrm{min}} =$$

$$\frac{1}{2\pi j}\oint \mathrm{tr}\{\boldsymbol{L}_*\boldsymbol{\beta}_*^{-1}\boldsymbol{\beta}^{-1}\boldsymbol{L} + \frac{1}{PP_*}\boldsymbol{H}(\mathbf{I}_n - \boldsymbol{C}_*\boldsymbol{B}_*\boldsymbol{N}_*\boldsymbol{\beta}_*^{-1}\boldsymbol{\beta}^{-1}\ \boldsymbol{NBC})\boldsymbol{H}_*\}\frac{dz}{z}\ . \tag{5.26}$$

The minimal criterion value consists of two terms. The first term represents the error caused by incomplete inversion of the system $\boldsymbol{A}^{-1}\boldsymbol{B}$. Only the use of an infinite smoothing lag will cause this term to vanish, unless the system is minimum phase and there is no noise. One can show that $\boldsymbol{L} \to \mathbf{0}$ when $m \to \infty$, see [58]. Thus, the second part of (5.26) represents the limit of performance approached by a noncausal Wiener filter.

There exists a very special case in which perfect input estimation is possible with finite smoothing lags. It is the case of minimum-phase systems without noise ($\boldsymbol{M} = \mathbf{o}$, $\boldsymbol{N} = \mathbf{I}$), with $q^m\boldsymbol{B}$ square and stably and causally invertible. Consider this situation and let $\boldsymbol{S} = \mathbf{I}_\ell$ and $T = 1$. Then, the direct inversion of the transducer dynamics

$$\boldsymbol{\mathcal{R}} = \boldsymbol{B}^{-1}\boldsymbol{A}\, q^{-m}$$

will result in a vanishing integrand in (5.21).

Adaptive and blind deconvolution

For scalar systems, the deconvolution problem has also been studied in an adaptive setting. An interesting feature here is that the spectral factor $\boldsymbol{\beta}$ need not

[11]For models with poles in $|z| > 1$, a second Diophantine equation will be required. See Ch. 3 and [6]. Such models are, however, of limited practical interest in open-loop filtering.

be calculated from the equation (5.15). Instead, it can be obtained directly from data, by estimating an appropriately parametrized innovations model

$$y(k) = \boldsymbol{\alpha}^{-1}(q^{-1})\boldsymbol{\beta}(q^{-1})\varepsilon(k)$$

of the measurement vector, cf. (5.14) and (5.3). See [73]. Based on this fact, multivariable adaptive deconvolution, for the special case of white input and noise, has been discussed in [71, 72] and [60]. Crucial for an adaptive algorithm to work in more general situations, with coloured input and noise, is that the model polynomials can be estimated from the output only. Algorithms which can be applied in the scalar case with signals and noises of unknown colour, but with a *known system*, have been presented in [73] and [74]. In [75], the identifiability properties of the scalar deconvolution problem are investigated and conditions for parameter identifiability are given, when $\mathcal{G}$ is known, while $\mathcal{F}$ and $\mathcal{H}$ have to be estimated. The conditions for identifiability in [75] are based on the use of second order moments only.

Another challenging problem is that of *blind deconvolution*, where both the input signal $u(k)$ and the transducer/channel $\mathcal{G}$ have to be estimated from data. The unique estimation of a possibly non-minimum phase system $\mathcal{G}$, based on only the second order statistics of its output is, in general, impossible. The second order statistics does not provide the appropriate phase information. If the input is non-Gaussian while the noise is Gaussian, then higher order statistics can be utilized [76]. Higher order statistics is used, directly or indirectly, in most proposed algorithms for blind deconvolution.

Algorithms based on higher order statistics require long convergence times, and the quality of the estimates may be sensitive to the assumption that the noise is Gaussian. A recent discovery is therefore of significant interest: Blind identification *is* possible, for cyclo-stationary inputs, if several output samples are available per input sample. The continuous-time baseband signals used in digital communications are cyclostationary; the result is therefore of interest for *blind equalization*. The number of received samples per symbol can here either be increased by oversampling (fractionally spaced equalization), or by the use of multiple antennas/sensors, so that $y(k)$ is a vector, while $u(k)$ is scalar [77, 78]. Radio systems with several receiver antennas are of increasing interest in particular for mobile applications, see section 5.3.4.

A duality to feedforward control

The set of problems for which the solution above is relevant can be enlarged further. The considered deconvolution problem turns out to be dual to the *LQG* (or $\mathcal{H}_2$)-feedforward control problem, with rational weights on control and output signals. See [79]. It is very simple to demonstrate this duality. Reverse all arrows, interchange summation points and node points and transpose all rational matrices in figure 5.3. Then, the block diagram for the LQG problem is obtained. The optimization problem remains the same for $\mathcal{H}_2$ problems, and indeed for the minimization for any transfer function norms that are invariant under transposition. Thus, the equations (5.15) and (5.23) can be used also to design disturbance measurement feedforward regulators, reference feedforward filters and feedforward decoupling filters.

5.3.4 Decision feedback equalizers

We now turn our interest to an important problem in digital communications and outline a polynomial solution, which was originally presented in [80] for the scalar case, and later in [6, 16]. The multivariable solution discussed here has been presented in [5, 81].

Digital communications in the presence of intersymbol interference and co-channel interference

When digital data are transmitted over multiple cross-coupled communication channels, *intersymbol interference*, *co-channel interference* and noise will prevent perfect detection.

The transmitted sequence $\{u(k)\}$ in question is white, and it may be real- or complex-valued.[12] It is to be reconstructed from a sampled received signal $y(k)$. Whenever the channel has an impulse response of length ≥ 1, we are said to encounter *intersymbol interference*: not only the symbol at time k, but also previous symbols contribute to the current received measurement $y(k)$. The channel is then said to be *dispersive*. Dispersive channels occur in digital mobile radio systems such as GSM. Other situations include modem connections over cable, transmission of data to and from hard discs in computers, as well as underwater acoustic communication. In radio communications and underwater acoustics, intersymbol interference is caused by *multipath propagation*: The transmitted signal travels along several paths with differing transmission delays.

A linear equalizer, for example the MSE-optimal design of section 5.3.3, can be used to retrieve the transmitted symbols. A linear MSE-optimal equalizer performs an approximate inversion of the channel. This inversion may result in noise amplification at the filter output, which severely limits the attainable performance.

The lowest error rate would be attained by maximum likelihood estimation of the entire transmitted sequence, implemented through the Viterbi algorithm [82], which constitutes forward dynamic programming. Close to optimal performance, at a much lower computational load, can often be obtained with a Decision Feedback Equalizer (DFE). A DFE is a nonlinear filter, which involves a decision circuit and a feedback of decisioned data through a linear filter to improve the current estimate. See, for example, [83, 84, 85], and the references therein. The attainable bit error rate of a DFE is in many cases several orders of magnitude

[12] One example is the = use of p-ary symmetric Pulse Amplitude Modulated (PAM) signals. Then, each component of $u(k)$ is a real, white, zero mean sequence which attains values $\{-p+1,\dots,-1,+1,\dots,p-1\}$ with some probability distribution. In other modulation schemes, such as Quadrature Amplitude Modulation (QAM), the signal $u(k)$ is complex-valued. For example, in 4-QAM, the symbols represented by the elements of $u(k)$ may attain the values $\{1+i,\ \ 1-i,\ \ -1+i\ \ -1-i\ \}$. See, for example, [45, 46].

lower than for a linear equalizer. We shall here consider DFE design as an application where noise free auxiliary information $a(k)$ is explicitely taken into account in the general structure (5.10). In order to make the discussion general, we will consider transmission and reception of several signals, that is, we will consider a Multivariable DFE. The design or adaptation of the DFE is here assumed to be based on an indirect approach, i.e. on an explicit multivariable channel model.

Let us describe a channel model structure which is appropriate for the purpose of mobile radio communications.

A FIR channel model

Consider a sampled data vector sequence $y(k)$, which represents measurements from a receiver in a radio communication system. The received signal is down-converted to the baseband [45, 46]. Under such circumstances, multiple cross-coupled communication channels are adequately modelled as FIR systems represented by polynomial matrices, with complex elements. The channel model includes pulse shaping, receiver filters and possible transmission delays. It will be described by the multivariable linear stochastic discrete-time model

$$\begin{pmatrix} y_1 \\ \vdots \\ y_{ny} \end{pmatrix} = \begin{pmatrix} B_{11}(q^{-1}) & \dots & B_{1n_u}(q^{-1}) \\ \vdots & \ddots & \vdots \\ B_{n_y1}(q^{-1}) & \dots & B_{n_yn_u}(q^{-1}) \end{pmatrix} \begin{pmatrix} u_1(k) \\ \vdots \\ u_{n_u}(k) \end{pmatrix} + \begin{pmatrix} v_1(k) \\ \vdots \\ v_{n_y}(k) \end{pmatrix}$$

or

$$\begin{aligned} y(k) &= \boldsymbol{B}(q^{-1})u(k) + v(k) \\ &= \mathbf{B}_0 u(k) + \ldots + \mathbf{B}_{n_b} u(k-n_b) + v(k) \quad = . \end{aligned} \tag{5.27}$$

Here, we have denoted by n_b the highest degree occurring in any matrix element $B_{ij}(q^{-1})$. The sequence $\{v(k)\}$ is assumed to be discrete-time white noise, which is zero mean, stationary and uncorrelated with $u(k)$.

The multivariable DFE

A detector designed to estimate only one message $u_i(k)$ would have to treat the co-channel interference from the remaining signals, $\sum_j B_{ij}u_j$, $i \neq j$, as noise. A multivariable detector, which estimates all components of $u(k)$ simultaneoulsy, can utilize the fact that the signals have discrete and known amplitude distributions. This knowledge is utilized in an efficient way by a multivariable DFE.

When the channel is adequately modelled by a polynomial matrix and the noise is white or autoregressive, the appropriate structure of a DFE, cf. [80], is described by

$$\hat{u}(k-m|k) = \boldsymbol{S}(q^{-1})y(k) - \boldsymbol{Q}(q^{-1})\tilde{u}(k-m-1) \quad . \tag{5.28}$$

Here, $\tilde{u}(k-m-1)$ denotes previously detected symbols, while $\hat{u}(k-m|k)$ represents the amplitude-continuous (soft) estimate, obtained with smoothing lag m.

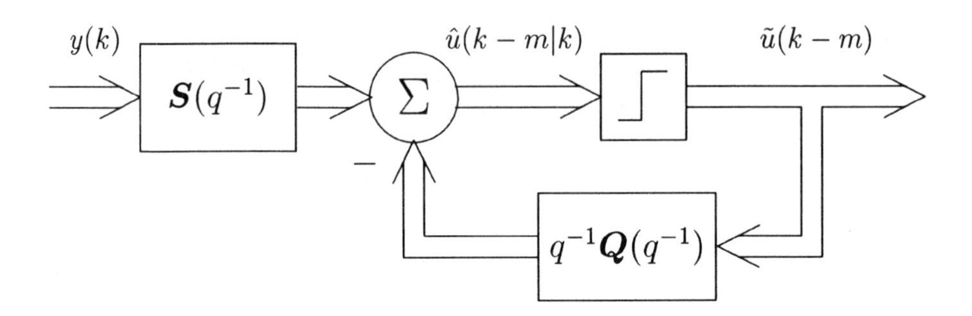

Figure 5.5: Structure of the multivariable DFE.

Detected symbols $\tilde{u}$ are obtained by feeding $\hat{u}$ through a decision device, which, in its simple form, is given by sign=C4($\hat{u}$) for binary PAM signals. The polynomial matrices $\boldsymbol{S}(q^{-1})$ and $\boldsymbol{Q}(q^{-1})$ have dimension $n_u|n_y$ and $n_u|n_u$, respectively. See figure 5.5.

Our multivariable DFE is obtained by minimizing the MSE criterion[13]

$$J = E \parallel u(k-m) - \hat{u}(k-m|k) \parallel_2^2 \quad . \tag{5.29}$$

An obstacle, preventing a direct minimization of J, is the presence of the nonlinear decision element. It is impossible to obtain closed form solutions, but the problem can be simplified by assuming *correct past decisions.* This is a common simplification, which allows us to replace previous decisions $\tilde{u}(j)$ by the correct values $u(j)$ for $j = k-m-1, k-m-2\ldots$ in (5.28).

If previous decisions are correct, they can be used to completely eliminate the contamination caused by past symbols, at the current time instant. In contrast to linear equalizers, this can be achieved without any noise amplification, since we need not invert the channel. Instead, a feedforward measurement of $a(k) = u(k-m-1)$ is used. Under the assumption of correct previous decisions, the MIMO-DFE can be included in the general structure (5.10), cf figure 5.2, by setting

$$\mathcal{G}_a \;=\; q^{-m-1}\mathbf{I}_{n_u} \;,\mathcal{F} \;=\; \mathbf{I}_{n_u} \;,\mathcal{D} \;=\; \mathbf{I}_{n_u} \;,\; \mathcal{R}_d \;=\; [\boldsymbol{S}\,,\boldsymbol{Q}] \;.$$

Thus, we have obtained an ordinary LQ-optimization problem. The resulting MIMO DFE, to be presented next, was derived in [81] and presented in [5].

Theorem 5.3. *Consider the DFE described by (5.28) and the channel model described by (5.27), where $E[v(k)v(l)^*] = \Psi\delta_{kl}$. Assuming correct past decisions,*

[13]It could be argued that a more relevant criterion is minimum probability of decision errors (MPE), which leads to a nonlinear optimization problem. However, Monsen [84] has concluded that consideration of MPE and MSE lead to essentially the same error probability. A more recent discussion of this issue can be found in [86].

the polynomial matrices $\boldsymbol{S}(q^{-1})$ and $\boldsymbol{Q}(q^{-1})$ of order m and $n_b - 1$ respectively, which minimize (5.29), are obtained as follows:

- *The feedforward filter $\boldsymbol{S}(q^{-1}) = \mathbf{S}_0 + \mathbf{S}_i q^{-1} + \ldots + \mathbf{S}_m q^{-m}$ is obtained by solving the system of linear equations*

$$\begin{pmatrix} \mathbf{B}_0^* & \ldots & \mathbf{0} & \mathbf{I}_{n_u} & \ldots & \mathbf{0} \\ \vdots & \ddots & \vdots & \vdots & \ddots & \vdots \\ \mathbf{B}_m^* & \ldots & \mathbf{B}_0^* & \mathbf{0} & \ldots & \mathbf{I}_{n_u} \\ -\Psi & \ldots & \mathrm{o} & \mathbf{B}_0 & \ldots & \mathbf{B}_m \\ \vdots & \ddots & \vdots & \vdots & \ddots & \vdots \\ \mathbf{0} & \ldots & -\Psi & \mathbf{0} & \ldots & \mathbf{B}_0 \end{pmatrix} \begin{pmatrix} \mathbf{S}_0^* \\ \vdots \\ \mathbf{S}_m^* \\ \mathbf{L}_{1m} \\ \vdots \\ \mathbf{L}_{10} \end{pmatrix} = \begin{pmatrix} \mathbf{0} \\ \vdots \\ \mathbf{0} \\ \mathbf{I}_{n_u} \\ \mathbf{0} \\ \vdots \\ \mathbf{0} \end{pmatrix} \tag{5.30}$$

  with respect to the matrices $\mathbf{S}_i^*$ and $\mathbf{L}_{1j}$. On the right-hand side, the zero matrices have dimension $n_u|n_u$ and $\mathbf{I}_{n_u}$ is positioned vertically in block number $m+1$.

- *The coefficients matrices of the feedback filter*

$$\boldsymbol{Q}(q^{-1}) = \mathbf{Q}_0 + \mathbf{Q}_1 q^{-1} + \ldots + \mathbf{Q}_{n_b - 1} q^{-n_b+1}$$

  are then given by

$$\mathbf{Q}_i = \sum_{j=0}^{m} \mathbf{S}_{m-j} \mathbf{B}_{j+i+1} \tag{5.31}$$

Proof. See [81].

Remarks. The derivation, which follows the same principles as outlined in section 5.3.3, leads to two coupled Diophantine equations. These equations can be transformed to the linear system of equations (5.30).

It can be shown that whenever the noise covariance matrix Ψ is nonsingular, the system of equations (5.30) will have a unique solution, regardless of the properties of the channel $\boldsymbol{B}(q^{-1})$.

The performance of the equalizer improves monotonically with an increased smoothing lag.

Adaptation and robustness

The equalizer coefficients can be calculated from data *directly*. They can also be adjusted *indirectly*, via model estimation and filter computation.[14] The *input*=80 MSE criterion (5.29) relevant for equalization is used also for filter adjustment by a direct algorithm. An estimator of the model (5.27) would instead typically minimize the *output* prediction error. For very long data windows, direct and

[14]When training data are available, they are used for identification. Otherwise, estimates $\tilde{u}(k)$ from the DFE can be used for so-called decision-directed adaptation.

indirect methods provide the same performance, if the channel is time-invariant. For *scalar* $y(k)$ and time-invariant channels, a nominal indirect design will tend to perform worse than a direct one if the data record is of medium lenght (40-300 data) [87]. A possible explanation is a higher sensitivity to model errors, due to the mismatch between the criteria used of identification and for filtering [42]. For *vector* measurements, the situation seems to be reversed; suitably parameterized indirect methods then outperform direct adaptation of DFE's [49], since a smaller number of parameters need to be adjusted.

For short data windows, indirect methods clearly outperform direct ones, for scalar as well as vector-valued received signals. For example, we advocate the indirect approach when tracking rapidly time-varying channels, a situation which is common in mobile radio communications due to the presence of fading. The reason for this is twofold. First, the time-variations of radio channel coefficients will tend to be smooth, while the corresponding optimal adjustments of the equalizer parameters will have strongly time-varying rates of change. This poses a difficult problem for the selection of the gains of a direct adaptive algorithm [54]. Second, the number of parameters in the equalizer will, in general, be larger than the number of channel coefficients, in particular if the smoothing lag is large. This makes it more difficult to directly adjust the filter to short data sets [49].

If an indirect multivariable adaptive DFE is applied to a time-varying channel, the filter coefficients will have to be recomputed periodically, using theorem 5.3. Note that the solution steps presented in theorem 5.3 require *no spectral factorization.* This reduces the computational complexity.

A major drawback with the DFE is that a single erroneous decision under unfortunate conditions may cause a whole sequence of errors, *an error burst.* This phenomenon is known as error propagation. It occurs, in particular, if the feedback filter $\boldsymbol{Q}(q^{-1})$ has a long impulse response.

In [38] and [88], *robust equalizers* are discussed for the scalar case. These algorithms are based on uncertain channel models and are optimized with respect to the averaged MSE criterion (5.8). They also provide means to control the error bursts and can basically trade shorter but more frequently occurring error bursts for a decreased frequency of long bursts. The system can then be designed with a coding scheme which gives rise to a smaller delay. The robust DFE is designed by assuming that the feedback signal is corrupted by white noise, with an adjustable variance. As this noise variance is increased, the gain of the feedback filter is reduced, until a (robust) linear equalizer, without decision feedback, is obtained as a limiting case. Channel uncertainty is taken into account as in section 5.4 below.

Next, we will present some problems of digital mobile radio communications in which multivariable DFE:s are applicable.

Example 5.1. (Combined temporal and spatial equalization) Consider a digital radio communications scenario where there is only one transmitted message $u(k)$ and n_y receiver antennas. Let the channel from the transmitter to receiver antenna i be described by

$$y_i(k) = B_i(q^{-1})u(k) + v_i(k) \quad . \tag{5.32}$$

The scenario for $n_y = 2$ is depicted in figure 5.6. The use of two antennas will improve the attainable performance significantly if the channels B_1 and B_2 differ. It will improve the performance moderately even for identical channels, if the noises v_1 and v_2 are not identical.

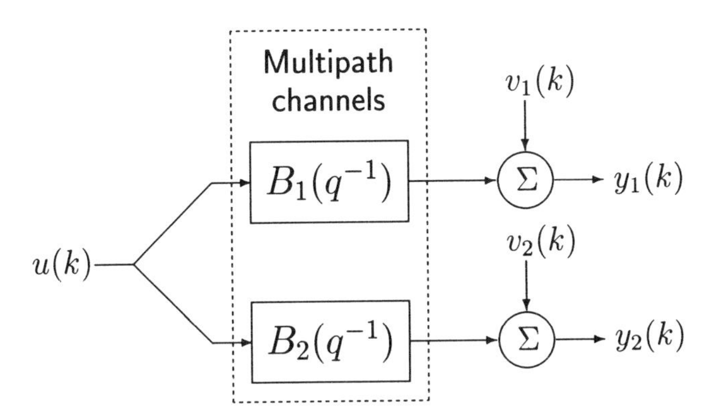

Figure 5.6: Multivariable channel model in a system with two receiver antennas.

Channel models of the type (5.32) occur not only in systems for digital mobile radio communications, but also in systems for acoustic underwater communication. The structure is of increasing interest, since it enables spatial filtering: The antenna may be adapted to have spatial nulls in the direction of interferers while maintaining high gain in the directions of arrival of $u(k)$. See, for example, [5, 47, 48, 49, 81] and [86].

The equalizer described in theorem 5.3 can be applied directly to the above scenario.

Example 5.2. (Narrowband multiuser detection in cellular digital mobile radio systems) The multivariable DFE can be used to detect multiple users on the same channel in the same cell simultaneously [5, 81, 89, 90]. See figure 5.7. The user u_2 could represent a second user in the same cell. Alternatively, u_2 could represent a co-channel interferer located in a nearby cell. The influence of this interferer on the detection of u_1 should then be minimized.

Consider a scenario with n_u transmitters and n_y receiver antennas. Let the channel from transmitter j to receiver antenna i be represented by $B_{ij}(q^{-1})$. The received signal at antenna i, $y_i(k)$, can in this case be expressed as

$$y_i(k) = \sum_{j=1}^{n_u} B_{ij}(q^{-1})u_j(k) + v_i(k) \quad .$$

By collecting the antenna signals in vector form, we obtain the model (5.27). Application of the DFE at the receiver can be seen as a way to utilize spatial diversity. If the transmitters are at different locations, then the transfer functions B_{ij} will differ for different transmitters j. It is here interesting to note that multipath propagation which leads to intersymbol interference will actually be of advantage. Propagations through different paths will tend to make the channels unequal, which improves the attainable performance.

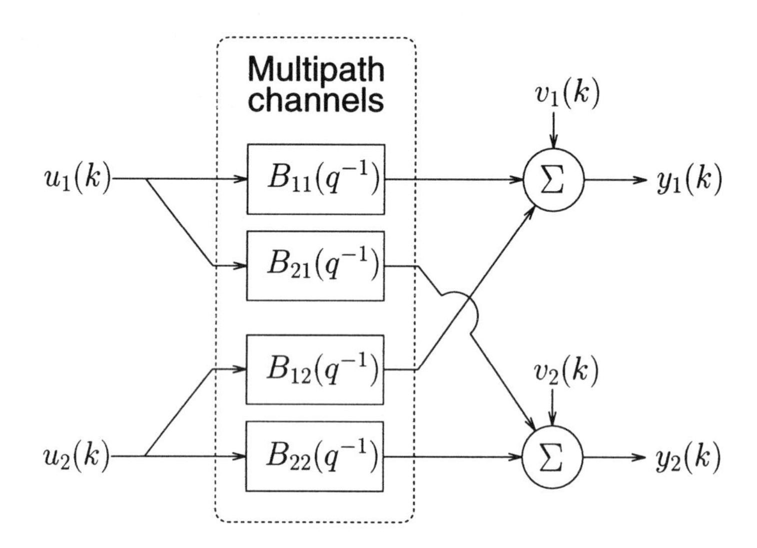

Figure 5.7: Multivariable channel model describing a situation where multiple users simultaneously share the same baseband channel.

Example 5.3. (Multiuser detection in asynchronous DS-CDMA (direct sequence-code division multiple access)) Multiple users within a cell of a cellular radio systems can share resources by utilizing different frequencies (frequency division), different time slots (time division) or different codes (code division).[15]

In systems using DS-CDMA, all active users within a cell transmit on the same frequency band at the same time [91]. In order to distinguish one message from another, each message is convolved with a different code (or signature) sequence at the transmitter. The sampling rate of the code sequence, the chip rate, is much higher than the symbol rate. The convolution with the code will transform the narrowband message into a rather broad band signal. A DS-CDMA system will thus constitute a spread spectrum transmission system.

A DS-CDMA radio system with n_u simultaneous users can be represented within the model structure (5.27). Define $u(k) = (u_1(k) \;\; u_2(k) \ldots u_{n_u}(k))^T$ as the vector of transmitted (scalar) symbols at time k. For simplicity, assume the channels to cause no intersymbol interference. The received scalar signal is assumed to be convolved with the appropriate code sequence and then sampled once per symbol. Following Verdú [92], the sampled outputs from a bank of n_u

[15]These strategies can be combined. They can also be complemented and enhanced by utilizing spatial diversity with multiple antennas, as indicated by examples 5.1 and 5.2.

filters, which are matched to the signature sequences of the individual users, may then be collected in vector form, as

$$y'(k) = \mathbf{R}(1)\mathbf{A}u(k+1) + \mathbf{R}(0)\mathbf{A}u(k) + \mathbf{R}^T(1)\mathbf{A}u(k-1) + v'(k) \ . \qquad (5.33)$$

Here, the matrices $\mathbf{R}(n)$, of dimension $n_u|n_u$, contain partial crosscorrelations between the signature sequences used to spread the data, whereas $v'(k)$ denotes noise. The square and diagonal matrix $\mathbf{A}$ contains channel coefficients associated with the different users. By redefining the measurement and the noise as $y(k) = y'(k-1)$ and $v(k) = v'(k-1)$ in (5.33), we obtain a causal multivariable channel model of the form (5.27), with $n_y = n_u$ outputs

$$y(k) = (\mathbf{R}(1)\mathbf{A} + \mathbf{R}(0)\mathbf{A}q^{-1} + \mathbf{R}^T(1)\mathbf{A}q^{-2})u(k) + v(k) \ . \qquad (5.34)$$

If the channel causes intersymbol interference, the constant matrix $\mathbf{A}$ would be substituted by a diagonal polynomial matrix. If multiple antennas are utilized, the dimensions of all matrices are increased correspondingly.

5.4 Design of robust filters in input-output form based on averaged $\mathcal{H}_2$ criteria

For any model-based filter, modelling errors will constitute a potential source of performance degradation. In this section, we propose a *cautious Wiener filter* for the prediction, filtering or smoothing of discrete-time signal vectors. The methodology has been presented in [4, 35] and [34]. A comprehensive exposition can be found in the thesis [37] by Öhrn.

The design of robust multivariable estimators, as it will be presented here, will be based on a stochastic description of model errors, related to the stochastic embedding concept of Goodwin and co-workers [12, 93].

To be more specific, the suggested approach is based on the following foundations:

- A set of (true) dynamic systems is assumed to be well described by a set of discrete-time, stable, linear and time-invariant transfer function matrices

$$\mathcal{F} = \mathcal{F}_o + \Delta\mathcal{F} \ . \qquad (5.35)$$

  We will call such a set an *extended design model.* Above, $\mathcal{F}_o$ represents a stable nominal model, while an *error model* $\Delta\mathcal{F}$ describes a set of stable transfer functions, parameterized by stochastic variables. The random variables enter linearly into $\Delta\mathcal{F}$.

- A single robust linear filter is to be designed for the whole class of possible systems. Robust performance is obtained by minimizing the averaged mean square estimation error criterion (5.8)

$$\bar{J} = \mathrm{tr}\bar{E}E(\varepsilon(k)\varepsilon(k)^*) \ . \qquad (5.36)$$

Here, $\varepsilon(k)$ is the weighted estimation error vector, E denotes expectation over noise and $\bar{E}$ is an expectation over the stochastic variables parameterizing the error model $\Delta\mathcal{F}$.

The averaged mean square error has been used previously in the literature by, for example, Chung and Bélanger [31], Speyer and Gustafson [32] and by Grimble [33, 94]. These works were based on assumptions of small parametric uncertainties and on series expansions of uncertain parameters. We suggest the use of the criterion (5.36), together with a particular description of the sets (5.35): Transfer function elements in $\Delta\mathcal{F}$ are postulated to have stochastic numerators and fixed denominators. Such models can describe non-parametric uncertainty and under-modelling as well as parametric uncertainty.

5.4.1 Approaches to robust $\mathcal{H}_2$ estimation

The most obvious *ad hoc* approach for increasing the robustness against model errors is perhaps to detune a filter, by increasing the measurement noise variance used in the design. This will work, in principle, if the transducer $\mathcal{G}$ and/or the noise model $\mathcal{H}$ is uncertain. It might, however, be very difficult to find an appropriate noise colour or covariance structure without a more systematic technique. When the signal model $\mathcal{F}$ is uncertain, a detuning approach will not work at all. The filter gain should in such situations instead be *increased*, in an appropriate way.

Most previous suggestions for obtaining robust filters in a systematic way have been based on some type of minimax approach [19, 95]. For example, a paper [96] by Martin and Mintz takes both spectral uncertainty and uncertainty in the noise distribution into account. The resulting filter will be of very high order. Minimax design of a filter $\mathcal{R}$ becomes very complex, unless there exists either a saddle point or a boundary point solution. A crucial condition here is that $\min_{\mathcal{R}} \max_{\mathcal{M}}$ equals $\max_{\mathcal{M}} \min_{\mathcal{R}}$. If so, instead of finding the worst case with respect to a set of models $\mathcal{M}$, one can search for models whose optimal filter gives the worst (nominal) performance, and use the corresponding filter. This is a much simpler task, but can still be computationally demanding. See [20, 97, 98, 99] and the survey paper [21] by Kassam and Poor. The condition $\min_{\mathcal{R}} \max_{\mathcal{M}} = \max_{\mathcal{M}} \min_{\mathcal{R}}$ is *not* fulfilled in numerous problems, which makes them very difficult to solve. See, for instance, example 5 in [34], and the example in [4].

Kalman filter-like estimators have recently been developed for systems with structured and possibly time-varying parametric uncertainty of the type

$$x(k+1) = (\mathbf{A} + \mathbf{D}\boldsymbol{\Delta}(k)\mathbf{E})\, x(k) + w(k)$$

where the matrix $\boldsymbol{\Delta}(k)$ contains norm-bounded uncertain parameters. See [27, 28, 30] and [100] for continuous-time results and [29] for the discrete-time one-step-ahead predictor. See also [101] for a related method. For systems which are stable for all $\boldsymbol{\Delta}(k)$, an upper bound on the estimation error covariance matrix can be minimized by solving two coupled Riccati equations, combined with a one-dimensional numerical search. This represents a computational simplification, as compared to previous minimax designs. Still, the resulting estimators are quite conservative, partly because they rest on worst case design. This conservatism is illustrated and discussed in [36, 37].

The method suggested in the present section and in section 5.5 is computationally simpler than any of the minimax schemes referred to above. It also avoids two drawbacks of worst case designs. First, the stochastic variables in the error model need not have compact support. Thus, the descriptions of model uncertainties may have *'soft' bounds*. These are more readily obtainable in a noisy environment than the hard bounds required for minimax design. Secondly, not only the range of the uncertainties, but also their likelihood is taken into account by using the expectation $\bar{E}(\cdot)$ of the MSE. Highly probable model errors will affect the estimator design more than do very rare 'worst cases'. Therefore, the performance loss in the nominal case, the price paid for robustness, becomes smaller than for a minimax design. In other words, conservativeness is reduced. There do, of course, exist applications where a worst case design is mandatory, e.g. for safety reasons. However, we believe that the average performance of estimators is often a more appropriate measure of performance robustness.

One of our goals will be to present transparent design equations, and to hold their number to a minimum, without sacrificing numerical accuracy. As in section 5.3.3, we therefore use matrix fraction descriptions with diagonal denominators and common denominator forms. This leads to a solution which is, in fact, significantly simpler, and more numerically well-behaved, than the corresponding nominal $\mathcal{H}_2$-designs (without uncertainty) presented in [17] or [24]. Somewhat surprisingly, taking model uncertainty into account does not require any new types of design equations. We end up with just two equations for robust estimator design: A polynomial matrix spectral factorization and a unilateral Diophantine equation. The solution has a strong formal similarity to the nominal design of section 5.3.3 and it provides structural insight; important properties of a robust estimator are evident by direct inspection of the filter expression.

5.4.2 The averaged $\mathcal{H}_2$ estimation problem

Consider the following extended design model

$$\begin{aligned} y(k) &= \boldsymbol{\mathcal{G}}(q^{-1})u(k) + \boldsymbol{\mathcal{H}}(q^{-1})v(k) \\ u(k) &= \boldsymbol{\mathcal{F}}(q^{-1})e(k) \\ z(k) &= \boldsymbol{\mathcal{D}}(q^{-1})u(k) \end{aligned} \tag{5.37}$$

where $\boldsymbol{\mathcal{G}}$, $\boldsymbol{\mathcal{H}}$, $\boldsymbol{\mathcal{F}}$ and $\boldsymbol{\mathcal{D}}$ are stable and causal, but possibly uncertain, transfer functions of dimension $p|s$, $p|r$, $s|n$ and $\ell|s$, respectively. The noise sequences $\{e(k)\}$ and $\{v(k)\}$ are mutually uncorrelated and zero mean stochastic sequences. To obtain a simple notation they are assumed to have unit covariance matrices, so scaling and uncertainty of the covariances are included in $\boldsymbol{\mathcal{F}}$ and $\boldsymbol{\mathcal{H}}$, respectively.

As before, an estimator

$$\hat{z}(k|k+m) = \mathcal{R}(q^{-1})y(k+m) \tag{5.38}$$

of $z(k)$ is sought. See figure 5.8. The transfer function $\mathcal{R}$ is designed to minimize the averaged mean square error (MSE) criterion (5.8).

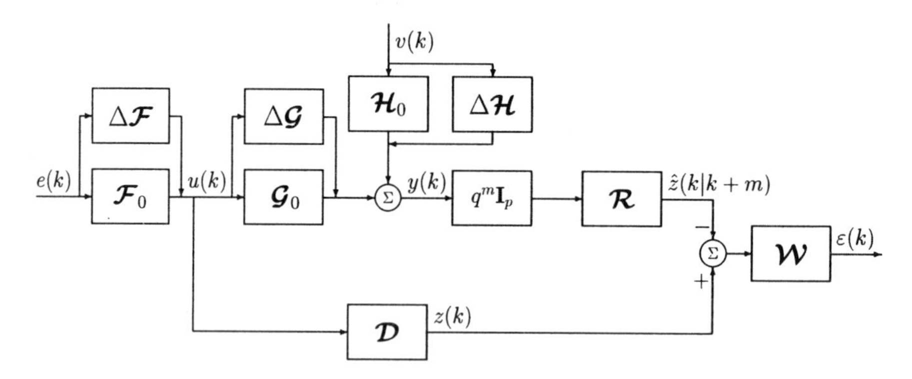

Figure 5.8: A general linear filtering problem formulation with uncertain linear models. Based on noisy measurements $y(k+m)$, the signal $z(k)$ is to be estimated, as in figure 5.3. Model errors in transfer functions are described by stochastic error models. Signals have the same dimension as in section 5.3.3.

Example 5.4. (Equalization based on an uncertain channel model) An application where uncertain dynamics in $\mathcal{G}$ is of interest is equalizer design for digital mobile radio communications. See section 5.3.4.

A signal $u(k)$ then propagates along multiple paths, with different time delays, represented by delays in $\mathcal{G}$. In present systems, both the transmitted symbol sequence $u(k)$ and the received baseband signal $y(k)$ are scalar. The use of multiple antenna elements ($p > 1$), are however of increasing interest, see example 5.1 in section 5.3.4. An appropriate model of $\mathcal{G}$ is then a column vector of FIR channels, (5.32), i.e. a vector of polynomials. The polynomial coefficients are normally estimated based on a short training sequence $\{u(k)\}$ which is known to the receiver. Estimation errors are inevitable. Furthermore, the channel coefficients will be slowly time-varying during the time intervals between training events in, for example, the GSM system [38].

The task of a (robust) equalizer is then to estimate $u(k)$, based on $y(k+m)$, a nominal model $\mathcal{G}_o$, and an estimate of the model uncertainty $\Delta\mathcal{G}$.

5.4.3 Parameterization of the extended design model

As in section 5.3.3, we choose to parameterize $\mathcal{G}$ and $\mathcal{H}$ as left MFDs having *diagonal* denominators, while $\mathcal{F}$, $\mathcal{D}$ and $\mathcal{W}$ are parameterized in common denominator form

$$\mathcal{G} = \boldsymbol{A}^{-1}\boldsymbol{B} \;\; ; \;\; \mathcal{H} = \boldsymbol{N}^{-1}\boldsymbol{M} \tag{5.39}$$

$$\mathcal{F} = \frac{1}{D}C \ ; \ \mathcal{D} = \frac{1}{T}S \ ; \ \mathcal{W} = \frac{1}{U}V \ .$$

It will be assumed that $\mathcal{G}$, $\mathcal{H}$ and $\mathcal{F}$ may be uncertain. The weighting matrix $\mathcal{W}$ is assumed to be exactly known. It is shown in [4] that uncertainty in $\mathcal{D}$ does not affect the optimal filter design, provided it is uncorrelated to uncertainties in other blocks. Therefore, uncertainty in $\mathcal{D}$ is not introduced.

The extended design models, cf. (5.35), which consist of a nominal model and an uncertainty model according to

$$\mathcal{G} = \mathcal{G}_o + \Delta\mathcal{G} \ , \ \mathcal{H} = \mathcal{H}_o + \Delta\mathcal{H} \ , \ \mathcal{F} = \mathcal{F}_o + \Delta\mathcal{F} \ .$$

are now expressed in polynomial matrix form. Using $\hat{\boldsymbol{B}}_o = \boldsymbol{A}_1\boldsymbol{B}_o$, $\hat{\boldsymbol{B}}_1 = \boldsymbol{A}_o\boldsymbol{B}_1$ etc. we introduce

$$\begin{aligned}
\mathcal{G} &= \boldsymbol{A}_o^{-1}\boldsymbol{B}_o + \boldsymbol{A}_1^{-1}\boldsymbol{B}_1\Delta\boldsymbol{B} = \boldsymbol{A}_o^{-1}\boldsymbol{A}_1^{-1}(\hat{\boldsymbol{B}}_o + \hat{\boldsymbol{B}}_1\Delta\boldsymbol{B}) \triangleq \boldsymbol{A}^{-1}\boldsymbol{B} \\
\mathcal{H} &= \boldsymbol{N}_o^{-1}\boldsymbol{M}_o + \boldsymbol{N}_1^{-1}\boldsymbol{M}_1\Delta\boldsymbol{M} = \boldsymbol{N}_o^{-1}\boldsymbol{N}_1^{-1}(\hat{\boldsymbol{M}}_o + \hat{\boldsymbol{M}}_1\Delta\boldsymbol{M}) \triangleq \boldsymbol{N}^{-1}\boldsymbol{M} \\
\mathcal{F} &= \frac{1}{D_o}\boldsymbol{C}_o + \frac{1}{D_1}\boldsymbol{C}_1\Delta\boldsymbol{C} = \frac{1}{D_oD_1}(\hat{\boldsymbol{C}}_o + \hat{\boldsymbol{C}}_1\Delta\boldsymbol{C}) \triangleq \frac{1}{D}\boldsymbol{C} \ .
\end{aligned} \tag{5.40}$$

Above, $\mathcal{G}_o = \boldsymbol{A}_o^{-1}\boldsymbol{B}_o$ represents the nominal model and $\Delta\mathcal{G} = \boldsymbol{A}_1^{-1}\boldsymbol{B}_1\Delta\boldsymbol{B}$ the error model. The same holds for $\mathcal{H}$ and $\mathcal{F}$. The diagonal polynomial matrices $\boldsymbol{A} = \boldsymbol{A}_o\boldsymbol{A}_1$, $\boldsymbol{N} = \boldsymbol{N}_o\boldsymbol{N}_1$ and the polynomials $D = D_oD_1, T$ and U are all assumed to be stable, with causal inverses. Denominator polynomials are assumed monic.

In the error models, the polynomial D_1, the diagonal polynomial matrices $\boldsymbol{A}_1$ and $\boldsymbol{N}_1$ and the polynomial matrices $\boldsymbol{C}_1$, $\boldsymbol{B}_1$ and $\boldsymbol{M}_1$ are fixed. They can be used to tailor the error models for specific needs. For example, if multiplicative error models are deemed appropriate, we use $\boldsymbol{A}_1 = \boldsymbol{A}_o$, $\boldsymbol{B}_1 = \boldsymbol{B}_o\boldsymbol{B}_m$ etc., with $\boldsymbol{B}_m$ to be specified.

The matrices $\Delta\boldsymbol{B}$, $\Delta\boldsymbol{C}$ and $\Delta\boldsymbol{M}$ contain polynomials, with jointly distributed random variables as coefficients. These coefficients are used to fit the model class to the set of true systems. One particular modelling error is represented by one particular realization of the random coefficients.[16] An element ij of a stochastic polynomial matrix $\Delta\boldsymbol{P}$ is denoted

$$\Delta P^{ij} \triangleq [\Delta\boldsymbol{P}]_{ij} = \Delta p_o^{ij} + \Delta p_1^{ij}q^{-1} + \ldots + \Delta p_{\delta p}^{ij}q^{-\delta p} \tag{5.41}$$

where δp is the degree of $\Delta\boldsymbol{P}$. All coefficients have zero means, so the nominal model is the average model in the set. Only the second order moments of the random coefficients need to be specified, since the type of distribution, and higher

[16] For a given system realization, the random coefficients are assumed time-invariant and independent of the time-series $e(k)$ and $v(k)$. This is in contrast to the approach of Haddad and Bernstein in [102], who represent the effect of uncertainties by multiplicative noises. For a given uncertainty variance, a noise representation would under-estimate the true effect of (time-invariant) parameter deviations on the dynamics.

order moments, will not affect the filter design. The parameter covariances are denoted $\bar{E}(\Delta p_r^{ij})(\Delta p_s^{\ell k})^*$ and are collected in covariance matrices $\mathbf{P}_{\Delta\mathbf{P}}^{(ij,\ell k)}$, discussed further in section 5.4.5.

We now introduce the following assumption.

Assumption 5.4. *The coefficients of all polynomial elements of $\Delta \boldsymbol{C}$ are independent of those of $\Delta \boldsymbol{B}$.*

It is possible to exclude assumption 5.4 but it does simplify the solution by eliminating the need of taking fourth order moments with respect to elements of $\Delta \boldsymbol{C}$ and $\Delta \boldsymbol{B}$ into account. The assumption is reasonable in most practical cases.

5.4.4 Obtaining error models

Error models can be obtained from ordinary identification experiments, provided the model structures match. SISO transfer function models can also be obtained in the presence of undermodelling, using a maximum likelihood approach [12]. We shall next outline various ways to adjust error models to the variability of the dynamics within sets of possible SISO-systems.

Obtaining extended design models from identification

Model error estimates are obtained from many types of identification algorithms, for example prediction error methods. In a Bayesian setting, a model error estimate can be said to represent the characteristics of a possible set of true systems, which might have generated the data used for identification. It is conventional to decompose the estimation error into a variance error, caused by noise in finite data sets, and a bias error, which would remain even for infinite data sets. The bias error is caused by the selection of an inappropriate model structure. Model structure selection can sometimes be difficult, but is aided by systematic procedures for model validation [104].

If the model error caused by bias is small, and if the data series is not too short, then the estimated parameter covariance matrix of an identified model provides acceptable estimates of the modelling errors.[17]

For model structures with denominator polynomials, such as AR, ARX and ARMAX models [103, 104], the estimated model uncertainty of denominator coefficients will have to be transformed into an additive error model by series expansion. Methods for series expansion are discussed in detail in [37]. In general, first or modified second order expansions will provide an approximation with sufficient accuracy. The estimation of error models for ARMAX structures, based on short data records generated by high order systems, has been investigated recently by Bigi in [42].

[17]For models of time-invariant systems, which pass standard validation tests, the variance error will, in general, dominate the bias error, see [39]. A reasonable, but perhaps somewhat conservative, estimate of the total model error is then obtained by doubling the parameter covariance matrix, which is a measure of the variance error.

For a system with measurable inputs, one way of directly obtaining extended design models of the type (5.40) is from identification experiments based on functional series expansions $\sum_{i=1}^{M} p_i \mathcal{B}_i(q^{-1})$. Here, $\mathcal{B}_i, i = 1 \dots M$ represents a set of predetermined rational basis functions, such as, e.g. discrete Laguerre functions. A functional series model is linear in the parameters $\{p_i\}$. The model structure has received increasing interest as a useful tool in system identification, see [105] or [106]. If an identification experiment provides parameter estimates $\{\bar{p}_{0i}\}$ and covariances for zero mean errors $\{\Delta\bar{p}_i\}$, we directly obtain the extended design model

$$\mathcal{P} \triangleq \sum_{i=1}^{M} \bar{p}_{0i}\mathcal{B}_i + \sum_{i=1}^{M} \Delta\bar{p}_i\mathcal{B}_i = \bar{\mathcal{P}}_0 + \Delta\bar{\mathcal{P}} .$$

Writing $\Delta\bar{\mathcal{P}}$ in common denominator form and using the covariance matrix for $\{\Delta\bar{p}_i\}$ gives the frequency domain variance $\bar{E}(\Delta\bar{\mathcal{P}}\Delta\bar{\mathcal{P}}_*)$, which will be needed in the robust design. See chapter 5 of [107].

Adjustment to sets of spectra or Nyquist plots

Error models representing nonparametric uncertainties can be adjusted directly to frequency domain data. In that context, a very useful concept is provided by the stochastic frequency domain theory of Goodwin and Salgado, see [93].

We will next very briefly recapitulate their stochastic embedding concept. An additive transfer function error $\Delta\mathcal{G}(e^{i\omega})$ is viewed as a realization of a stochastic process in the frequency domain, with zero mean and covariance function

$$\bar{E}\{\Delta\mathcal{G}(e^{i\omega_1})\Delta\mathcal{G}_*(e^{i\omega_2})\} \triangleq \Gamma(e^{i\omega_1}, e^{i\omega_2}) \geq 0 .$$

For stationary processes, the covariance depends only on the difference in frequency, $\Gamma(e^{i\omega_1}, e^{i\omega_2}) = \Gamma_s(e^{i(\omega_1-\omega_2)})$. The shape of Γ_s is a measure of the assumed frequency domain smoothness of realizations of the model error. The variance $(\omega_1 - \omega_2 = 0)$ is a scale factor for the uncertainty.

The frequency domain stochastic process $\Delta\mathcal{G}(e^{i\omega})$ corresponds to a time-domain filter with stochastic, zero mean, impulse response coefficients

$$\Delta\mathcal{G}(q^{-1}) = \sum_{j=0}^{\infty} g_j q^{-j} \; ; \qquad \bar{E}(g_j, g_\ell) = \gamma(j, \ell) . \tag{5.42}$$

Here, $\gamma(j,\ell)$ can be calculated from the inverse two-dimensional discrete Fourier transform of $\Gamma(e^{i\omega_1}, e^{i\omega_2})$. For stationary stochastic processes in the frequency domain, the corresponding time-domain process will be white, with

$$\bar{E}(g_j, g_\ell) = \gamma_j \delta_{j,\ell} . \tag{5.43}$$

For example, consider a frequency domain stochastic process $\mathcal{H}(e^{i\omega})$, with a zero mean Gaussian distribution and with covariance function

$$\bar{E}\{\mathcal{H}(e^{i\omega_1})\mathcal{H}_*(e^{i\omega_2})\} = \frac{\alpha e^{i(\omega_1-\omega_2)}}{e^{i(\omega_1-\omega_2)} - \lambda} . \tag{5.44}$$

This process corresponds to the time domain model (5.42), (5.43), with Gaussian distributed independent parameters with variances $\gamma_j = \alpha\lambda^j$. See [12] by Goodwin *et al.* By truncating at some $j = M$ for which λ^M is small, we obtain

$$\mathcal{H}(q^{-1}) \approx h_0 + \ldots + h_M q^{-M}$$

with $\bar{E}(h_j)^2 = \alpha\lambda^j$ and $\bar{E}(h_j) = 0$.

A priori information may be available about the frequency domain distribution of the unmodelled dynamics. It can be incorporated by using a fixed prefilter, to obtain the total model

$$\Delta\mathcal{G}(q^{-1}) = \mathcal{M}(q^{-1})\mathcal{N}'_\Delta(q^{-1}) \ . \tag{5.45}$$

Here, $\mathcal{M}(q^{-1})$ is a known shaping filter and $\mathcal{N}'_\Delta(q^{-1})$ is a stationary process in the frequency domain, with covariance function $\Gamma_s(e^{i(\omega_1-\omega_2)})$. Examples of the use of this modelling procedure can be found in [34, 37] and [93].

Example 5.5. (A frequency-shaped error model) Assume that the variance in the frequency domain of the model error can be described by a squared magnitude response of a first order filter. Then we may use a model of the type (5.45), with

$$\mathcal{M}(q^{-1}) = \frac{1+\eta q^{-1}}{1+a_{11}q^{-1}} \ . \tag{5.46}$$

Also, assume that the parameter λ in (5.44) can be tuned to give a reasonable description of the degree of smoothness (in the frequency domain) of the most probable model errors. The process in (5.44) can then be used to represent the stationary part $\Gamma_s(e^{i(\omega_1-\omega_2)})$ of the frequency-domain process. Truncation of its corresponding time-domain impulse response gives a model (5.45). The error model has the structure introduced in (5.40)

$$\begin{aligned}\Delta\mathcal{G}(q^{-1}) &= \frac{1+\eta q^{-1}}{1+a_{11}q^{-1}}(h_o + h_1 q^{-1} + \ldots + h_M q^{-M}) \\ &= \frac{B_1(q^{-1})\Delta B(q^{-1})}{A_1(q^{-1})} \ .\end{aligned} \tag{5.47}$$

The covariance matrix of $\{h_j\}$ is $\mathbf{P}_{\Delta B} = \text{diag}(\alpha\lambda^j)$. Note that the model is characterized by only five parameters: α, λ, a_{11}, η and the truncation length M.

Pragmatic tuning of covariances

Even if the statistics are hard to obtain, one could still use the elements of covariance matrices pragmatically, as robustness 'tuning knobs'. They are then used similarly as when weighting matrices are adjusted in LQG controller design. One objective could be to obtain as good a performance as possible, under the constraint of a prespecified level of degradation in the nominal case. Another objective could be to limit the maximal error within a specified range of model dynamics. The error models may also be used to account for a slowly time-varying dynamics, see [38].

5.4.5 Covariance matrices for the stochastic coefficients

In order to represent the uncertainties of the system in a natural way, covariance matrices will be organized as follows. The ij-th element of a stochastic polynomial matrix $\Delta \boldsymbol{P}$ can be expressed as

$$\Delta P^{ij}(q^{-1}) = \varphi^T(q^{-1})\bar{p}_{ij} \tag{5.48}$$

where

$$\varphi^T(q^{-1}) = (1 \;\; q^{-1} \ldots q^{-\delta p}) \quad ; \quad \bar{p}_{ij} = (\Delta p_o^{ij} \;\; \Delta p_1^{ij} \ldots \Delta p_{\delta p}^{ij})^T \quad . \tag{5.49}$$

The cross covariance matrix $\mathbf{P}_{\Delta \mathbf{P}}^{(ij,\ell k)}$, of dimension $\delta p + 1 | \delta p + 1$, between elements of $\Delta P^{ij}(q^{-1})$ and $\Delta P^{\ell k}(q^{-1})$, is given by

$$\mathbf{P}_{\Delta \mathbf{P}}^{(ij,\ell k)} = \bar{E}\,\bar{p}_{ij}\bar{p}_{\ell k}^* = \begin{bmatrix} \bar{E}(\Delta p_o^{ij})(\Delta p_o^{\ell k})^* & \ldots & \bar{E}(\Delta p_o^{ij})(\Delta p_{\delta p}^{\ell k})^* \\ \vdots & \ddots & \vdots \\ \bar{E}(\Delta p_{\delta p}^{ij})(\Delta p_o^{\ell k})^* & \ldots & \bar{E}(\Delta p_{\delta p}^{ij})(\Delta p_{\delta p}^{\ell k})^* \end{bmatrix} \tag{5.50}$$

where $\mathbf{P}_{\Delta \mathbf{P}}^{(ij,ij)}$ is Hermitian and positive semidefinite, while $\mathbf{P}_{\Delta \mathbf{P}}^{(ij,\ell k)} = (\mathbf{P}_{\Delta \mathbf{P}}^{(\ell k,ij)})^*$. Thus,

$$\bar{E}(\Delta P^{ij}\Delta P_*^{\ell k}) = \bar{E}(\varphi^T(q^{-1})\bar{p}_{ij}\bar{p}_{\ell k}^*\varphi_*^T(q)) = \varphi^T \mathbf{P}_{\Delta \mathbf{P}}^{(ij,\ell k)}\varphi_*^T \quad . \tag{5.51}$$

With autocovariances, $(ij) = (\ell k)$, we model the uncertainty within each input–output pair. Cross-dependencies between different transfer functions may also be known. For example, uncertainty in one single physical parameter may very well enter into several transfer functions between inputs and outputs. Such effects are captured by cross covariances, $(ij) \neq (\ell k)$.

We collect all matrices of type (5.50) into one large covariance matrix, organized as

$$\mathbf{P}_{\Delta \mathbf{P}} = \begin{pmatrix} \begin{bmatrix} \mathbf{P}_{\Delta \mathbf{P}}^{(11,11)} & \ldots & \mathbf{P}_{\Delta \mathbf{P}}^{(11,1m)} \\ \vdots & \ddots & \vdots \\ \mathbf{P}_{\Delta \mathbf{P}}^{(1m,11)} & \ldots & \mathbf{P}_{\Delta \mathbf{P}}^{(1m,1m)} \end{bmatrix} & \ldots & \begin{bmatrix} \mathbf{P}_{\Delta \mathbf{P}}^{(11,n1)} & \ldots & \mathbf{P}_{\Delta \mathbf{P}}^{(11,nm)} \\ \vdots & & \vdots \\ \mathbf{P}_{\Delta \mathbf{P}}^{(1m,n1)} & \ldots & \mathbf{P}_{\Delta \mathbf{P}}^{(1m,nm)} \end{bmatrix} \\ \vdots & \ddots & \vdots \\ \begin{bmatrix} \mathbf{P}_{\Delta \mathbf{P}}^{(n1,11)} & \ldots & \mathbf{P}_{\Delta \mathbf{P}}^{(n1,1m)} \\ \vdots & & \vdots \\ \mathbf{P}_{\Delta \mathbf{P}}^{(nm,11)} & \ldots & \mathbf{P}_{\Delta \mathbf{P}}^{(nm,1m)} \end{bmatrix} & \ldots & \begin{bmatrix} \mathbf{P}_{\Delta \mathbf{P}}^{(n1,n1)} & \ldots & \mathbf{P}_{\Delta \mathbf{P}}^{(n1,nm)} \\ \vdots & \ddots & \vdots \\ \mathbf{P}_{\Delta \mathbf{P}}^{(nm,n1)} & \ldots & \mathbf{P}_{\Delta \mathbf{P}}^{(nm,nm)} \end{bmatrix} \end{pmatrix} . \tag{5.52}$$

If $\Delta \boldsymbol{P}$ has dimension $n|m$, then $\mathbf{P}_{\Delta \mathbf{P}}$ has nm by nm blocks $\mathbf{P}_{\Delta \mathbf{P}}^{(ij,\ell k)}$. The structure of (5.52) is useful from a design point of view. If, for example, a multivariable moving average model, or FIR model, is to be identified, then (5.52) is the natural way of representing the covariance matrix. If we instead prefer to use the blocks $\mathbf{P}_{\Delta \mathbf{P}}^{(ij,\ell k)}$ of (5.52) as multivariable 'tuning knobs', a given amount of uncertainty can be assigned to a specific input-output pair.

5.4.6 Design of the cautious Wiener filter

Nominal Wiener filter design involves two key elements, namely a spectral factorization and a Diophantine equation. These two elements will remain crucial also when models are uncertain. However, they will now be based on the average behaviour with respect to an underlying set of systems.

Averaged spectral factorization

We define an averaged spectral factor $\boldsymbol{\beta}(q^{-1})$ as the numerator polynomial matrix of an averaged innovations model. It constitutes a key element of the robust filter. The average, over the set of models, of the spectral density matrix $\boldsymbol{\Phi}_y(e^{i\omega})$ of the measurement $y(k)$ in (5.37) is given by

$$\bar{E}\{\boldsymbol{\Phi}_y(e^{i\omega})\} = \frac{1}{DD_*}\boldsymbol{A}^{-1}\boldsymbol{N}^{-1}\boldsymbol{\beta}\boldsymbol{\beta}_*\boldsymbol{N}_*^{-1}\boldsymbol{A}_*^{-1} \ .$$

The square polynomial matrix $\boldsymbol{\beta}(z^{-1})$ is given by the stable solution to

$$\boldsymbol{\beta}\boldsymbol{\beta}_* = \bar{E}\{\boldsymbol{N}\boldsymbol{B}\boldsymbol{C}\boldsymbol{C}_*\boldsymbol{B}_*\boldsymbol{N}_* + D\boldsymbol{A}\boldsymbol{M}\boldsymbol{M}_*\boldsymbol{A}_*D_*\} \ . \tag{5.53}$$

Note that $\boldsymbol{N}^{-1}$ and $\boldsymbol{A}^{-1}$ are diagonal, and will thus commute. The averaged second order statistics of $y(k)$ is thus described by the same spectral density as for a vector-ARMA model

$$\bar{y}(k) = \frac{1}{D}\boldsymbol{A}^{-1}\boldsymbol{N}^{-1}\boldsymbol{\beta}\,\varepsilon(k) \tag{5.54}$$

where $\varepsilon(k)$ is white with a unit covariance matrix. This model is denoted the *averaged innovations model.* (Note that $\bar{y}(k) \neq y(k)$, but $\boldsymbol{\Phi}_{\bar{y}}(e^{i\omega}) = \bar{E}\{\boldsymbol{\Phi}_y(e^{i\omega})\}$). When constructing the right-hand side of (5.53), the following results are useful.

Lemma 5.5. *Let $\boldsymbol{H}(q,q^{-1})$ be an $m|m$ polynomial matrix with double-sided polynomial elements having stochastic coefficients. Also, let $\boldsymbol{G}(q^{-1})$ be an $n|m$ polynomial matrix with polynomial elements having stochastic coefficients, independent of all those of $\boldsymbol{H}$. Then,*

$$\bar{E}[\boldsymbol{G}\boldsymbol{H}\boldsymbol{G}_*] = \bar{E}[\boldsymbol{G}\bar{E}(\boldsymbol{H})\boldsymbol{G}_*] \tag{5.55}$$

Proof. See [4].

Now, introduce the double-sided polynomial matrices

$$\tilde{\boldsymbol{C}}\tilde{\boldsymbol{C}}_* \triangleq \bar{E}(\boldsymbol{C}\boldsymbol{C}_*) \ ; \ \tilde{\boldsymbol{B}}_C\tilde{\boldsymbol{B}}_{C*} \triangleq \bar{E}(\boldsymbol{B}\tilde{\boldsymbol{C}}\tilde{\boldsymbol{C}}_*\boldsymbol{B}_*) \ ; \ \tilde{\boldsymbol{M}}\tilde{\boldsymbol{M}}_* \triangleq \bar{E}(\boldsymbol{M}\boldsymbol{M}_*) \ . \tag{5.56}$$

Invoking (5.40) and using the fact that the stochastic coefficients are assumed to be zero mean, gives

$$\tilde{\boldsymbol{C}}\tilde{\boldsymbol{C}}_* = \hat{\boldsymbol{C}}_o\hat{\boldsymbol{C}}_{o*} + \hat{\boldsymbol{C}}_1\bar{E}(\Delta\boldsymbol{C}\Delta\boldsymbol{C}_*)\hat{\boldsymbol{C}}_{1*}$$

$$\tilde{\boldsymbol{B}}_C\tilde{\boldsymbol{B}}_{C*} = \hat{\boldsymbol{B}}_o\tilde{\boldsymbol{C}}\tilde{\boldsymbol{C}}_*\hat{\boldsymbol{B}}_{o*} + \hat{\boldsymbol{B}}_1\bar{E}(\Delta\boldsymbol{B}\tilde{\boldsymbol{C}}\tilde{\boldsymbol{C}}_*\Delta\boldsymbol{B}_*)\hat{\boldsymbol{B}}_{1*} \tag{5.57}$$

$$\tilde{\boldsymbol{M}}\tilde{\boldsymbol{M}}_* = \hat{\boldsymbol{M}}_o\hat{\boldsymbol{M}}_{o*} + \hat{\boldsymbol{M}}_1\bar{E}(\Delta\boldsymbol{M}\Delta\boldsymbol{M}_*)\hat{\boldsymbol{M}}_{1*} \ .$$

Factorizations to obtain $\tilde{C}$, $\tilde{B}_C$ etc. need not be performed. The double-sided polynomial matrices are expressed as $\tilde{C}\tilde{C}_*$ etc. merely to simplify the notation.

Lemma 5.6. *Let assumption 5.4 hold. By using (5.56), (5.57) and invoking lemma 5.1, the averaged spectral factorization (5.53) can be expressed as*

$$\boldsymbol{\beta}\boldsymbol{\beta}_* = \boldsymbol{N}\tilde{\boldsymbol{B}}_C\tilde{\boldsymbol{B}}_{C*}\boldsymbol{N}_* + D\boldsymbol{A}\tilde{\boldsymbol{M}}\tilde{\boldsymbol{M}}_*\boldsymbol{A}_*D_* \tag{5.58}$$

Proof. See [4].

With a given right-hand side, equation (5.58) is just an ordinary polynomial matrix left spectral factorization, of the type encountered in (5.15). It is solvable under the following mild assumption.

Assumption 5.7. *The averaged spectral density matrix $\bar{E}\{\boldsymbol{\Phi}_y(e^{i\omega})\}$ is nonsingular for all ω.*

This assumption is equivalent to the right-hand side of (5.58) being nonsingular on $|z| = 1$. Then, the solution to (5.58) is unique, up to a right orthogonal factor. Under assumption 5.7, a solution exists, with $\boldsymbol{\beta}$ having nonsingular leading coefficient matrix $\boldsymbol{\beta}(0)$. Its degree, $n\beta$, will be determined by the maximal degree of the two right-hand side terms in (5.58).

To obtain the right-hand side of (5.58), averaged polynomial matrices like $\bar{E}(\Delta\boldsymbol{P}\boldsymbol{H}\Delta\boldsymbol{P}_*)$ have to be computed, where $\boldsymbol{H}(q,q^{-1}) = \tilde{\boldsymbol{C}}\tilde{\boldsymbol{C}}_*$ or $\mathbf{I}$. It is shown in [4] that the ij-th element of $\bar{E}(\Delta\boldsymbol{P}\boldsymbol{H}\Delta\boldsymbol{P}_*)$ is given by

$$\bar{E}[\Delta\boldsymbol{P}\boldsymbol{H}\Delta\boldsymbol{P}_*]_{ij} = \operatorname{tr}\boldsymbol{H}\begin{bmatrix}\varphi^T & & 0\\ & \ddots & \\ 0 & & \varphi^T\end{bmatrix}\begin{bmatrix}\mathbf{P}_{\Delta\mathbf{P}}^{(i1,j1)} & \cdots & \mathbf{P}_{\Delta\mathbf{P}}^{(im,j1)}\\ \vdots & \ddots & \vdots\\ \mathbf{P}_{\Delta\mathbf{P}}^{(i1,jm)} & \cdots & \mathbf{P}_{\Delta\mathbf{P}}^{(im,jm)}\end{bmatrix}\begin{bmatrix}\varphi_*^T & & 0\\ & \ddots & \\ 0 & & \varphi_*^T\end{bmatrix} \tag{5.59}$$

where φ^T is defined in (5.49). The block covariance matrix in (5.59) is obtained by taking the block-transpose of the ij-th block $[\,\cdot\,]$ of $\mathbf{P}_{\Delta\mathbf{P}}$ in (5.52). Average factors in (5.57) are readily obtained by substituting $\Delta\boldsymbol{C}$, $\Delta\boldsymbol{B}$ and $\Delta\boldsymbol{M}$ for $\Delta\boldsymbol{P}$ in (5.59).

We are now ready to present the solution to the robust $\mathcal{H}_2$ filter design problem.

The cautious multivariable Wiener filter

Theorem 5.8. *Assume an extended design model (5.37), (5.39), (5.40), to be given, with known covariance matrices (5.52). Under assumptions 5.4 and 5.7, a realizable estimator of $z(k)$ then minimizes the averaged MSE (5.8), among all linear time-invariant estimators based on $y(k+m)$, if and only if it has the same coprime factors as*

$$\hat{z}(k|k+m) = \boldsymbol{\mathcal{R}}\,y(k+m) = \frac{1}{T}\boldsymbol{V}^{-1}\boldsymbol{Q}\boldsymbol{\beta}^{-1}\boldsymbol{N}\boldsymbol{A}\,y(k+m) \quad . \tag{5.60}$$

Here, $\boldsymbol{\beta}(q^{-1})$ is obtained from (5.58), while $\boldsymbol{Q}(q^{-1})$ together with $\boldsymbol{L}_*(q)$, both of dimensions $\ell|p$, is the unique solution to the unilateral Diophantine equation

$$q^{-m}\boldsymbol{V}\boldsymbol{S}\tilde{\boldsymbol{C}}\tilde{\boldsymbol{C}}_*\hat{\boldsymbol{B}}_{o*}\boldsymbol{N}_* = \boldsymbol{Q}\boldsymbol{\beta}_* + q\boldsymbol{L}_*UTD\,\mathbf{I}_p \tag{5.61}$$

with generic[18] *degrees*

$$\begin{aligned} nQ &= \max(n\tilde{v} + ns + n\tilde{c} + m, nu + nt + nd - 1) \\ nL_* &= \max(n\tilde{c} + n\hat{b}_o + nn - m, n\beta) - 1 \end{aligned} \tag{5.62}$$

where $ns = \deg \boldsymbol{S}$ etc. When applying the estimator (5.60) on an ensemble of systems, the minimal criterion value becomes

$$\mathrm{tr}\bar{E}E(\varepsilon(k)\varepsilon(k)^*)_{\min} = \mathrm{tr}\frac{1}{2\pi j}\oint_{|z|=1}\{\boldsymbol{L}_*\boldsymbol{\beta}_*^{-1}\boldsymbol{\beta}^{-1}\boldsymbol{L} +$$

$$+\frac{1}{UTDD_*T_*U_*}\boldsymbol{V}\boldsymbol{S}\tilde{\boldsymbol{C}}\left[\mathbf{I}_n - \tilde{\boldsymbol{C}}_*\hat{\boldsymbol{B}}_{o*}\boldsymbol{N}_*\boldsymbol{\beta}_*^{-1}\boldsymbol{\beta}^{-1}\boldsymbol{N}\hat{\boldsymbol{B}}_o\boldsymbol{C}\right]\tilde{\boldsymbol{C}}_*\boldsymbol{S}_*\boldsymbol{V}_*\Big\}\frac{dz}{z} \tag{5.63}$$

Proof. See [4], where the variational approach outlined in section 5.3.3 is utilized, with $\bar{E}E(\cdot)$ substituted for $E(\cdot)$. In the case considered here, with a known filter $\mathcal{W}$, it is straightforward to show that the cautious Wiener filter minimizes not only the scalar averaged MSE, but also the average covariance matrix $\bar{E}E\varepsilon(k)\varepsilon(k)^*$.

Remarks. Note the very close formal similarity of (5.60), (5.61) and (5.63) to their nominal counterparts (5.22), (5.23) and (5.26), respectively. The only new required type of computation, as compared to the nominal case described in section 5.3.3, is the calculation of averaged polynomials in (5.57) performed by using (5.59). Design examples can be found in [4] and in [37].

Since both $\boldsymbol{V}$ and $\boldsymbol{\beta}$ are stable, the estimator $\mathcal{R}$ will be stable[19]. Since $\boldsymbol{V}(0)$ and $\boldsymbol{\beta}(0)$ are nonsingular, $\mathcal{R}$ will be causal.

Note that the diagonal matrix $\boldsymbol{NA} = \boldsymbol{N}_o\boldsymbol{N}_1\boldsymbol{A}_o\boldsymbol{A}_1$ appears explicitly in the filter (5.60). Important properties of the robust estimator are evident by direct inspection. For example, assume some diagonal elements of $\boldsymbol{N}_1^{-1}$ or $\boldsymbol{A}_1^{-1}$ in the error models to have resonance peaks, indicating large uncertainty at the corresponding frequencies. Then, the filter will have notches, so the filter gain from the uncertain components of $y(k+m)$ will be low at the relevant frequencies.

The nominal Wiener filter (5.22) has a whitening filter as a right factor. The robust estimator has a similar structure. After multiplying $\mathcal{R}$ by the stable common factor D/D, the cautious filter (5.60) will contain $\boldsymbol{\beta}^{-1}\boldsymbol{NA}D$ as right

[18] In special cases, the degrees may be lower.

[19] Stable common factors may exist in (5.60). They could be detected by calculating invariant polynomials of the involved matrices. If such factors have zeros close to the unit circle, it is advisable to cancel them before the filter is implemented. Otherwise, slowly decaying (initial) transients may deteriorate the filtering performance.

factor. This averaged counterpart of a whitening filter is the inverse of the averaged innovations model (5.54).

The model structure (5.39)-(5.40) was selected to obtain a few simple design equations. Other choices are possible, but lead to various complications. For example, if stochastic polynomials had been introduced in the denominators, no exact analytical solution could have been obtained. Also stability would have been a problem. The use of general left MFD representations, instead of forms with diagonal denominators or common denominators, would have led to a solution involving five coprime factorizations. Such a solution is presented in [37], but it provides less physical insight. It does also exhibit worse numerical behaviour, since algorithms for coprime factorization are numerically sensitive.

Robust design also makes the solution less numerically sensitive. Almost common factors of $\det \boldsymbol{\beta}_*$ and UTD with zeros close to $|z| = 1$ would make the solution of (5.61) numerically sensitive. The averaged spectral factor $\boldsymbol{\beta}$ will, in general, have its zeros more distant from the unit circle than the nominal spectral factor, given by (5.15). This reduces the numerical difficulty of solving both (5.58) and (5.61).

The equivalent-noise interpretation

For every cautious Wiener filter, there exists a system (without uncertainty) for which this estimator is the optimal Wiener filter, see [37]. For example, if $\mathcal{G} = \mathbf{I}_p$, then we can utilize the modified signal and noise spectral density matrices

$$\mathcal{F}_o\mathcal{F}_{o*} + \bar{E}(\Delta\mathcal{F}\Delta\mathcal{F}_*) \quad ; \quad \mathcal{H}_o\mathcal{H}_{o*} + \bar{E}(\Delta\mathcal{H}\Delta\mathcal{H}_*) \tag{5.64}$$

to obtain the averaged innovations model. The spectral densities (5.64) may be obtained directly from frequency domain data if such are available, by simply averaging the spectral densities of the model sets $\mathcal{F}$ and $\mathcal{H}$.

It is also possible to represent model uncertainties by coloured noises, and then to design a Wiener filter for the corresponding system. This correspondence provides a way of understanding the structure of the above design equations. However, we do not recommend the use of an equivalent noise-approach in the actual design, for two reasons:

- It is far from trivial to obtain an equivalent noise representation of the uncertainties in the block $\mathcal{G}$, with appropriate colour and covariance structure. This is true in particular if the block $\mathcal{F}$ is also uncertain, and if the problem is multivariable.

- It is an advantage from a design point of view to have separate tools which handle different aspects of the design: Error models to represent the effect of modelling uncertainty; noise models to represent disturbances; criterion weighting functions to reflect the priorities of the user. A method which does not distinguish between these aspects will tend to confuse the designer.

The attainable limit of estimator performance

The attainable performance improves monotonically with an increasing smoothing lag m. The following result gives the lower bound of the averaged estimation error. This bound can be approached pointwise in the frequency domain for $m < \infty$, by using a criterion filter $\mathcal{W}$ with a high resonance peak.

Corollary 5.9. *The limiting estimator for $m \to \infty$, the non-realizable cautious Wiener filter, can be expressed as*

$$\lim_{m\to\infty} q^m \mathcal{R} = \frac{1}{T} \boldsymbol{S}\tilde{\boldsymbol{C}}\tilde{\boldsymbol{C}}_* \hat{\boldsymbol{B}}_{o*} \boldsymbol{N}_* \boldsymbol{\beta}_*^{-1} \boldsymbol{\beta}^{-1} \boldsymbol{N} \boldsymbol{A} \ . \tag{5.65}$$

Its average performance is given by (5.63) with $\boldsymbol{L} = \boldsymbol{0}$. If $\mathcal{W} = \mathbf{I}_\ell$, the trace of the spectral density of the lower bound of the estimation error $z(k) - \hat{z}(k|k+m)$ is

$$\lim_{m\to\infty} \mathrm{tr}\bar{E}E\ \boldsymbol{\Phi}_{z-\hat{z}}(e^{i\omega})$$

$$= \frac{1}{TDD_*T_*} \mathrm{tr} \left\{ \boldsymbol{S}\tilde{\boldsymbol{C}} \left[\mathbf{I}_n - \tilde{\boldsymbol{C}}_* \hat{\boldsymbol{B}}_{o*} \boldsymbol{N}_* \boldsymbol{\beta}_*^{-1} \boldsymbol{\beta}^{-1} \boldsymbol{N} \hat{\boldsymbol{B}}_o \tilde{\boldsymbol{C}} \right] \tilde{\boldsymbol{C}}_* \boldsymbol{S}_* \right\} \ . \tag{5.66}$$

The bound can be attained at a frequency ω_1 by an estimator with finite smoothing lag, if the estimator is designed using a weighted criterion where

$$U(e^{-i\omega_1}) \approx 0 \tag{5.67}$$

Proof. In a similar way as in appendix A.3 of [59], it is straightforward to show that $\boldsymbol{L} \to \boldsymbol{0}$ as $m \to \infty$ in (5.61). Thus, (5.61) gives

$$\lim_{m\to\infty} q^m \boldsymbol{Q} = \boldsymbol{V}\boldsymbol{S}\tilde{\boldsymbol{C}}\tilde{\boldsymbol{C}}_* \hat{\boldsymbol{B}}_{o*} \boldsymbol{N}_* \boldsymbol{\beta}_*^{-1} \ . \tag{5.68}$$

The substitution of this expression into (5.60) gives (5.65). The use of $\boldsymbol{L} = \boldsymbol{0}$, $\boldsymbol{V} = \mathbf{I}_\ell$ and $U = 1$ in the integrand of (5.63) gives (5.66). When $U(e^{-i\omega_1}) \approx 0$, we obtain the same effect on the Diophantine equation (5.61) at the frequency ω_1 as if $\boldsymbol{L} \to \boldsymbol{0}$: the rightmost term vanishes. Thus, at ω_1, the gain and the phase of the elements of the polynomial matrix $q^m \boldsymbol{Q}$ are approximately equal to those of (5.68) and the estimation error approaches the lower bound (5.66).

Remark. Note that for realizable estimators (m finite) the lower bound (5.66) is only attainable at distinct frequencies ω_i by means of frequency weighting. For frequencies outside the bandwidth of $\mathcal{W}$, the estimate may be severely degraded.

5.5 Robust $\mathcal{H}_2$ filter design based on state-space models with parametric uncertainty

Let us now consider state space models with parametric uncertainty obtained, for example, by physical modelling. The section illustrates the utility of combining state space and polynomial representations. It summarizes results from [36] and

[37]. Extended design models of the type (5.35) will be obtained by a series expansion methodology outlined in section 5.5.1 below, and discussed in more detail in [37]. The method is an improvement upon a similar suggestion by Speyer and Gustafsson [32], in that ℓ'th order expansions will lead to modified state space models of order $n(\ell+1)$, rather than $n(\ell+1)^2$. The subsequent design of an averaged robust $\mathcal{H}_2$ estimator can be performed either by using the cautious Wiener estimator of theorem 5.8 or by designing a robustified Kalman estimator, as outlined in section 5.5.2.

The results of [36] are here generalized slightly to include time-varying measurement equations. The resulting estimators are then directly applicable as robust adaptive algorithms, which can be applied for tracking the parameters of linear regression models, as discussed briefly in section 5.6.

5.5.1 Series expansion

Assume a set of stable discrete-time models

$$
\begin{aligned}
x(k+1) &= (\mathbf{A}_0 + \Delta\mathbf{A}(\rho))x(k) + (\mathbf{B}_0 + \Delta\mathbf{B}(\rho))e(k) \\
y(k) &= \mathbf{C}(k)x(k) + (\mathbf{M}_0 + \Delta\mathbf{M}(\rho))v(k) \\
z(k) &= \mathbf{L}x(k)
\end{aligned}
\tag{5.69}
$$

where $x(k) \in \mathbf{R}^n$ is the state vector and $e(k) \in \mathbf{R}^{n_e}$ is zero mean process noise with unit covariance matrix. The output $y(k) \in \mathbf{R}^p$ is the measurement signal, with $v(k) \in \mathbf{R}^p$ being white zero mean noise having unit covariance matrix. The vector $z(k) \in \mathbf{R}^l$ is to be estimated. The nominal model is

$$
x_0(k+1) = \mathbf{A}_0 x_0(k) + \mathbf{B}_0 e(k) \tag{5.70}
$$

$$
y_0(k) = \mathbf{C}(k)x_0(k) + \mathbf{M}_0 v(k) \quad .
$$

We assume the matrices $\Delta\mathbf{A}, \Delta\mathbf{B}$ and $\Delta\mathbf{M}$ to be known functions of the unknown parameter vector ρ. The vector ρ may, for example, contain uncertain physical parameters of a continuous-time model. The robust estimation of $z(k)$ will be founded on the following assumptions:

- The uncertain parameters ρ are treated as if they were stochastic variables. Their realizations represent particular models in the set.
- All models (5.69) are assumed internally and BIBO stable. In other words, the eigenvalues of $\mathbf{A}_0 + \Delta\mathbf{A}(\rho)$ are located in $|z| < 1$ for all admissible ρ, and the elements of $\mathbf{C}(k)$ are bounded.
- The effect of ρ on the set of models (5.69) is described by known covariances between elements of the matrices $\Delta\mathbf{A}, \Delta\mathbf{B}$ and $\Delta\mathbf{M}$. The nominal model (5.70) is selected as the average model of the set; $\Delta\mathbf{A}, \Delta\mathbf{B}$ and $\Delta\mathbf{M}$ have mean value zero.

The aim is to obtain an approximate modified Kalman estimator which minimizes the criterion (5.8), i.e. the average, over the set of models, of the mean square estimation error.

In order to apply the framework of section 5.4, a model with uncertainties in the system matrix must be approximated by a new model, in which the uncertainties appear only in the input matrix. One way of doing this is to use series expansion, based on the denominator terms of a transfer function representation of (5.69). See section 5 of [34] or chapter 3 of [37]. Here we shall, instead, perform the expansion directly in the state space representation, by augmenting the nominal state vector $x_0(k+1)$ by additional vectors. These vectors correspond to sets of perturbations caused by the different powers of $\Delta\mathbf{A}$ occurring in a series expansion.

Introduce the set of possible state trajectory variations $\delta x(k+1)$, caused by $\Delta\mathbf{A}(\rho)$ and $\Delta\mathbf{B}(\rho)$, such that

$$x(k+1) = x_0(k+1) + \delta x(k+1) \tag{5.71}$$

where x_0 is the nominal state vector given by (5.70) and

$$\delta x(k+1) = \Delta\mathbf{A}x_0(k) + \mathbf{A}_0\delta x(k) + \Delta\mathbf{B}e(k) + \Delta\mathbf{A}\delta x(k)\ . \tag{5.72}$$

The equality (5.72) is an exact expression derived from (5.69). We now express $\delta x(k)$ in (5.72) as

$$\delta x(k) = x_1(k) + x_2(k) + \ldots + x_d(k)$$

for a given expansion order d. The terms $x_m(k), m < d$ are defined as being affected by powers of $\Delta\mathbf{A}$ up to m only. Specifying state equations for the additional state vectors, $x_m(k)$, is now a matter of pairing terms $x_m(k+1)$ on the left-hand side of (5.72) with appropriate terms on the right-hand side. The choice

$$\begin{aligned} x_1(k+1) &= \Delta\mathbf{A}x_0(k) + \mathbf{A}_0x_1(k) + \Delta\mathbf{B}e(k) \\ x_2(k+1) &= \mathbf{A}_0x_2(k) + \Delta\mathbf{A}x_1(k) \\ &\ \vdots \\ x_d(k+1) &= \mathbf{A}_0x_d(k) + \Delta\mathbf{A}x_{d-1}(k) + \Delta\mathbf{A}x_d(k) \end{aligned}$$

yields the augmented state space model

$$\begin{bmatrix} x_0(k+1) \\ x_1(k+1) \\ \vdots \\ x_d(k+1) \end{bmatrix} = \underbrace{\begin{bmatrix} \mathbf{A}_0 & \mathbf{0} & \ldots & \mathbf{0} \\ \Delta\mathbf{A} & \mathbf{A}_0 & & \vdots \\ & \ddots & \ddots & \mathbf{0} \\ \mathbf{0} & & \Delta\mathbf{A} & \mathbf{A}_0 + \Delta\mathbf{A} \end{bmatrix}}_{\bar{\mathbf{A}}} \begin{bmatrix} x_0(k) \\ x_1(k) \\ \vdots \\ x_d(k) \end{bmatrix} + \begin{bmatrix} \mathbf{B}_0 \\ \Delta\mathbf{B} \\ \mathbf{0} \\ \vdots \\ \mathbf{0} \end{bmatrix} e(k)$$

$$x(k) = x_0(k) + x_1(k) + \ldots x_d(k). \tag{5.73}$$

So far, no approximation has been made. The term $\Delta\mathbf{A}$ in the lower right corner of $\bar{\mathbf{A}}$ represents the effect of $(d+1)$'th and higher powers of $\Delta\mathbf{A}$ on $x(k)$. We *neglect this term* from now on, and thus discard terms of higher order than d. The characteristic polynomial is then given by

$$\det(z\mathbf{I}_{(d+1)n} - \bar{\mathbf{A}}) = \det(z\mathbf{I}_n - \mathbf{A}_0)^{d+1}$$

so perturbations will no longer affect any transfer function denominator.

To keep the notation simple, we shall in the sequel specialize to first order expansions. By transforming to an input-output model in forward shift operator form, we then obtain

$$x(k) = [\ \mathbf{I}_n \ \ \mathbf{I}_n\ ] \begin{bmatrix} q\mathbf{I}_n - \mathbf{A}_0 & \mathbf{0} \\ -\Delta\mathbf{A} & q\mathbf{I}_n - \mathbf{A}_0 \end{bmatrix}^{-1} \begin{bmatrix} \mathbf{B}_0 \\ \Delta\mathbf{B} \end{bmatrix} e(k)\ . \tag{5.74}$$

Now, introduce the $n|n$ polynomial matrices $\widetilde{\boldsymbol{D}}(q)$ and $\widetilde{\Delta\boldsymbol{A}}(q)$ as a solution to the coprime factorization

$$\widetilde{\boldsymbol{D}}(q)\Delta\mathbf{A} = \widetilde{\Delta\boldsymbol{A}}(q)(q\mathbf{I}_n - \mathbf{A}_0) \tag{5.75}$$

where $\widetilde{\boldsymbol{D}}(q)$ should contain no stochastic coefficients[20] and $\deg\det\widetilde{\boldsymbol{D}} = n$. Then, (5.74) can be written as a left matrix fraction description

$$\begin{aligned} x(k) &= (q\mathbf{I}_n - \mathbf{A}_0)^{-1}(\mathbf{B}_0 + \Delta\mathbf{B} + \Delta\mathbf{A}(q\mathbf{I}_n - \mathbf{A}_0)^{-1}\mathbf{B}_0)e(k) \\ &= (q\mathbf{I}_n - \mathbf{A}_0)^{-1}\widetilde{\boldsymbol{D}}^{-1}(q)(\widetilde{\boldsymbol{D}}(q)(\mathbf{B}_0 + \Delta\mathbf{B}) + \widetilde{\Delta\boldsymbol{A}}(q)\mathbf{B}_0)e(k) \\ &\triangleq \boldsymbol{D}^{-1}(q)\boldsymbol{C}(q)e(k) \end{aligned} \tag{5.76}$$

where $\deg\det\boldsymbol{D}(q) = 2n$. This representation is of the form (5.35), with

$$\begin{aligned} \mathcal{F}_o &= (q\mathbf{I}_n - \mathbf{A}_0)^{-1}\mathbf{B}_0 \\ \Delta\mathcal{F} &= (q\mathbf{I}_n - \mathbf{A}_0)^{-1}\widetilde{\boldsymbol{D}}^{-1}(q)(\widetilde{\boldsymbol{D}}(q)\Delta\mathbf{B} + \widetilde{\Delta\boldsymbol{A}}(q)\mathbf{B}_0). \end{aligned}$$

It can easily be converted to backward shift operator form. The input-output representation could be complemented by stochastic additive error models which represent unmodelled higher-order dynamics.

5.5.2 The robust linear state estimator

The model (5.76) can form the basis of a robust Wiener filter design, as described in section 5.4, in which also uncertainty in a time-invariant matrix $\mathbf{C}$ of (5.69) can be handled. If we prefer to work with state space estimators, the set of models (5.76) can be realized on observable state space form [55], with $2n$ states:

[20] This step is superfluous if the original system is realized in diagonal form. As explained in [37], the factorization actually corresponds to a polynomial matrix spectral factorization. A d'th order expansion will require d factorizations of the type (5.75).

$$\xi(k+1) = \mathbf{F}\xi(k) + (\mathbf{G}_0 + \Delta\mathbf{G})e(k) \; ; \; x(k) = \mathbf{H}\xi(k) \tag{5.77}$$

where $\Delta\mathbf{G}$ has zero mean. Note that since the denominator matrix $\boldsymbol{D}(q)$ of (5.76) contains no uncertain coefficients, neither will $\mathbf{F}$ in (5.77). The covariance matrix of the uncertain elements of $\Delta\mathbf{G}$ in (5.77) can be calculated straightforwardly from the covariances of the elements of $\Delta\mathbf{A}$ and $\Delta\mathbf{B}$ in (5.69).

Let us restrict attention to linear estimators[21]. The model (5.77) can now be utilized for designing robust Kalman predictors, filters and smoothers, using well-known techniques [56]. For example, it can be shown, see [37] chapter 7, that if $\Delta\mathbf{M}$ is independent of $\Delta\mathbf{A}, \Delta\mathbf{B}$ and if $e(k)$ is uncorrelated to $v(s)$ for all k, s, then the one-step predictor minimizing (5.36) is given by

$$\begin{aligned}\hat{\xi}(k+1) &= \mathbf{F}\hat{\xi}(k) + \mathbf{K}(k)(y(k) - \mathbf{C}_1(k)\hat{\xi}(k)) \\ \hat{z}(k+1) &= \mathbf{L}\mathbf{H}\hat{\xi}(k+1)\end{aligned}$$

with $\mathbf{K}(k)$ calculated from

$$\begin{aligned}\mathbf{K}(k) &= \mathbf{F}\mathbf{P}(k)\mathbf{C}_1^T(k)(\mathbf{C}_1(k)\mathbf{P}(k)\mathbf{C}_1^T(k) + \mathbf{R}_2)^{-1} \\ \mathbf{P}(k+1) &= \mathbf{F}\mathbf{P}(k)\mathbf{F}^T + \mathbf{R}_1 \\ &\quad -\mathbf{F}\mathbf{P}(k)\mathbf{C}_1^T(k)(\mathbf{C}_1(k)\mathbf{P}(k)\mathbf{C}_1^T(k) + \mathbf{R}_2)^{-1}\mathbf{C}_1(k)\mathbf{P}(k)\mathbf{F}^T\end{aligned} \tag{5.78}$$

where $\mathbf{C}_1(k) \triangleq \mathbf{C}(k)\mathbf{H}$, with initial values

$$\begin{aligned}\hat{\xi}(0) &= \bar{E}E\xi(0) \triangleq \xi^0 \\ \mathbf{P}(0) &= \bar{E}E(\xi(0) - \xi^0)(\xi(0) - \xi^0)^T \; .\end{aligned}$$

The robustifying modified covariance matrices are given by

$$\begin{aligned}\mathbf{R}_1 &= \mathbf{G}_0\mathbf{G}_0^T + \bar{\mathbf{E}}(\Delta\mathbf{G}\Delta\mathbf{G}^T) \\ \mathbf{R}_2 &= \mathbf{M}_0\mathbf{M}_0^T + \bar{\mathbf{E}}(\Delta\mathbf{M}\Delta\mathbf{M}^T)\end{aligned} \tag{5.79}$$

with $\Delta\mathbf{G}$ and $\Delta\mathbf{M}$, both with zero mean, introduced in (5.77) and (5.69), respectively.

Numerical illustrations can be found in [36] and in chapter 7 of [37].

5.6 Parameter tracking

In digital communication and, in particular, in digital mobile radio applications, the problem of adjusting filters to a rapidly time-varying dynamics is encountered. If models cannot be re-adjusted with sufficient speed and accuracy, sophisticated tools for model-based filtering will be of little utility.

[21]The variable $\Delta Ge(k)$ will, in general, not be Gaussian. Robustified Kalman estimators are, however, the optimal *linear* estimators for arbitrary noise and uncertainty distributions.

Models of time-varying communication channels, can often be represented in a linear regression form

$$y(k) = \varphi^*(k)\hat{\theta}(k) + \varepsilon(k) \tag{5.80}$$

where $\varphi^*(k)$ is a regressor matrix containing known signals, $\hat{\theta}(k)$ is a column parameter vector and $\varepsilon(k)$ is a vector of residuals.[22] If the model structure is correct, it becomes meaningful to formulate the problem of model adjustment as a problem of following (tracking) a vector $\theta(k)$ which parameterizes a system

$$y(k) = \varphi^*(k)\theta(k) + w(k) \tag{5.81}$$

where the disturbance $w(k)$ is assumed independent of both $\varphi^*(k)$ and $\theta(k)$.

Parameter tracking is normally performed by utilizing the LMS algorithm or exponentially windowed RLS [7, 8, 10, 11, 103]. The initial convergence of the Newton-based RLS algorithm is much faster than that of the gradient-based LMS, but their performance when tracking continuously changing parameters is about equal [108]. In general, both LMS and windowed RLS algorithms turn out to be structurally mismatched to the tracking problem at hand [50, 54].

If the dynamics of the elements of $\theta(k)$ and the noise have known second order moments, then a Kalman-based adaptation algorithm represents the optimal scheme, which attains the best tracking performance. In many problems, the dynamic properties of time-varying parameters are approximately known. They are, however, rarely known exactly. Methods for *robust* model-based parameter tracking would therefore be of interest. The robust Kalman scheme outlined in section 5.5 can be used in such cases.

Let $z(k) = \mathbf{L}x(k)$ in (5.69) represent the parameter vector $\theta(k)$, to be tracked. The state update equation will then represent the set of possible parameter dynamics, whereas the measurement equation corresponds to the linear regression (5.81)

$$y(k) = \varphi^*(k)\theta(k) + w(k) = \varphi^*(k)\mathbf{L}x(k) + (\mathbf{M}_o + \Delta\mathbf{M})v(k) \ . \tag{5.82}$$

Thus, $\mathbf{C}(k) = \varphi^*(k)\mathbf{L}$. The cautious Kalman predictor outlined in section 5.5.2 can now be used directly for tracking the parameters of linear regression models.

Kalman-based adaptation algorithms require a Riccati update step at each sample, and they will therefore have an unacceptable complexity in many high-speed applications. Algorithms of lower computational complexity, which can still take *a priori* knowledge about the statistical properties of the parameter variations into account, have therefore been sought [50, 109, 110].

By using the polynomial approach, a family of algorithms with low computational complexity, but still close to optimal performance, has been derived by Lindbom in [54]. The algorithms within this family have a *constant* adaptation

[22] In an adaptive equalizer which runs in decision-directed mode, the elements of $\varphi^*(k)$ contain estimates of transmitted symbols. These estimates are obtained from an equalizer or Viterbi detector which, in its turn, is adjusted based on the the changing channel model.

gain, and will require no Riccati update. They are characterized by the general recursive structure

$$\varepsilon(k) = y(k) - \boldsymbol{\varphi}^*(k)\hat{\theta}(k|k-1) \tag{5.83}$$
$$\hat{\theta}(k+m|k) = \boldsymbol{F}(q^{-1})\hat{\theta}(k+m-1|k-1) + \boldsymbol{\mathcal{G}}_m(q^{-1})\varphi(k)\varepsilon(k)$$

which includes predictors, filters and fixed-lag smoothers. Above, the polynomial matrix $\boldsymbol{F}$ is obtained directly from the assumed stochastic model for the parameter variations, while the stable rational matrix $\boldsymbol{\mathcal{G}}_m$ is obtained via spectral factorization and Diophantine design equations, which need to be re-computed only if the model describing the parameter variations changes. In simple but useful special cases, no design equations at all need to be solved. The complexity of the algorithm (5.83) will increase only linearly with the dimension of $\hat{\theta}(k)$, if the regressor correlation matrix is known.

The thesis [54] also includes an analysis of tracking algorithms and a case study, describing adaptive equalization in the US D-AMPS standard (IS-54). For these two purposes, as well as for the design of constant-gain tracking algorithms, the use of a polynomial equations approach has turned out to be very fruitful.

5.7 Acknowledgement

We thank Claes Tidestav for many useful comments on the manuscript for the present chapter.

5.8 References

[1] Kučera V. Discrete Linear Control. The Polynomial Equation Approach. Wiley, Chichester, 1979

[2] Kučera V. Analysis and Design of Discrete Linear Control Systems. Academia, Prague and Prentice Hall International, London, 1991

[3] Kwakernaak H, Sivan R. Linear Optimal Control Systems. Wiley, New York, 1972.

[4] Öhrn K, Ahlén A, Sternad M. A probabilistic approach to multivariable robust filtering and open-loop control. IEEE Transactions on Automatic Control,1995; 40:405-418

[5] Tidestav C, Ahlén A, Sternad M. Narrowband and broadband multiuser detection using a multivariable DFE. IEEE PIMRC'95. Toronto, 1995, pp 732-736

[6] Ahlén A, Sternad M. Derivation and design of Wiener filters using polynomial equations. In: Leondes CT (ed) Control and Dynamic Systems, vol 64: Stochastic Techniques in Digital Signal Processing Systems. Academic Press, New York, 1994

[7] Honig ML, Messerschmidt DG. Adaptive Filters: Structures, Algorithms and Applications. Kluwer Academic, Boston, 1984

[8] Haykin S. Adaptive Filter Theory. Prentice-Hall, Englewood Cliffs, Third ed. 1996.

[9] Oppenheim AV, Shafer RW. Digital Signal Processing. Prentice-Hall, Englewood Cliffs, 1975

[10] Widrow B, Stearns SD. Adaptive Signal Processing. Prentice Hall, Englewood Cliffs, 1985

[11] Treichler JR, Johnson Jr CR, Larimore MG. Theory and Design of Adaptive Filters. Wiley, New York, 1987

[12] Goodwin GC, Gevers M, Ninness B. Quantifying the error in estimated transfer functions with application to model order selection. IEEE Transactions on Automatic Control. 1992; 27:913-928

[13] Bode HW, Shannon CE. A simplified derivation of linear least square smoothing and prediction theory. Proceedings of the I.R.E. 1950; 38:417-425

[14] Kailath T. Lectures on Wiener and Kalman Filtering. Springer, Wien, 1981

[15] Cadzow JA. Foundations of Digital Signal Processing and Data Analysis. Macmillan, New York, 1987

[16] Ahlén A, Sternad M. Optimal Filtering Problems. In: Hunt K (ed) Polynomial Methods in Optimal Control and Filtering, chapter 5. Control Engineering Series, Peter Peregrinus, London, 1993

[17] Ahlén A, Sternad M. Wiener filter design using polynomial equations. IEEE Transactions on Signal Processing. 1991; 39:2387-2399

[18] Grimble MJ. Polynomial systems approach to optimal linear filtering and prediction. International Journal of Control. 1985; 41:1545-1564

[19] Leondes CT, Pearson JO. A minimax filter for systems with large plant uncertainties. IEEE Transactions on Automatic Control. 1972; 17:266-268

[20] Kassam SA, Lim TL. Robust Wiener filters. Journal of The Franklin Institute. 1977; 171-185

[21] Kassam SA, Poor HV. Robust techniques for signal processing: a survey. Proceedings of the IEEE. 1985; 73:433-481

[22] Francis B, Zames G. On $\mathcal{H}_\infty$ optimal sensitivity theory for SISO feedback systems. IEEE Transactions on Automatic Control, 1984;29:9-16

[23] Doyle J, Glover K, Khargonekar P, Francis B. State space solutions to standard $\mathcal{H}_\infty$ and $\mathcal{H}_\infty$ control problems. IEEE Transactions on Automatic Control, 1989;34:831-847

[24] Grimble MJ, ElSayed A. Solution to the $\mathcal{H}_\infty$ optimal linear filtering problem for discrete-time systems. IEEE Transactions on Signal Processing. 1990; 38:1092-1104

[25] Shaked U, Theodor Y. A frequency domain approach to the problems of $\mathcal{H}_\infty$-minimum error state estimation and deconvolution. IEEE Transactions on Signal Processing. 1992; 40:3001-3011.

[26] Theodor Y, Shaked U, de Souza C E. A game theory approach to robust discrete-time $\mathcal{H}_\infty$-estimation. IEEE Transactions on Signal Processing. 1994; 42:1486-1495.

[27] Bolzern P, Colaneri P, De Nicolao G. Optimal robust filtering for linear systems subject to time-varying parameter perturbations In: Proceedings of the IEEE 32nd Conference on Decision and Control. San Antonio, 1993, pp 1018-1023

[28] Xie L, Soh YC. Robust Kalman filtering for uncertain systems. Systems & Control Letters. 1994; 22:123-129

[29] Xie L, Soh YC, de Souza CE. Robust Kalman filtering for uncertain discrete-time systems. IEEE Transactions on Automatic Control. 1994; 39:1310-1314

[30] Shaked U, de Souza C. Robust minimum variance filtering. IEEE Transactions on Signal Processing. 1995; 43:2474-2483

[31] Chung RC, Bélanger PR. Minimum-sensitivity filter for linear time-invariant stochastic systems with uncertain parameters. IEEE Transactions on Automatic Control. 1976; 21:98-100

[32] Speyer JL, Gustafson DE. An approximation method for estimation in linear systems with parameter uncertainty. IEEE Transactions on Automatic Control. 1975; 20:354-359

[33] Grimble MJ. Wiener and Kalman filters for systems with random parameters. IEEE Transactions on Automatic Control. 1984; 29:552-554

[34] Sternad M, Ahlén A. Robust filtering and feedforward control based on probabilistic descriptions of model errors. Automatica. 1993; 29:661-679

[35] Öhrn K, Ahlén A, Sternad M. A probabilistic approach to multivariable robust filtering, prediction and smoothing. In: Proceedings of the 32nd IEEE Conference on Decision and Control. San Antonio, 1993, pp 1227-1232

[36] Sternad M, Öhrn K, Ahlén A. Robust $\mathcal{H}_2$ filtering for structured uncertainty: the performance of probabilistic and minimax schemes. European Control Conference. Rome, 1995, pp 87-92

[37] Öhrn K. Design of Multivariable Cautious Wiener Filters: A Probabilistic Approach. PhD thesis, Uppsala University, 1996

[38] Lindskog E, Sternad M, Ahlén A. Designing decision feedback equalizers to be robust with respect to channel time variations. Nordic Radio Society Symposium on interference resistant radio and radar. Uppsala, 1993

[39] Guo L, Ljung L. The role of model validation for assessing the size of the unmodelled dynamics. In: Proceedings of the 33rd IEEE Conference of Decision and Control. Lake Buena Vista, 1994, pp 3894-3899

[40] Gevers M. Towards a joint design of identification and control. In Trentelman HL and Willems JC ed. Essays on Control: Perspectives in the Theory and itn Applications. Birkhauser, 1993,111-151.

[41] Kosut R L, Goodwin G C and Polis, eds. Special issue on system identification for robust control design. IEEE Transactions on Automatic Control, 1992;37(7)

[42] Bigi S. On the use of system identification for design purposes and parameter estimation. Licentiate Thesis, Uppsala University, 1995

[43] Lee W. Mobile Communications Engineering. McGraw Hill, New York, 1982

[44] Lee W. Mobile Cellular Telecommunications Systems McGraw Hill, New York, 1989

[45] Proakis JG. Digital Communications. McGraw Hill, New York, 3rd ed. 1995

[46] Lee EA, Messerschmitt DG. Digital Communication. Kluwer Academic Publisher, Boston, Second ed. 1994

[47] Lindskog E. On Equalization and Beamforming for Mobile Radio Applications. Licentiate Thesis, Uppsala University, 1995

[48] Lindskog E, Ahlén A, Sternad M. Combined spatial and temporal equalization using an adaptive antenna array and a decision feedback equalization scheme. In: Proceedings of the IEEE International Conference on Acoustics, Speech and Signal Processing. Detroit, 1995

[49] Lindskog E, Ahlén A, Sternad M. Spatio-temporal equalization for multipath environments in mobile radio application. In: Proceedings of the 45th IEEE Vehicular Technology Conference, Chicago, 1995, pp 399-403

[50] Benveniste A. Design of adaptive algorithms for tracking of time-varying systems. International Journal of Adaptive Control and Signal Processing. 1987; 2:3-29

[51] Kubin G. Adaptation in rapidly time-varying environments using coefficient filters. In: Proceedings of the IEEE International Conference on Acoustics, Speech and Signal Processing, vol 3. Toronto, 1991,pp 2097-2100

[52] Lindbom L. Simplified Kalman estimation of fading mobile radio channels: high performance at LMS computational load. In: Proceedings of the IEEE International Conference on Acoustics, Speech and Signal Processing, vol 3. Minneapolis, 1993, pp 352-355

[53] Lindbom L. Adaptive equalizers for fading mobile radio channels. Licenciate Thesis, Uppsala University, 1992

[54] Lindbom L. A Wiener filtering approach to the design of tracking algorithms: with applications in mobile radio communications. PhD Thesis, Uppsala University, 1995

[55] Kailath T. Linear Systems. Prentice-Hall, Englewood Cliffs, New Jersey, 1980

[56] Anderson BDO, Moore JB. Optimal Filtering. Prentice-Hall, Englewood Cliffs, 1979

[57] Ahlén A, Sternad M. Optimal deconvolution based on polynomial methods. IEEE Transactions on Acoustics, Speech and Signal Processing. 1989; 37:217-226

[58] Carlsson B, Ahlén A, Sternad M, Optimal differentiation based on stochastic signal models. IEEE Transactions on Signal Processing. 1991; 39:341-353

[59] Carlsson B, Sternad M, Ahlén A. Digital differentiation of noisy data measured through a dynamic system. IEEE Transactions on Signal Processing. 1992; 40:218-221

[60] Moir TJ. A polynomial approach to optimal and adaptive filtering with application to speech enhancement. IEEE Transactions on Signal Processing. 1991; 39:1221-1224

[61] Roberts AP, Newmann MM. Polynomial approach to Wiener filtering. International Journal of Control. 1988; 47:681-696

[62] Grimble MJ. Time-varying polynomial systems approach to multichannel optimal linear filtering. Proceedings of the American Control Conference. WA6-10:45, Boston, 1985, pp 168-174

[63] Fitch SM, Kurz L. Recursive equalization in data transmission - A design procedure and performance evaluation. IEEE Transactions on Communication. 1975; 23:546-550

[64] Qureshi SUH. Adaptive equalization. Proceedings of the IEEE. 1985; 73:1349-1387

[65] Nelson PA, Hamada H and Elliot SJ. Adaptive inverse filters for stereophonic sound reproduction. IEEE Transactions on Signal Processing. 1992; 40:1621-1632

[66] Mendel JM. Optimal Seismic Deconvolution. An Estimation-Based Approach. Academic Press, New York, 1983

[67] Ahlén A, Sternad M. Filter design via inner-outer factorization: Comments on "Optimal deconvolution filter design based on orthogonal principle". Signal Processing, 1994; 35:51-58

[68] Kučera V. Factorization of rational spectral matrices: a survey. In: Preprints of IEE Control'91. Edinburgh, 1991, pp 1074-1078

[69] Ježek J, Kučera V. Efficient algorithm for matrix spectral factorization. Automatica. 1985; 21:663-669

[70] Demeure CJ, Mullis CT. A Newton-Raphson method for moving average spectral factorization using the Euclid algorithm. IEEE Transactions on Acoustics, Speech and Signal Processing. 1990; 38:1697-1709

[71] Deng ZL. White-noise filter and smoother with application to seismic data deconvolution. In: Preprints 7th IFAC/IFORS Symp. Identification Syst. Parameter Estimation. York, 1985, pp 621-624

[72] Deng Z-L, Zhang H-S, Liu S-J, Zhou L. Optimal and self-tuning white noise estimator with applications to deconvolution and filtering problems. Automatica. 1996; 32:199-216

[73] Ahlén A, Sternad M. Adaptive input estimation. IFAC Symposium ACASP-89, Adaptive Systems in Control and Signal Processing. Glasgow, 1989, pp 631-636

[74] Ahlén A, Sternad M. Adaptive deconvolution based on spectral decomposition. SPIE Annual Symposium. vol 1565. San Diego, 1991, pp 130-142

[75] Ahlén A. Identifiability of the deconvolution problem. Automatica. 1990; 26:177-181

[76] Nikias CL, Petropulu AP. Higher Order Spectral Analysis: A Nonlinear Signal Processing Framework. Prentice Hall, Englewood Cliffs, 1993.

[77] Tong L, Xu G, Kailath T. Fast blind equalization via antenna arrays. In: Proceedings of the IEEE International Conference on Acoustics, Speech and Signal Processing. Minneapolis, 1993, 4:272-275.

[78] Moulines E, Duhamel P, Cardoso J-F, Mayrargue S. Subspace methods for the blind identification of multichannel FIR filters. IEEE Transactions on Signal Processing. 1995; 43(2):516-525.

[79] Bernhardsson B, Sternad M. Feedforward control is dual to deconvolution. International Journal of Control. 1993; 47:393-405

[80] Sternad M, Ahlén A. The structure and design of realizable decision feedback equalizers for IIR-channels with coloured noise. IEEE Transactions on Information Theory. 1990; 36:848-858

[81] Tidestav C. Optimum diversity combining in multi-user digital mobile telephone systems. Master's Thesis, Uppsala University,1993

[82] Forney Jr GD. The Viterbi algorithm. In: Proceedings of IEEE. 1973; 61:268-278

[83] Belfiore CA, Park JH. Decision feedback equalization. Proceedings of the IEEE. 1979;67:1143-1156

[84] Monsen P. Feedback equalization for fading dispersive channels. IEEE Transactions on Information Theory. 1971;17:56-64

[85] Salz J. Optimum mean-square decision feedback equalization. Bell System Technical Journal. 1973; 52:1341-1374

[86] Balaban P, Salz J. Optimum diversity combining and equalization in digital data transmission with applications to cellular mobile radio - part I: theoretical considerations. IEEE Transactions on Communications 1992; 40:885-894

[87] Lo NWK, Falconer DD, Sheikh AUH. Adaptive equalizer MSE performance in the presence of multipath fading, interference and noise. In: Proceesings of the IEEE Vehicular Technology Conference, Rosemont, 1995, 409-413

[88] Sternad M, Ahlén A, Lindskog E. Robust decision feedback equalization. In: Proceedings of the IEEE International Conference on Acoustics, Speech and Signal Processing, Minneapolis, 1993;3:555-558.

[89] Winters J. On the capacity of radio communication systems with diversity in a Rayleigh fading environment. IEEE Journal on Selected Areas in Communication. 1987; 5(5):871-878

[90] Falconer DD, Abdulrahman M, Lo NWK, Petersen BR, Sheikh AUH. Advances in equalization and diversity for portable wireless systems. Digital Signal Processing. 1993; 3(3):148-162

[91] Gilhousen KS, Jacobs IW, Padovani R, Viterbi A, Weaver LA, Wheatly CE. On the capacity of a cellular CDMA system. IEEE Transactions on Vehicular Technology. 1991; 40:303-312

[92] Verdú S. Optimum Multi-user Signal Detection. PhD thesis, University of Illinois, 1984.

[93] Goodwin GC. Salgado ME. A stochastic embedding approach for quantifying uncertainty in the estimation of restricted complexity models. International Journal of Adaptive Control and Signal Processing. 1989; 3:333-356

[94] Grimble MJ. Generalized Wiener and Kalman filters for uncertain parameters. In: Proceedings of the 21th IEEE Conference on Decision and Control. Orlando, 1982, pp 221-227

[95] D'Appolito JA, Hutchinson CE. A minimax approach to the design of low sensitivity state estimators. Automatica. 1972; 8:599-608

[96] Martin CJ, Mintz M. Robust filtering and prediction for linear systems with uncertain dynamics: a game-theoretic approach. IEEE Transactions on Automatic Control. 1983; 28:888-896

[97] Moustakides G. Kassam SA. Robust Wiener filters for random signals in correlated noise. IEEE Transactions on Information Theory. 1983; 29:614-619

[98] Poor HV. On robust Wiener filtering. IEEE Transactions on Automatic Control. 1980; 25:531-536

[99] Vastola KS, Poor HV. Robust Wiener-Kolmogorov theory. IEEE Transactions on Information Theory. 1984; 30:316-327

[100] Petersen IR, McFarlane DC. Robust state estimation for uncertain systems. In: Proceedings of the 30th Conference on Decision and Control. Brighton, 1991, pp 2630-2631

[101] Haddad WF, Bernstein DS. Robust, reduced-order, nonstrictly proper state estimation via the optimal projection equations with guaranteed cost bounds. IEEE Transactions on Automatic Control. 1988; 33:591-595

[102] Haddad WF, Bernstein DS. The optimal projection equations for reduced-order discrete-time state estimation for linear systems with multiplicative noise. System & Control Letters. 1987; 8:381-388

[103] Ljung L, Söderström T. Theory and Practice of Recursive Identification. MIT Press, Cambridge, 1983

[104] Ljung L. System Identification: Theory for the User. Prentice Hall, Englewood Cliffs, 1987

[105] Wahlberg B. System identification using Laguerre models. IEEE Transactions on Automatic Control. 1991; 36:551-562

[106] P. Lindskog and B. Walhberg, Applications of Kautz models in system identification. In: Proceedings of the IFAC World Congress, Sydney, 1993(5):309-312

[107] Hakvoort R. System identification for robust process control. PhD Thesis, Delft University of Technology, The Netherlands, 1994.

[108] Widrow B, Walach E. On the statistical efficiency of the LMS algorithm with nonstationary inputs. IEEE Transactions on Information Theory. 1984; 30:211-221

[109] Benveniste A, Métivier M, Priouret P. Adaptive Algorithms and Stochastic Approximations. Springer, Berlin, 1990

[110] Clark AP. Adaptive Detectors for Digital Modems. Pentech Press, London, 1989

6

Polynomial Solution of H_2 and H_∞ Optimal Control Problems with Application to Coordinate Measuring Machines

M. J. Grimble and M. R. Katebi

6.1 Abstract

The H_2 and H_∞ optimal control problems are considered for the case where the output to be controlled is different from the signal available for feedback. To improve the tracking properties of the controller, a two degree-of-freedom structure is employed. This has the advantage of allowing a lower optimal tracking cost to be achieved. The solution involves polynomial spectral-factorization and diophantine-like equations.

The general system model employed enables a wide class of industrial problems to be considered. The proposed technique is applied to the position control of a coordinate measuring machine (CMM) which is a very high accuracy application.

6.2 Introduction

In most conventional control problems the output to be controlled is also measured and used for feedback control purposes. However, there are also applications where it is physically impossible to measure the signal which must follow a given reference trajectory. This control problem is referred to as inferential control, since the output to be controlled must be inferred from the measured output signal.

The Linear Quadratic Gaussian (LQG) and H_∞ solutions of inferential control problems, when the plant is represented in polynomial form and the controlled output must follow a reference trajectory, have important applications. Different control design methods can of course be applied to the inferential control problem but LQG and H_∞ controllers are particularly relevant when robust self- tuning features may be needed [1]. The continuous-time problem has been discussed in [2] from a Wiener-Hopf transfer-function solution viewpoint. This solution is not so practical for use in adaptive control where polynomial based algorithms are normally employed.

Recently a *standard system* model description has been introduced (described in chapter 1) which encompasses the above class of inferential control problems. Unusual controller structures have also been employed providing advantages in certain tracking problems. The background to these developments is introduced in the next two subsections.

The proposed technique is applied to the position control of a Coordinate Measuring Machine (CMM). CMMs are widely used in manufacturing processes for obtaining rapid measurements of micrometer accuracy on machined products. The basic types of CMM are moving bridge gantry and horizontal arm. A horizontal arm CMM, as shown in figure 6.1, is considered in the following. The main components are the three orthogonal arms, represented by links 1, 2, and 3, for movement in the Cartesian coordinate system and a two degree-of-freedom (DOF) probe attached at the end of the horizontal arm, represented by links 4 and 5, which rotate about the vertical and horizontal axes. Knowing the dimensions of the arms and the probe, the three arm positions and the probe orientation completely define the tip position of the probe in space relative to a given reference position.

The main control design objective is to accurately position the tip of the probe at a given point in a minimum time. However, speed and accuracy are two conflicting design requirements due to the existence of the interaction between the arms and the probe. For example, the movement of the arm introduces a force on the probc and causc it to deviate from its nominal angular position, thus resulting in inaccuracies. The feedback control problem is illustrated in one degree-of-freedom form in figure 6.2.

Unlike the conventional control design problems where the controlled variables are also the measured variables the probe positions which are to be controlled are not measurable. This problem is solved by formulating H_2 and H_∞ control design problems which contain the spectrum of the probe positions as terms in the cost functions. The resulting function is then minimised using a polynomial approach.

Simulation and experimental results are presented to demonstrate the effectiveness of the proposed control scheme. Proportional + Integral + Derivative (PID) control is also used for comparison purposes.

The chapter is organised as follows : H_2 control design is discussed in section

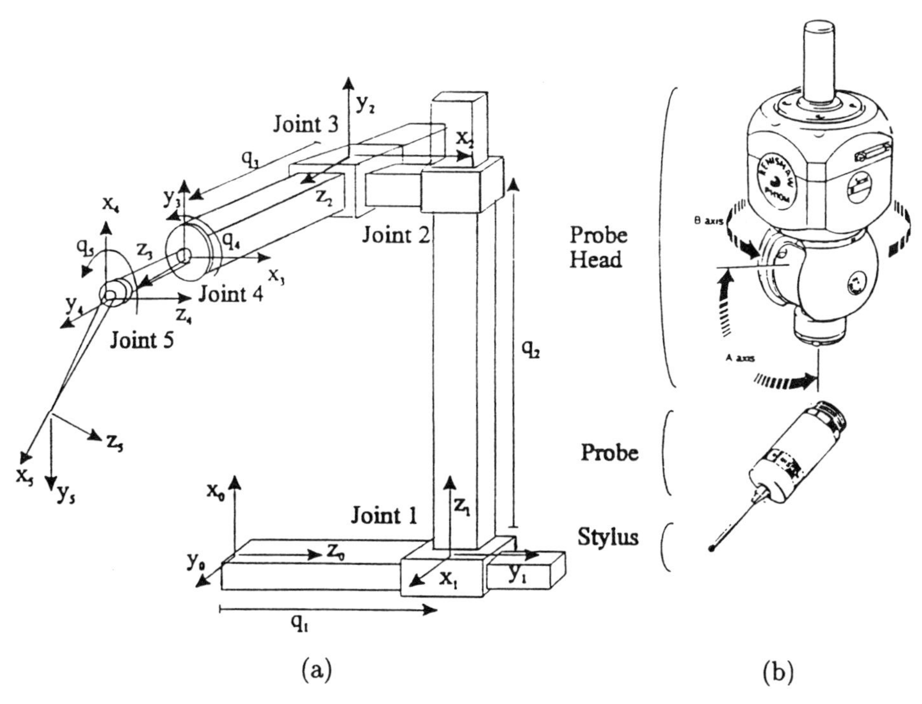

(a) (b)

Figure 6.1: (a) The coordinate measuring machine. (b) The probe.

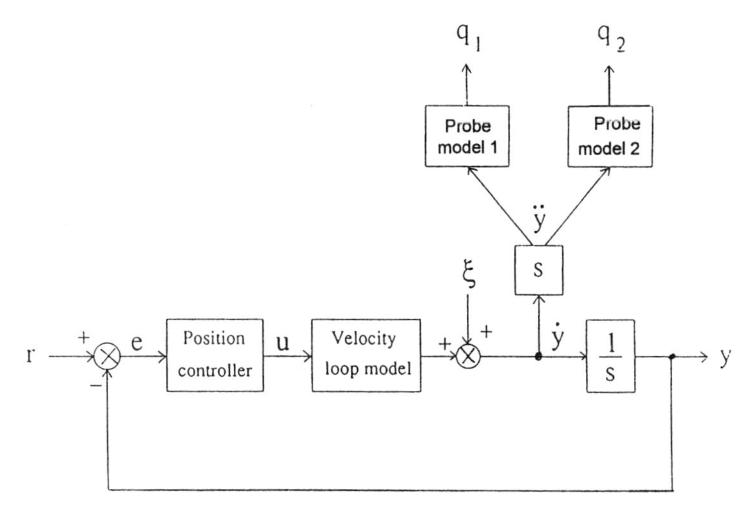

Figure 6.2: Position loop block diagram.

6.3. The H_∞ controller is derived in section 6.4. The CMM model is developed in section 6.5. Simulation and experimental results are presented in section 6.6. Finally, conclusions are drawn in section 6.6.

6.3 H_2 control design

Motivation for the use of the robust H_2 approach will now be provided. The CMM control problem is also introduced. The control design procedure is based on discrete polynomial system approach of Kucera [5] and Grimble [6] (a continuous-time approach gives almost identical results).

A two DOF feedback controllers structure is employed. It has the advantage of allowing a lower optimal cost to be achieved [7]. The solution involves polynomial spectral factorisations and the solution of diophantine equations (see the appendix for the derivation).

As there is little interaction between the arms of the CMM, the control of the three position loops is assumed to be independent, hence a single-input single-output system is considered here.

6.3.1 System model

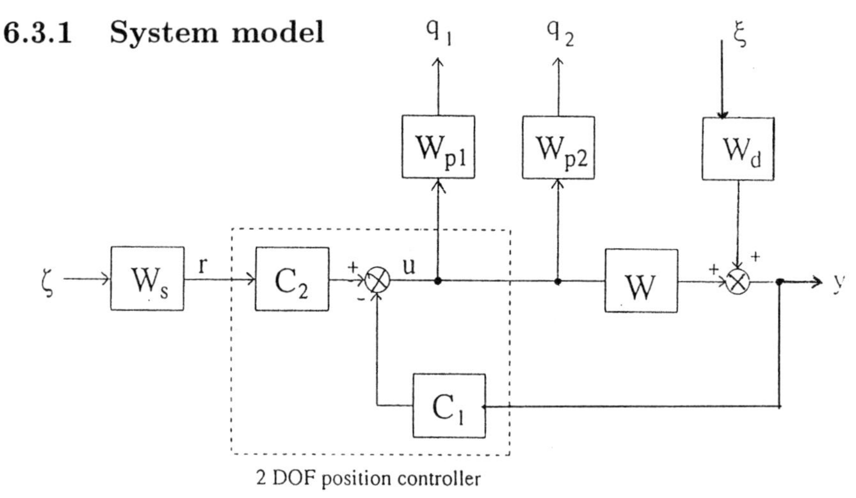

Figure 6.3: Two-degree-of-freedom controller structure.

Rearranging figure 6.2, and utilizing a two DOF controller structure the position loop is as shown in figure 6.3, where

ζ is the white noise with variance $\Phi_{\zeta\zeta}$,

ξ is the white process noise with variance $\Phi_{\xi\xi}$,

r is the stochastic reference signal with variance Φ_{rr} and $r = W_s\zeta$,

q_1, q_2 are the deviations of the two probe positions from nominal positions,

u is the control signal,

y is the arm position,

and the subsystems (all expressed in ratios of two polynomials in z^{-1}) are as follows:

$W = B/A$ is the plant,

$H_d = C/A$ is the process noise colouring filter,

$W_s = B_s/A_s$ is the stochastic reference model,

$W_{p1} = B_{p1}/A_{p1}$, $W_{p2} = B_{p2}/A_{p2}$ are the models describing the probe model and the velocity loop dynamics,

$C_1 = C_{1n}/C_{1d}$ is the feedback controller,

$C_2 = C_{2n}/C_{2d}$ is the reference controller.

The reason for modifying the probe models so that they are at the output of the controller will be apparent after the dynamic weights are introduced in the cost function. The control weighting function can in fact represent the probe signal costing, as described later.

6.3.2 Assumptions

1 The white noise sources ζ and ξ have zero means and are mutually independent.

2 The plant W is free of unstable hidden modes and the probe models W_{p1}, W_{p2} are asymptotically stable.

3 The plant W and the process noise colouring filter share the same common denominator polynomial A.

6.3.3 The H_2 cost function

In the time domain, the LQG cost function to be minimised is quadratic in the tracking error e (defined as $e = r - y$) and control signal u, this is given by

$$J = E\left[(H_q e)(t)^2 + (H_{r1} u)(t)^2 + (H_{r2} u)(t)^2\right] \tag{6.1}$$

where $E[\ldots]$ denotes the unconditional expectation operator, t is the discrete time index 1, 2, 3, ... ∞ and H_q, H_{r1}, H_{r2} are the weighting terms on the tracking error and control signals, respectively. The two control terms in the cost function describe the two probe positions in this problem.

Using Parseval's theorem, the cost function in the frequency domain is transformed as follows:

$$J = \frac{1}{2\pi j} \oint_{|z|=1} \left(Q\Phi_{ee}(z^{-1}) + R_1 \Phi_{uu}(z^{-1}) + R_2 \Phi_{uu}(z^{-1})\right) \frac{dz}{z} \tag{6.2}$$

where Φ_{ee} and Φ_{uu} denote the rational spectral densities of the error and control signals respectively.

The dynamic frequency dependent weights Q, R_1 and R_2 are expressed as:

$$Q = H_q^* H_q, \quad H_q = \frac{B_q}{A_q}$$

$$R_1 = H_{r1}^* H_{r1}, \quad H_{r1} = \frac{B_{r1}}{A_{r1}}$$

$$R_2 = H_{r2}^* H_{r2}, \quad H_{r2} = \frac{B_{r2}}{A_{r2}}$$

with H_q, H_{r1}, H_{r2} being minimum-phase, strictly stable transfer functions and $(\ldots)^*$ denoting the adjoint of a transfer function or polynomial. A constant variable term K will now be introduced on the error spectrum. The constant K is a positive tuning parameter which determines the relative importance of the dynamic weights Q, R_1 and R_2. The effect of the dynamic weighting is to allow the error and control variations to be penalised in different frequency ranges, whilst a constant weighting places equal importance at all frequencies.

Combining the two dynamic weights R_1 and R_2, the cost function of equation (6.2) has the more familiar LQG form:

$$J = \frac{1}{2\pi j} \oint_{|z|=1} \left(K^2 Q \Phi_{ee}(z^{-1}) + R\Phi_{uu}(z^{-1})\right) \frac{dz}{z} \tag{6.3}$$

where

$$R = R_1 + R_2 = H_r^* H_r \quad \text{and} \quad H_r = \frac{B_r}{A_r}$$

6.3.4 Dynamic weightings

In this section methods of choosing appropriate frequency responses for the dynamic weights Q, R_1 and R_2 will be discussed [13].

Choice of Q

As zero steady-state error is required, it is desirable to have infinite cost weighting on Φ_{ee} at zero frequency by introducing integral action. Since H_q has been defined to be strictly stable, the integral action can be approximated by choosing $Q = (B_q/A_q)^*(B_q/A_q)$ such that (B_q/A_q) behaves like an integrator in the low frequency region and a constant at high frequencies. Hence B_q/A_q takes on the form of a lag-lead transfer-function $(1 - \alpha z^{-1})/(1 - \rho z^{-1})$, where ρ is close to unity, giving a high cost in the low frequency region.

Choice of R_1 and R_2

The main objective of the control system is to control the probe positions. Although they are not measurable, the known probe models W_{p1} and W_{p2} can be used. In order to minimise the variations of the probe positions q_1 and q_2, the gains from the control signal point u to q_1 and q_2 are to be minimised. This can be achieved by introducing dynamic weightings R_1 and R_2 on the control signal such that $H_{r1} = B_{r1}/A_{r1}$ and $H_{r2} = B_{r2}/A_{r2}$ have the same frequency responses of W_{p1} and W_{p2} respectively. In other words the control signal is penalised more at the frequencies where the gains between u and q_1, q_2 are high. The weightings R_1 and R_2 are then combined to give a single dynamic- weighting term R on the control signal.

Closed-loop system bandwidth

The bandwidth of the closed-loop system is directly related to the plant W and the dynamic weights Q, R. A rule of thumb is that the frequency at which the gains of KH_qW and H_r are the same will be close to the closed-loop system bandwidth. Hence, by plotting the frequency-responses of KH_qW and H_r the closed-loop system bandwidth can be determined by noting the frequency at which the two curves cross.

6.3.5 The H_2 controller

The optimal LQG controller which minimises the cost function of equation (6.3) for the CMM control design problem is obtained by solving the following spectral-factor and diophantine equations.

Spectral factors

Control

$$D_c D_c^* = K^2 B^* B_q^* A_r^* B B_q A_r + A^* B_r^* A_q^* A B_r A_q \tag{6.4}$$

Filter

$$D_{f1} D_{f1}^* = \Phi_{\xi\xi} C^* C \tag{6.5}$$

$$D_{f2} D_{f2}^* = \Phi_{\zeta\zeta} B_s^* B_s \tag{6.6}$$

where the control spectral factor D_c and filter spectral factors D_{f1}, D_{f2} are strictly-Schur polynomials.

Diophantine equations

The following coupled diophantine equations, corresponding to the feedback controller, must be solved for the minimum-degree solution (H_0, G_0, F_0), with respect to F.

$$G D_c^* D_{f1}^* z^{-g} + F A_q A = \Phi_{\xi\xi} K^2 B_q^* C^* B^* A_r^* B_1 C z^{-g} \tag{6.7}$$

$$H D_c^* D_{f1}^* z^{-g} - F A_r B = \Phi_{\xi\xi} A^* A_q^* B_r^* C^* C B_r z^{-g} \tag{6.8}$$

It can be shown that if A is stable, only the solution of the first diophantine equations is required. The following single diophantine equation, corresponding to the reference controller, must also be solved for the minimum-degree solution (M_0, N_0), with respect to N.

$$MD_c^* z^{-g_1} + NA_s A_q = K^2 B_q^* B^* A_r^* B_q D_{f2} z^{-g_1} \tag{6.9}$$

where g, g_1 are the smallest positive integers which make the above diophantine equations polynomials in z^{-1}.

Optimal controller

Feedback controller

$$C_1 = \frac{C_{1n}}{C_{1d}} = \frac{G_0 A_r}{H_0 A_q} \tag{6.10}$$

Reference controller

$$C_2 = \frac{C_{2n}}{C_{2d}} = \frac{M_0 A_r D_{f1}}{D_{f2} H_0 A_q} \tag{6.11}$$

Implied equation

Manipulation of the coupled diophantine equations (6.7) and (6.8) results in an implied equation which determines the stability of the closed-loop system:

$$H_0 A A_q + G_0 A_r B = D_c D_{f1} \tag{6.12}$$

The proof of these results is given in the appendix.

6.3.6 Properties of the controller

1 On substitution for the feedback controller, it is easily seen that the characteristic polynomial is the same as the implied equation. Hence, even though the open-loop plant is unstable, the system is guaranteed to be closed-loop stable, since D_c and D_{f1} are defined to be strictly stable.

2 The feedback controller is independent of the reference controller and the reference model W_s. However, the reference controller is dependent upon the feedback controller and reference model.

3 The effect of the dynamic weights Q and R on the controller are apparent. The feedback and reference controllers have poles due to the A_q weighting term and zeros due to the A_r weighting term.

6.3.7 Design procedure

The LQG control design procedure is summarised as follows:

1 Obtain model polynomials, A, B, C.
Obtain reference model polynomials A_s, B_s.
Obtain noise variances $\Phi_{\zeta\zeta}$, $\Phi_{\xi\xi}$.

2 Determine the dynamic weights Q, R_1, R_2.
Determine K.

3 Obtain spectral factors D_c, D_{f1}, D_{f2} from equations (6.4), (6.5), and (6.6).

4 Solve diophantine equations for G, H, M from equations (6.7) and (6.8).

5 Calculate the controllers according to equations (6.10) and (6.11).

6.4 H_∞ Robust control problem

An H_∞ robust feedback controller C_1 can be derived directly from the H_2 polynomial system results by computing a particular cost-function weighting W which links the two problems. This method of solving H_∞ control problems is referred to as H_∞ embedding [4] and invokes a lemma due to Kwakernaak [2].

$$J = \frac{1}{2\pi j} \oint_{|z|=1} \left(X(z^{-1}) \Sigma(z^{-1}) \right) \frac{dz}{z} \tag{6.13}$$

Lemma 6.1. *Consider the auxiliary problem of minimising the criterion (6.13) and suppose that for some real rational $\Sigma(z^{-1}) = \Sigma^*(z^{-1}) > 0$ the criterion J is minimised by a function $X(z^{-1}) = X^*(z^{-1})$ for which $X(z^{-1}) = \lambda^2$ (a real constant) on $|z| = 1$. Then this function also minimises $\sup_{|z|=1} \left(X(z^{-1}) \right)$.*

The function X to be minimised involves the cost terms in the integrand of equation (6.13) which depend upon the feedback controller C_1. The H_∞ norm of the following function is therefore to be minimised:

$$X \underline{\underline{\Delta}} \left[K^2 Q \Phi_{ee}(z^{-1}) + R\Phi_{uu}(z^{-1}) \right]$$

The integrand of the cost index at the optimum must satisfy:

$$F_1^* F_1 / (D_{fc}^* D_{fc}) + W_\sigma^* T_0 W_\sigma = \lambda^2 W_\sigma^* W_\sigma \tag{6.14}$$

where $\Sigma = W_\sigma W_\sigma^*$. It follows from lemma 6.1 that the weighting $W_\sigma = A_\sigma^{-1} W_\sigma$ must satisfy:

$$\begin{aligned} W_\sigma^* W_\sigma &= \frac{B_\sigma^* B_\sigma}{A_\sigma^* A_\sigma} \\ &= \frac{F_1^* F_1}{(\lambda^2 - T_0)(D_{fc}^* D_{fc})} \\ &= \frac{F_{1s}^* F_{1s}}{\lambda^2 D_{fc}^* D_{fc} - T_{0n}/(T_d^* T_d)} \end{aligned} \tag{6.15}$$

where F_{1s} is Schur and satisfies $F_{1s}^* F_{1s} = F_1^* F_1$ and after cancellation of common factors define:

$$T_{0n}/(T_d T_d^*) = T_0 D_{fc}^* D_{fc}$$

where T_d is Schur. Clearly B_σ and A_σ can be obtained as:

$$B_\sigma = F_{1s} T_d \quad and \quad A_\sigma^* A_\sigma = \lambda^2 D_{fc}^* D_{fc} T_d^* T_d - T_{0n} \tag{6.16}$$

The following theorem is a direct consequence of these results.

Theorem 6.2 (H_∞ optimal controller). *Consider the system defined in section 6.3 and assume that the feedback controller C_1 is to minimise the following H_∞ cost-function:*

$$J_\infty = \sup_{|z|=1} \left| K^2 Q \Phi_{ee}(z^{-1}) + R \Phi_{uu}(z^{-1}) \right| \tag{6.17}$$

Let the minimum-phase robustness weighting function $W_\sigma = A_\sigma^{-1} B_\sigma$ be defined (after cancellation of common factors) to satisfy (6.15), using the smallest value of λ. The feedback controller C_1 can then be computed from the results in section 6.3.

6.4.1 Generalised H_2 and H_∞ controllers

The generalised H_2 controller minimizes the following cost function:

$$J = E\left\{\phi(t)^2\right\} = \frac{1}{2\pi j} \oint_{|z|=1} \left\{\Phi_{\phi\phi}(z^{-1})\right\} \frac{dz}{z} \tag{6.18}$$

where

$$\phi(t) = P_c(z^{-1}) e(t) + F_c(z^{-1}) u(t)$$

The weighting elements can be dynamical and can be presented in terms of polynomials as:

$$P_c(z) = P_d^{-1} P_n \quad \text{and} \quad F_c(z) = P_d^{-1} F_n$$

where P_d is a monic, strictly Schur common denominator polynomial $P_n(0) \neq 0$ and often $F_n(z^{-1}) = z^{-k}(F_k(z^{-1}))$. The weighting polynomials P_n and F_n are chosen to ensure that the polynomial:

$$L_c \triangleq (P_n B - F_n A)$$

is free of zeros on the unit-circle of the z-plane or $L_c L_c^* > 0$ on $|z| = 1$. It may therefore be assumed that the spectral-factors D_c, D_f defined in the following are strictly-Schur polynomials. To determine the expression for the corresponding GH_∞ controller the lemma 6.1 can again be invoked. Note that since an adaptive controller is of interest the simplest possible solution is important and hence attention will return to the one DOF problem structure.

Theorem 6.3 (**GH_∞ optimal controller**). *The GH_∞ optimal controller which minimises $\left\|\Phi_{\phi\phi}(z^{-1})\right\|_\infty$, for the single DOF system shown in figure 6.2, when the measurement noise is zero, can be computed from the strictly Schur spectral factors D_0 and F_{1s}:*

$$D_0 D_0^* = E_r E_r^* + C_d C_d^* \quad \text{and} \quad F_{1s} F_{1s}^* = F_1 F_1^* \tag{6.19}$$

where the disturbance and reference models are denoted here as $A^{-1}C_d$ and $A^{-1}E_r$, respectively. The solution (G_1, H_1, F_1) of the following equations is also required:

$$F_1 A P_d + L_2 G_1 = P_n D_0 F_{1s}/\lambda \tag{6.20}$$

$$F_1 B P_d - L_2 H_1 = F_n D_0 F_{1s}/\lambda \tag{6.21}$$

Controller:

$$C_1 = H_1^{-1} G_1 \tag{6.22}$$

Sensitivities:

$$S = A H_1 \lambda/(L_1 D_0 F_{1s})$$
$$T = B G_1 \lambda/(L_1 D_0 F_{1s})$$
$$M = A G_1 \lambda/(L_1 D_0 F_{1s})$$

Implied equation:

$$\rho = G_1 B + H_1 A_0 = L_1 D_0 F_{1s}/\lambda$$

Minimal cost:

$$J_\infty = \left\|\Phi_{\phi\phi}(z^{-1})\right\|_\infty = \lambda^2$$

Signal ϕ:

$$\phi(t) = F_{1s}^{-1} F_1 \lambda \varepsilon(t)$$

where $L_c = P_n B - F_n A$ and L_c is factorised as: $L_c = L_1 L_2$ where L_1 is strictly Schur, and $L_{2s} L_{2s}^* = L_2 L_2^*$ for $L_{2s} = F_{1s}/\lambda$.

Proof. The proof is given in Grimble (1994 [13]).

6.5 System and disturbance modelling

The CMM consists of three orthogonal arms in the x, y and z axes as shown in figure 6.1 and a two DOF probe capable of rotating about the A axis and 0° to 105° about the B axis. The arms are controlled by independent servo loops consisting of a Pulse Width Modulated (PWM) current amplifier, three phase brushless DC Motors with incremental shaft encoders to detect the velocity, a preamp component for signal conditioning, a velocity loop control and a position controller implemented in software and downloaded into a transputer. The position transducer is a combination of an optical reading head and grating. The output of the transducer is fed to the transputer for decoding the position. A 20 lines/mm optical grating is used for position measurement. The interface between the system and the controller is through a commercially available input/output card programmed through high-level software.

6.5.1 System modelling

Velocity loop modelling

The velocity loop consists of a pre-amp, an amplifier, a servo motor and a velocity transducer. The pre-amp and amplifier, which are implemented in hardware electronics, act as the velocity controller of a PI (proportional + integral) controller [2]. When the position loop is open, the potentiometer gains of the pre-amp are tuned manually to give a good velocity response. The velocity loop is then modelled after conducting step input tests.

Stepper motor modelling

The two stepper motors driving the probe can each be modelled by four non-linear differential equations [3] with the input being the disturbance torque, due to interaction from the movement of the arms, and the output the angular position of the stepper motor.

CMM modelling

For modelling purposes the arms and the probe are considered as a set of rigid links connected together in series by three prismatic and two revolute joints. Using the Euler-Lagrange equation, a model of the CMM containing five non-linear equations was derived [4] which relates the five disturbance joint forces/torques of the arms/probe with the positions, velocities and accelerations of the arms and problem.

These disturbance forces/torques can be interpreted as interactions between the arms and the probe. Replacing the two disturbance torques of the probe by the stepper motor model, the CMM model is linearised about the nominal probe positions and then its order is reduced by discarding the fast modes.

Probe modelling

The linearised CMM model can be further simplified by ignoring the interaction of the probe and the arms and this results in two linear second order probe models which have the acceleration of the arms as the inputs and the two probe positions as the outputs. This simplification is justified because the weight of the arms is very much heavier than that of the probe and the interaction effect of the movement of the probe to the arms is negligible [5].

6.5.2 Disturbance modelling

Process noise

It is assumed that the process noise comes from the hardware electronics of the velocity loop and enters the system at the output of the velocity loop.

Measurement noise

A measure of the position of the arm is required to control the probe. The position transducer contains an optical grating and reading head which provides highly accurate position measurement. Thus, the measurement noise is assumed to be negligible and this assumption also simplifies the LQG control design procedures described later.

6.5.3 Overall system model

The input and outputs selected for the CMM are the three input voltages to the servo systems, the output velocities corresponding to the five joints and the three vibration signals. The joint positions can be obtained by integrating the velocities. The non-linear model has 24 state variables Fortunately, many of the poles and zeros of the system are stable and well damped. Thus, they can be removed from the linear dynamics. The linear model order is reduced to five by freezing the states which are not required for control design studies. A study of the rank of these matrices indicates that the system is both stabilizable and detectable despite the fact that the probe positions are not measured.

The total reduced linear continuous-time model for each arm may be described in the frequency-domain by the following relationships:

- *System model:* $X(s) = G(s)U(s); \quad G(s) = (sI - A)^{-1}B$
- *Measurement models:* $Z(s) = HX(s)$
- *Output model:* $Y(s) = CX(s)$

where $x = [q_1, q_1, p_1, q_2, q_2, p_2, q_3, q_3, p_3, q_4, q_5]^T$ is the vector of the state variables with Laplace transform $X(s); U(s) = L\{u(t)\}$ is the vector of the Laplace transform of the three input voltages to the DC servos; $Z(s)$ is the vector of measurements which includes three servo velocities, three arm positions and three vibration signals; $Y(s)$ is the vector of the outputs to be controlled and includes the position of the three arms and the probe position. Here $G(s)$ is defined as:

$$G(s) = \begin{bmatrix} \frac{k_1}{\tau_1 s+1} & 0 & 0 & 0 & 0 & 0 & 0 & 0 & 0 & 0 & 0 \\ \frac{1}{s} & 0 & 0 & 0 & 0 & 0 & 0 & 0 & 0 & 0 & 0 \\ 0 & 0 & G_1^p(s) & 0 & 0 & 0 & 0 & 0 & 0 & 0 & 0 \\ 0 & 0 & 0 & \frac{k_2}{\tau_2 s+1} & 0 & 0 & 0 & 0 & 0 & 0 & 0 \\ 0 & 0 & 0 & \frac{1}{2} & 0 & 0 & 0 & 0 & 0 & 0 & 0 \\ 0 & 0 & 0 & 0 & 0 & G_2^p(s) & 0 & 0 & 0 & 0 & 0 \\ 0 & 0 & 0 & 0 & 0 & 0 & \frac{k_3}{\tau_3 s+1} & 0 & 0 & 0 & 0 \\ 0 & 0 & 0 & 0 & 0 & 0 & \frac{1}{s} & 0 & 0 & 0 & 0 \\ 0 & 0 & 0 & 0 & 0 & 0 & 0 & 0 & G_3^p(s) & 0 & 0 \\ \frac{k_4}{\tau_4 s+1} & 0 & 0 & \frac{k_5}{\tau_5 s+1} & 0 & 0 & 0 & 0 & 0 & 0 & 0 \\ \frac{k_6}{\tau_6 s+1} & 0 & 0 & \frac{k_7}{\tau_7 s+1} & 0 & 0 & \frac{k_8}{\tau_8 s+1} & 0 & 0 & 0 & 0 \end{bmatrix}$$

To be more specific $Y(s)$ is the vector of the Laplace transform of the five outputs to be controlled, i.e. the arm positions $x_2 = q_1$, $x_4 = q_2$, $x_6 = q_3$ and the probe positions $x_{10} = q_4$ and $x_{11} = q_5$. The current and velocity loops are assumed to be closed. The measurements $Z(s)$ are the three arm positions and the three strain gauge outputs i.e. $x_3 = p_1$, $x_6 = p_2$, $x_9 = p_3$ and $x_1 = q_1$, $x_4 = q_2$, $x_6 = q_3$. The H and C matrices are constants of appropriate dimensions and represent the outputs and measurements. The parameters of the transfer-function matrices may be found by transforming the state-space model into the frequency-domain.

6.6 Simulation and experimental studies

Simulation studies were first used to examine the performance of the LQG optimal controller. The use of an integral error weighting and a constant weighting on the control signal will first be considered. Then consider a dynamic weighting on the control signal and finally consider the use of a classical PID controller.

6.6.1 System definition

Plant model

The plant model W is composed of two subsystems, the velocity loop subsystem and an integrator. The velocity loop model was obtained experimentally and can be accurately represented by a linear 4th order transfer function. The plant is therefore of 5th order.

Probe models

The models W_{p1} and W_{p2} are defined for the nominal probe positions of 45°. After linearisation, the reduced models are of 6th order.

Reference model

When step responses are considered the reference model is $W_s = 1/(1 - z^{-1})$, i.e. $B_s = A_s = 1$ and $B_s = 1$, $A_s = 1 - z^{-1}$.

Dynamic weighting

The dynamic weighting $Q = H_q^* H_q$ is chosen such that H_q has the form of a lead transfer function having high gain at low frequencies.

The dynamic weights $R_1 = H_{r1}^* H_{r1}$ and $R_2 = H_{r2}^* H_{r2}$ are chosen such that H_{r1}, H_{r2} have the same frequency responses as the probe models, i.e. $H_{r1} = W_{p1}$, $H_{r2} = W_{p2}$. The frequency responses of the two probe models are fairly similar and thus both H_{r1} and H_{r2} are chosen to be equal to W_{p1}. Furthermore, to reduce the complexity of R $(= R_1 + R_2)$ and hence the order of the controller, H_{r1} and H_{r2} are approximated by a 3rd order transfer function whose frequency response is shown in figure 6.4 alongside that of W_{p1} and W_{p2}.

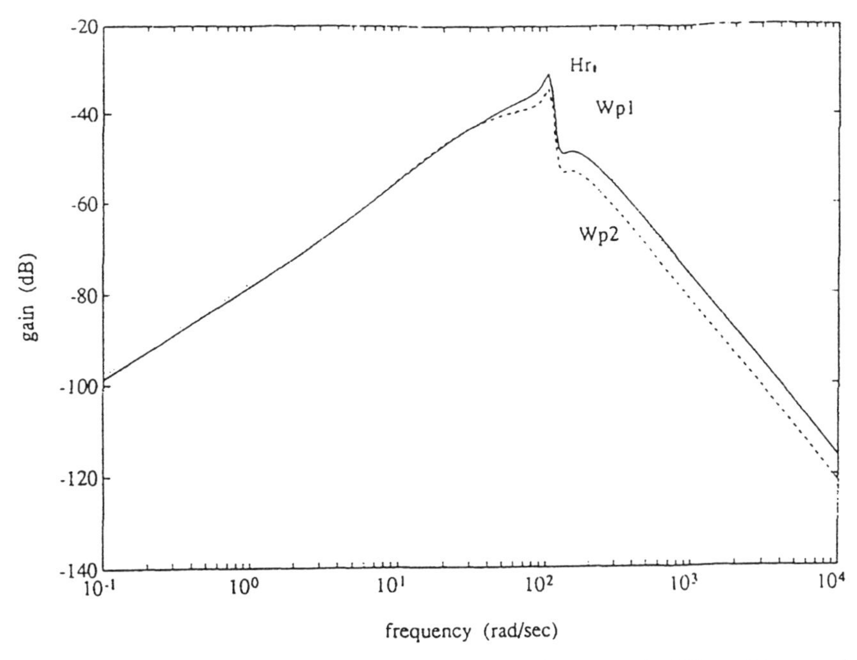

Figure 6.4: The frequency response of the probe models W_{p1}, W_{p2}, and the dynamic weight R_1.

Noise statistics

The process noise ξ is assumed to have a standard deviation of 0.01 and a variance $\Phi_{\xi\xi} = 0.0001$.

6.6.2 Simulation studies

Non-dynamic weighting on the control

With a non-dynamic weighting on the control signal u, i.e. constant R, frequency responses of the closed loop system were evaluated for several values of K (the constant weight on the error term in the cost function) and are shown in figure 6.5. The corresponding frequency responses of the control sensitivity function are shown in figure 6.6. As expected the closed loop bandwidth becomes higher and the gain of the control sensitivity function increases as K increases. The figure 6.7 shows the step- responses of the arm. The response becomes faster as K increases and the system is slightly underdamped with a maximum overshoot of about 3%. The corresponding step responses of the probe are shown in figure 6.8. Only one probe position is shown because the two probe models are almost the same. It is clear that the control signal is more active as K increases and the transient deviation of the probe from its nominal position becomes larger.

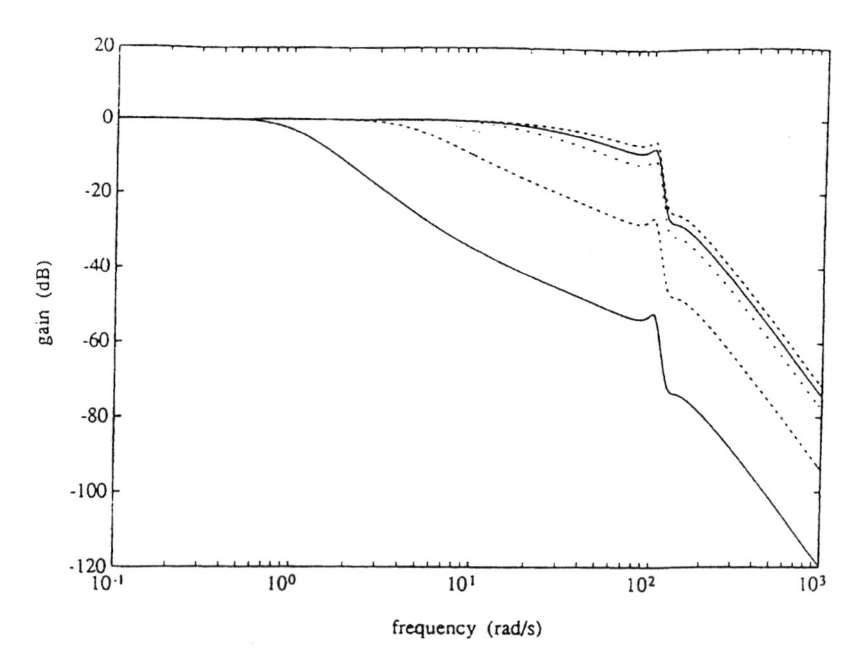

Figure 6.5: The frequency responses of the closed-loop system (non-dynamic weighting on control).

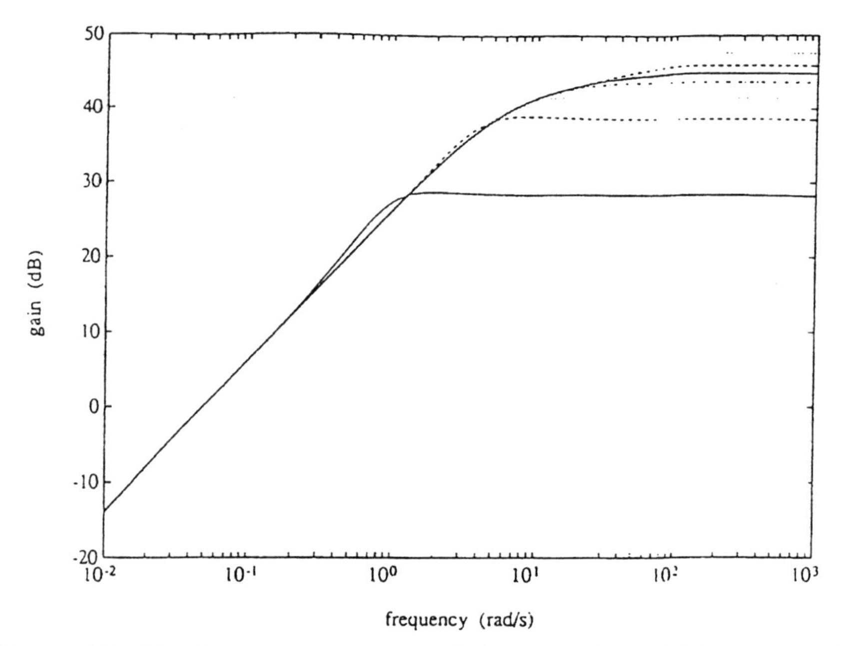

Figure 6.6: The frequency responses of the control sensitivity system (non-dynamic weighting on control).

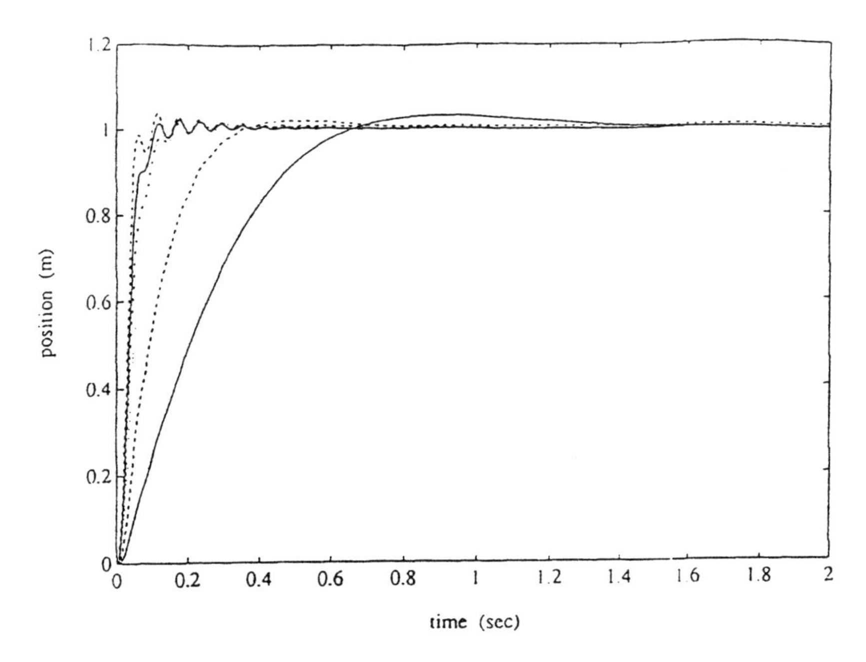

Figure 6.7: The step responses of the arm (non-dynamic weighting on control).

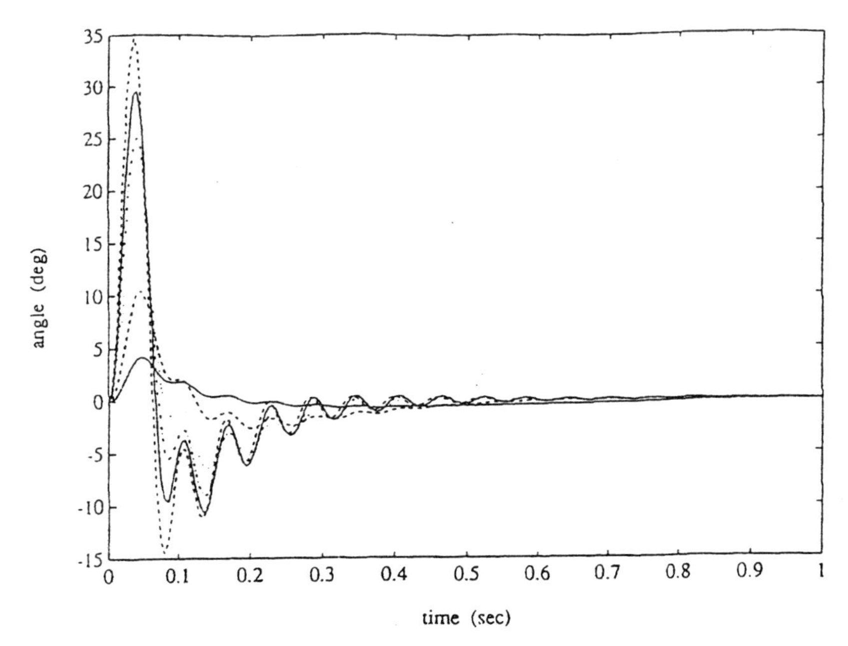

Figure 6.8: The step responses of the probe (non-dynamic weighting on control).

Dynamic weighting on the control

Consider a dynamic weighting on the control signal u. The frequency-response of the closed- loop system and the control-sensitivity function, for differing values of K, are shown in figure 6.9 and 6.10 respectively. The values of K are selected such that the closed-loop bandwidths are the same as those in the case of non-dynamic R so that a comparison can be made. As in the case of non-dynamic R, the closed-loop bandwidth increases with K. However, the response of the control-sensitivity function is quite different, there is large drop in gain at frequencies around 100 rad/sec which coincides with the peak gain of H_{r1} (see figure 6.4). This is a result of the choice of the dynamic weighting R as it tries to counterbalance the high gain of the probe model at these frequencies. The corresponding step-responses of the arm and the probe are shown in figures 6.11 and 6.12 respectively. In comparison with figures 6.7 and 6.8, the arm responses are less oscillatory, with slightly larger overshoot, otherwise they are similar. The probe responses are less oscillatory with a slightly shorter settling time and they exhibit a smaller deviation from the nominal position.

PID control

To compare with the effectiveness of LQG control, the conventional PID controller was also used. The design of the PID controller requires knowledge of the open-loop plant but not that of the probe models. The design method is based on moving a point in the Nyquist plot of the open-loop plant to a certain location to achieve the desired closed-loop bandwidth and phase or gain margin [8]. The step responses of the arm and probe for various closed-loop bandwidths and a phase margin of 60° are shown in figures 6.13 and 6.14 respectively. Comparison with those for the LQG controller reveals that the responses are more oscillatory with much larger overshoot and they also take longer to settle.

Comparison of the simulation results

To summarise the simulation results, the maximum deviation of the probe from the nominal position is plotted as a function of the bandwidth of the closed-loop system in figure 6.15 for the three cases. That is, using LQG control with and without dynamic control weighting and PID control.

It is clear that with PID control the maximum deflection of the probe is the largest of the three for all closed-loop bandwidths, especially below 13 rad/sec, and the corresponding time responses of the arm and the probe are much worse than those of the LQG controller (see figures 6.13 and 6.14).

For LQG control with dynamic control weighting, the maximum deflection of the probe is smaller than that with constant weighting for all closed-loop bandwidths simulated. In addition, the time responses of the arm are comparable and the time responses of the probe and are better in both cases, (see figures 6.7, 6.8, 6.11 and 6.12).

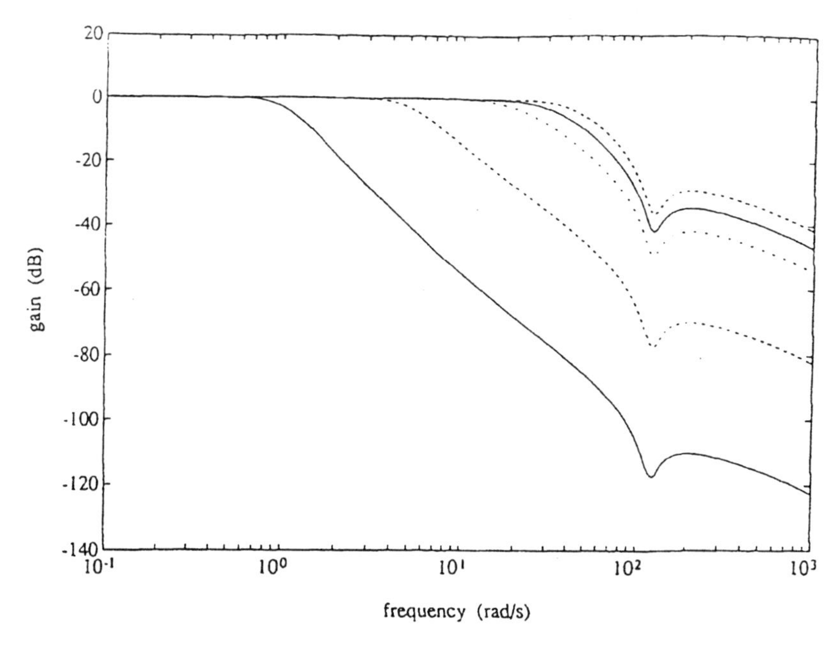

Figure 6.9: The frequency responses of the closed-loop system (dynamic weighting on control).

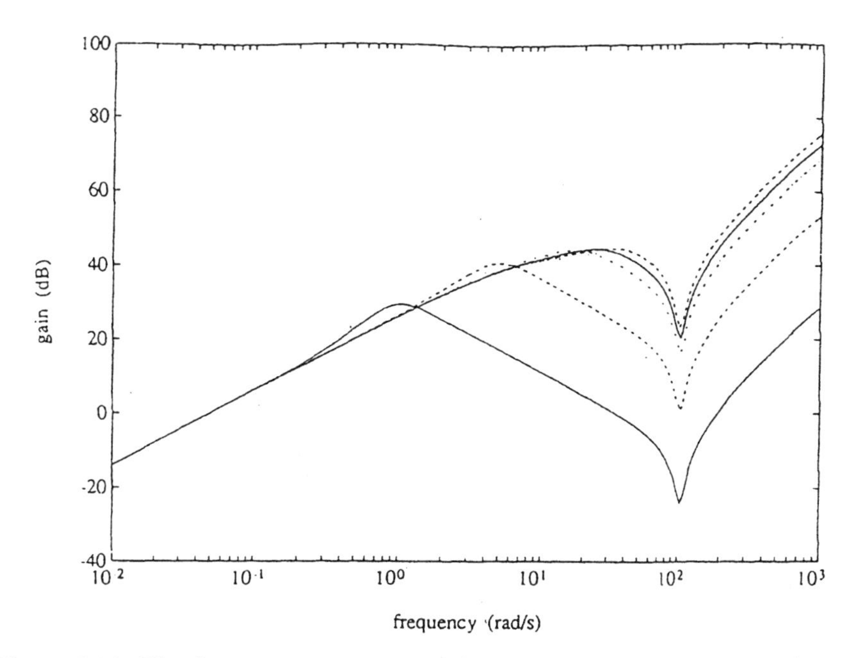

Figure 6.10: The frequency responses of the control sensitivity system (dynamic weighting on control).

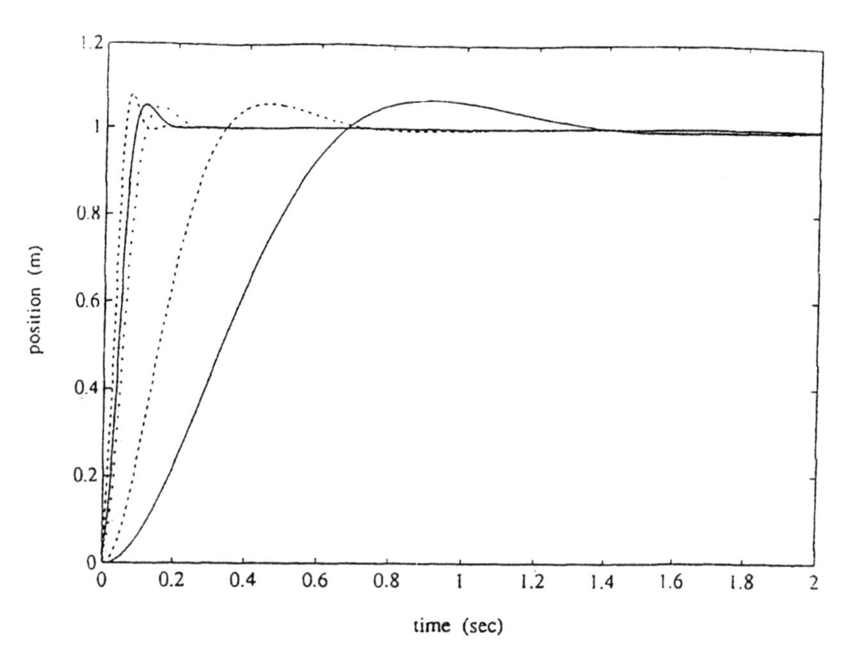

Figure 6.11: The step responses of the arm (dynamic weighting on control).

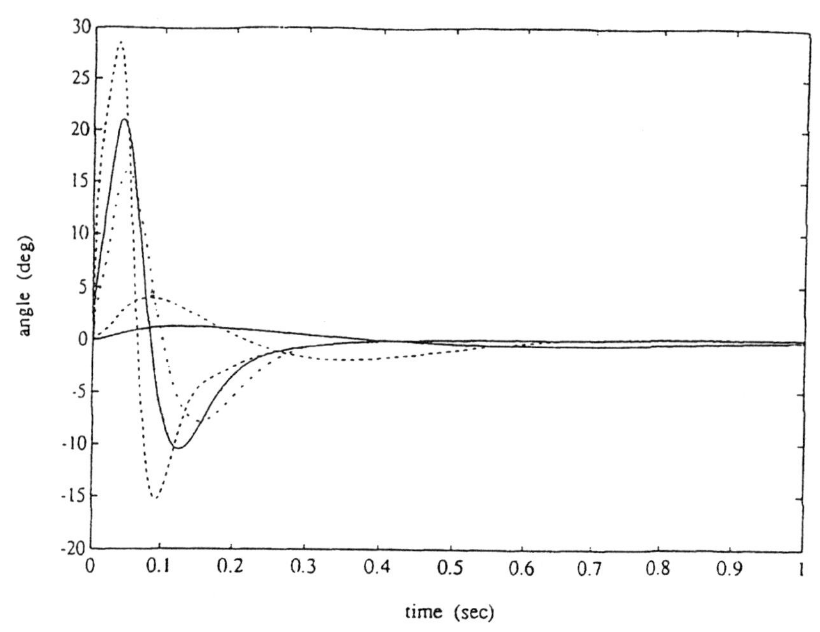

Figure 6.12: The step responses of the probe (dynamic weighting on control).

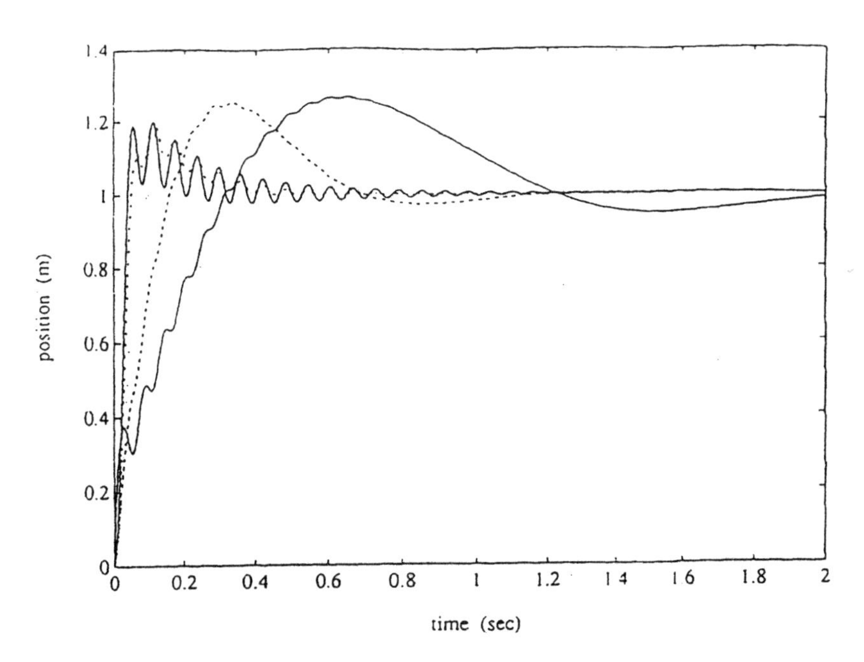

Figure 6.13: The step responses of the arm (PID control).

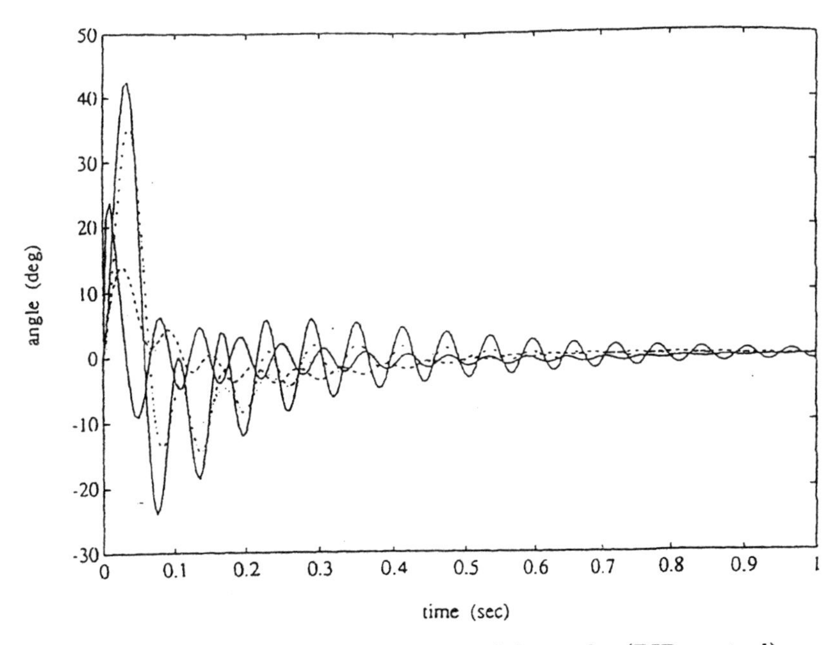

Figure 6.14: The step responses of the probe (PID control).

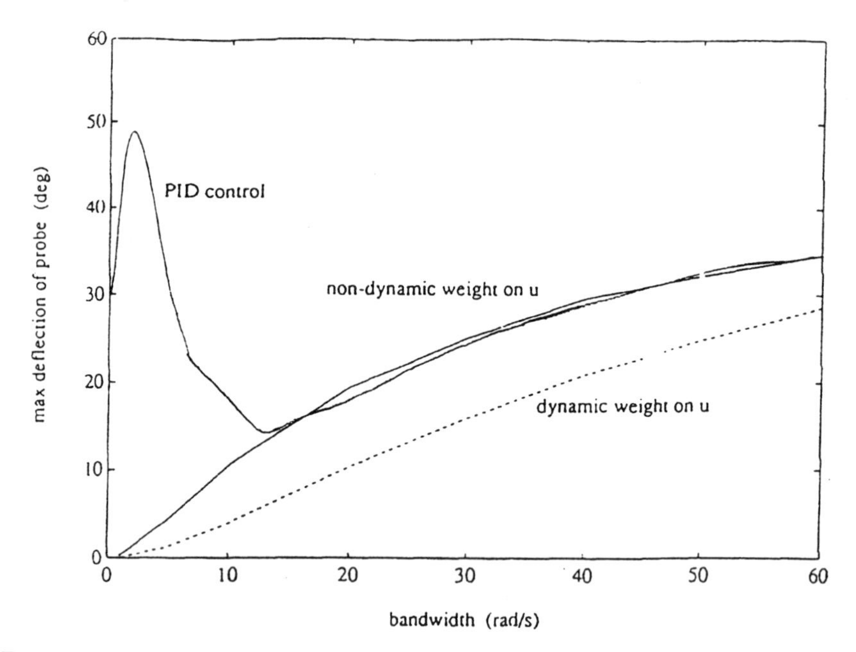

Figure 6.15: The variation of the maximum deflection of the probe as a function of the closed loop system bandwidth.

6.6.3 Experimental studies

A single axis test rig representing an arm was built to test and develop the control algorithm [9]. It included a servo system consisting of a pre-amp, an amplifier, a servo motor and a position transducer. The motor was connected to a worm screw and a trolley was mounted upon it to give the linear position. The position controller was implemented on a personal computer capable of a sampling frequency of 20 Hz, giving a maximum closed-loop bandwidth of 2 Hz (assuming the sampling frequency is at least 10 times the bandwidth).

The LQG controllers for the case of non-dynamic weighting R were tested on the test rig. The step-responses of the arm to a step demand of 0.2 m are shown in figure 6.16 for various K designed for closed-loop bandwidths of 1 to 5 rad/sec. This reveals that as K gets larger, the slope of the response has a maximum. This corresponds with the maximum speed of the motor resulting from saturation of the controller. In figure 6.17 a comparison between the simulation and experimental step responses are shown . For a small K the simulation and experimental results are almost the same. As K increases the system moves into a non-linear region and the responses become very different.

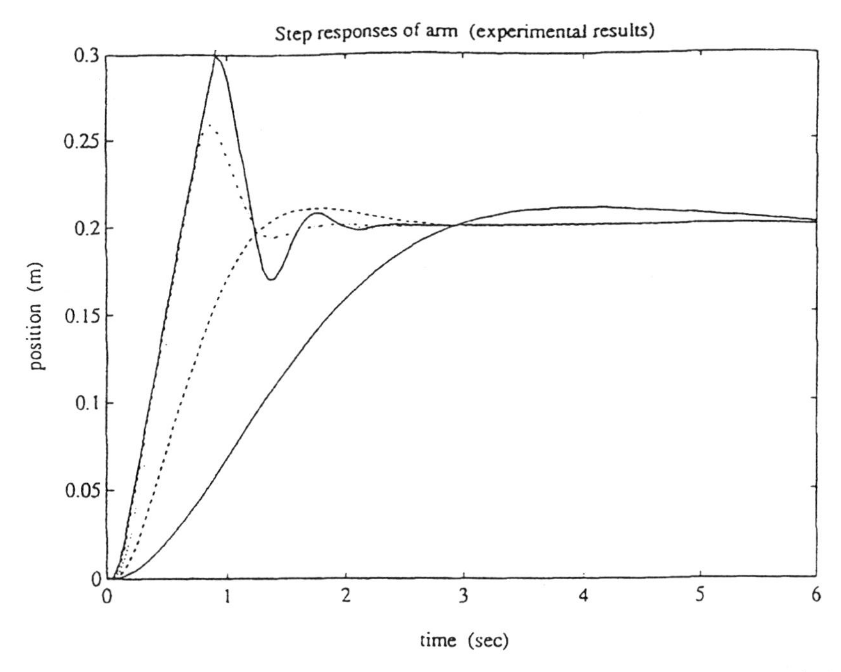

Figure 6.16: The step responses of the arm (experimental results).

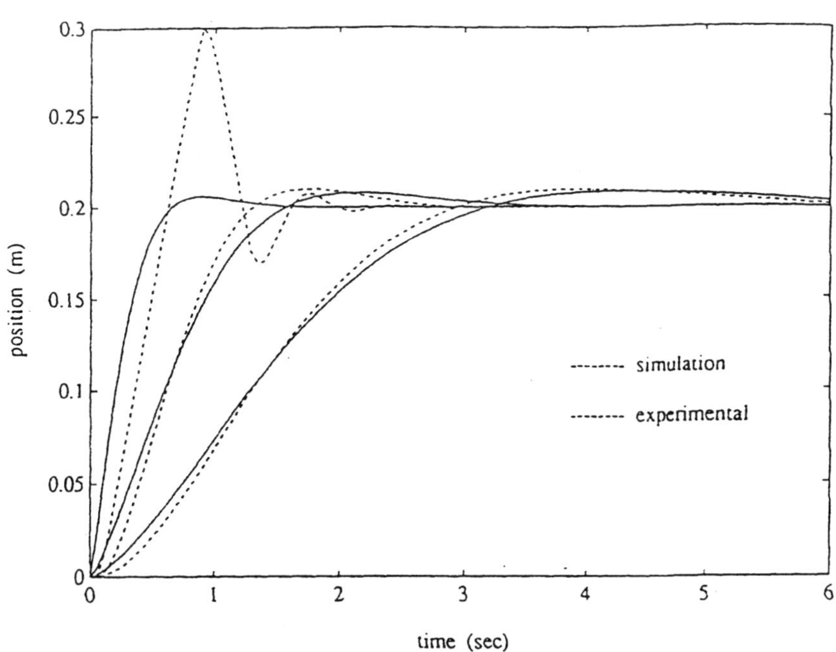

Figure 6.17: The step responses of the arm (simulation and experimental results).

6.6.4 H_∞ control

The H_∞ problem design utilized a simplified plant model, that is a discrete integrator. The generalized H_∞ solution which is particularly appropriate for this case, can be found through solving two linear equations [1]:

$$F_0^- A P_{cd} \lambda + L_k G_0 = P_{cn} D_f F_{0s}^-$$

$$F_0^- B F_{cd} \lambda + L_k H_0 F_{cn} D_f F_{0s}^-$$

The controller C_1 is found as:

$$C_1 = (H_0 P_{cd})^{-1} G_0 F_{cd}$$

where

$$D_f D_f^* = \Phi_{\xi\xi} C C^* \quad (\text{but } \Phi_{\xi\xi} \underline{\underline{\Delta}} 1)$$

$$P = P_{cn}/P_{cd} \quad (\text{error weighting})$$

$$F = F_{cn}/F_{cd} \quad (\text{control weighting})$$

where L_k is the closed-loop polynomial defined by:

$$L_k = P_{cn} F_{cd} B - F_{cn} P_{cd} A = (\iota_{k0} + l_{k0} z^{-1})(\iota_{k1} + l_{k1} z^{-1})$$

For the specific problem of the servo system, the following polynomials may be defined:

$$B = k_b z^{-1}, \quad A = (1 - \lambda z^{-1})$$

$$C = (1 - \beta z^{-1}) \Rightarrow D_f = (d_{f0} + d_{f1} z^{-1}) = (1 - \beta z^{-1})$$

The weighting functions are assumed as:

$$P_{cn} = P_0, \quad P_{cd} = (1 - \alpha z^{-1}), \quad F_{cn} = r, \quad F_{cd} = 1$$

Solution of equation (6.8) for the feedback controller gives:

$$C_1(z^{-1}) = \frac{(g_0 + g_1 z^{-1})}{h_0 (1 - \alpha z^{-1})}$$

where

$$g_0 = \frac{P_0 (d_{f1} l_1 + d_{f0} L_1 (\gamma + \alpha) + d_{f0} \iota_0 \gamma \alpha)}{(l_1 + \gamma \iota_0 L_1)(\iota_1 + \alpha \iota_0)}$$

$$g_1 = \frac{P_0 \gamma \alpha (d_{f1} l_1 + d_{f0} L_1)}{(l_1 + \gamma \iota_0)(\iota_1 + \alpha \iota_0)}$$

$$h_0 = \frac{-r d_{f0}}{l_0}$$

This controller is in the form of a PI control law with gains:

$$K_p = \frac{g_0}{h_0}, \quad K_1 = \frac{g_1 + \alpha g_0}{h_0}$$

There are two related design approaches to the H_∞ problem presented above. Either the weights P_{cn} and F_{cn} can be chosen and the controller calculated, or the closed-loop pole given by $(l_{k0} + \iota_{k1} z^{-1})$ can be placed at $(l + z_1 z^{-1})$. The value of either P_{cn} or F_{cn} can then be chosen and the second weighting calculated as follows:

$$P_{cn} = P_0 = \frac{-r(z_1 + \gamma)(z_1 \alpha)}{k_b z_1}$$

or

$$F_{cn} = r = \frac{-P_0 k_b z_1}{(z + \gamma)(z + \alpha)}$$

The two parameters l_0 and l_1 are found as follows:

$$l_0 = -r; \quad l_1 = -r\gamma\alpha / z_1$$

and equation (6.9) can be solved for the controller gains.

Choosing $\alpha = 0.999$, $P_{cn} = 0.2$ leads to the PI controller $K_p = 128.019$, $K_1 = 16.345$ for the plant model of $K_b = 0.05$, $\gamma = 1$ and $t_2 = 80$ms. The response to the pulse input for this controller is illustrated in figure 6.18. The rise-time and the overshoot are very similar to that for the manually tuned PI controller. At larger step sizes the overshoot and settling time are also larger caused by the high level of integral action. The choice of the weighting functions must be investigated to improve the plant response. The transfer-function for the H_∞ controller is given as:

$$C_1(z^{-1}) = \frac{0.564z^{-1}(1 - 0.9864z^{-1})}{(1 - 0.4489z^{-1})(1 - 0.9861z^{-1})}$$

The controller was simulated with the integrator model for the position loop. The response of the simulation is illustrated in figures 6.18 and 6.19 and is very close to that of the real system response for both small and large step inputs.

6.7 Conclusions

A H_∞ controller was developed based on the polynomial approach and applied to a CMM control problem where one of the outputs to be controlled was not measurable. Integral action was included in the dynamic weighting on the error to eliminate the steady-state error. To achieve optimal performance a compromise between speed of the arm and the position accuracy of the probe was made by tuning a scalar parameter K. This determined the relative importance of the error and control terms in the cost function and hence the bandwidth of the closed-loop system. By taking into account knowledge of the probe model an appropriate dynamic weighting on the control signal was introduced. The simulation results show that the accuracy of the probe is much improved even though its position is not measurable.

The controllers was tested on an experimental rig and the results verify the simulation within the linear region when the weighting on the error is relatively small. The PID controller was also used and the simulation results indicate that H_∞ control is superior in several respects.

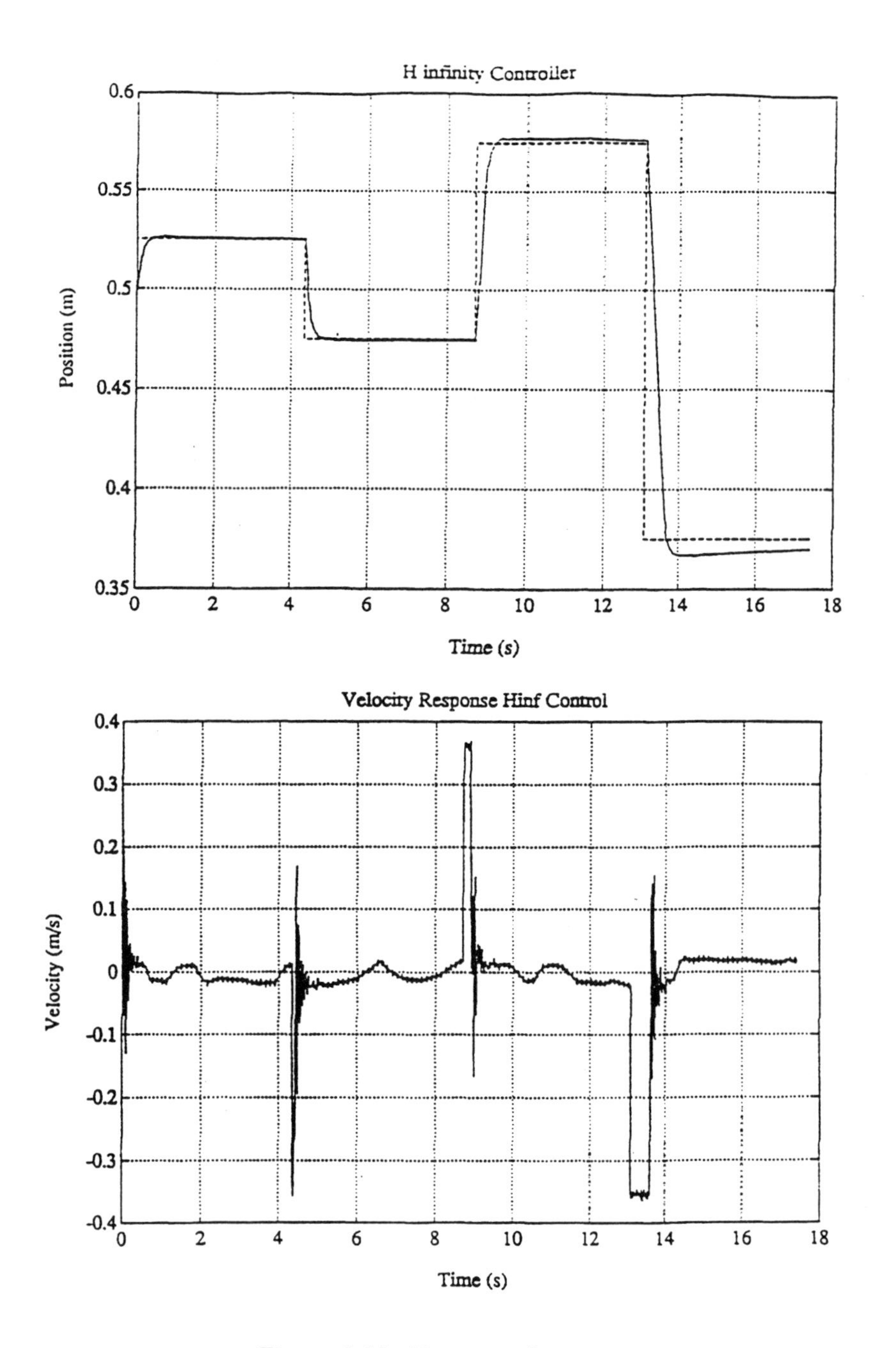

Figure 6.18: H_∞ control response.

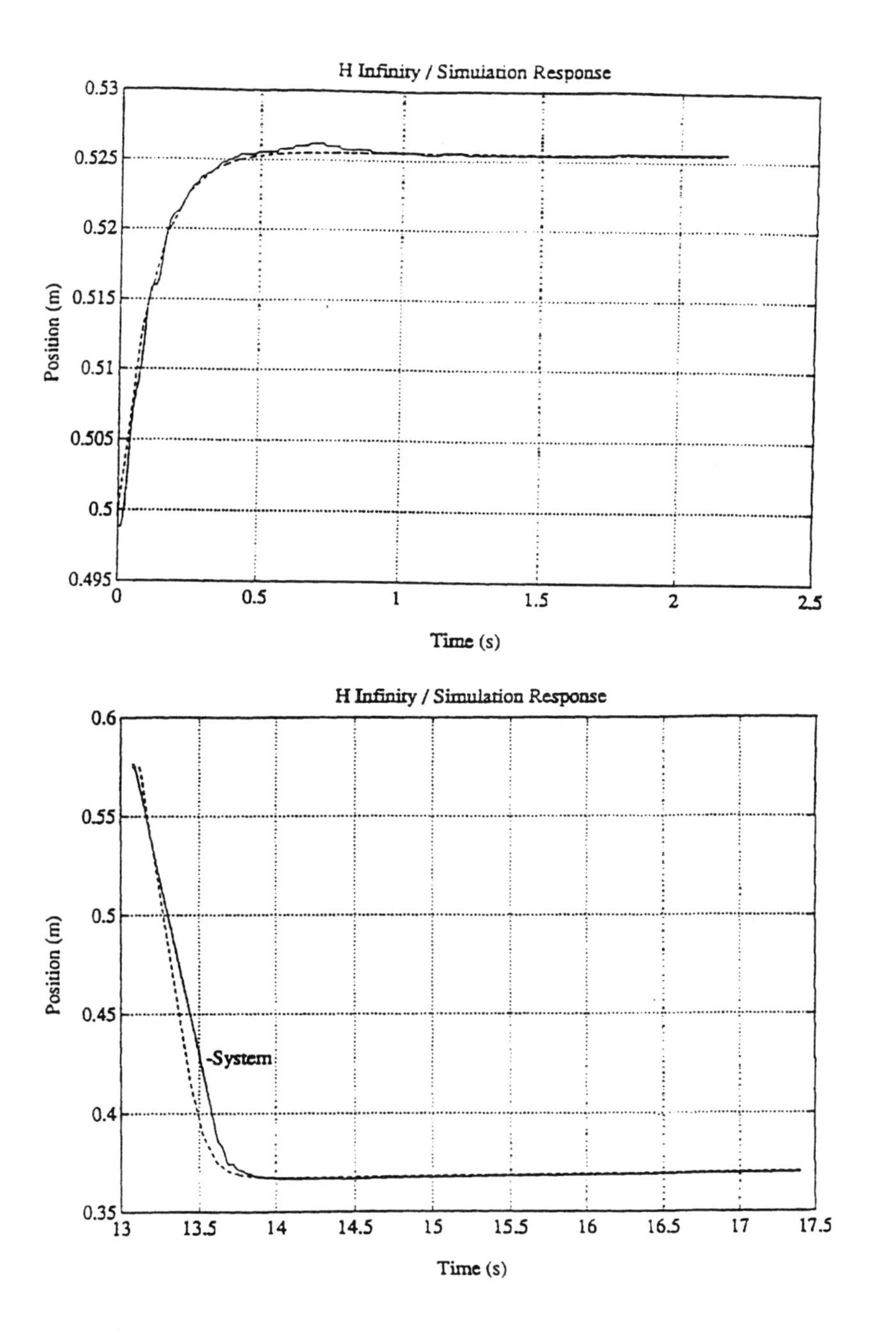

Figure 6.19: H_∞ simulation and actual response.

6.8 Acknowledgements

The authors are grateful to Mr Thomas Lee and Mr John McIntyre for the simulation results in this paper. Support from Ferranti Industrial Electronics, LK Ltd and the SERC's ACME Directorate are gratefully acknowledged.

6.9 References

[1] Grimble, M. J., 1987 *H_∞ robust controller for self-tuning control application; Part I: Controller design*, Int. J. Cont., 46, 4.

[2] McIntyre, J., M. R. Katebi and M. J. Grimble, 1990 *The modelling and control design analysis of the pre-amp component of the CMM servo systems*, ICU Report 315.

[3] Zribi, M., and J. Chiasson, 1991 *Position control of a PM stepper motor by exact linearisation*, IEEE Trans. On Automatic Control. Vol. 36, No. 5.

[4] Lee, T., M. R. Katebi and M. J. Grimble, 1991 *The modelling and control analysis of a co-ordinate measuring machine*, ICU Report 316.

[5] Lee, T., M. R. Katebi and M. J. Grimble, 1991 *Nonlinear and linear model for a co-ordinate measuring machine*, ICU Report 338.

[6] Kučera, V., 1979 *Discrete linear control*, John Wiley and Sons.

[7] Grimble, M. J., 1991 *Polynomial solution of the standard H_∞ optimal control problem for machine control systems applications*, ICU Report.

[8] Gawthrop, P. J., 1978 *Developments in optimal and self-tuning control theory*, PhD thesis, OUEL Report 1239/78, Oxford University.

[9] Åström, K. J., and T. Hägglund, 1989 *Automatic tuning of PID regulators* Instrument Society of America.

[10] Katebi, M. R., J. McIntyre, T. Lee, and M. J. Grimble, 1992 *Integrated process and control design for fast CMM*, International Conference on Mechatronics, Dundee.

[11] Kwakernaak, H., 1986 *A polynomial approach to minimax frequency domain optimisation of multivariable feedback systems*, Int. J. Control, Vol. 44, (1), pp. 117–156.

[12] Grimble, M. J., 1986 *Optimal H_∞ robustness and the relationship to LQG design problems*, Int. J. Control, Vol. 43, (2), pp. 351–372.

[13] Grimble, M. J., 1994 *Robust Industrial Control*, Prentice Hall, Hemel Hempstead.

6.10 Appendix: two-DOF H_2 optimal control problem

A proof of the results in section 6.3.5 is given in the following.

According to figure 6.2, the control signal u is given by:

$$u = C_2 r - C_1(W_d\xi + W_u) \tag{6.23}$$

$$= \frac{C_2}{1 + C_1 W} r - \frac{C_1 W_d}{1 + C_1 W}\xi \tag{6.24}$$

Let

$$S = \frac{1}{1 + C_1 W} \tag{6.25}$$

$$u = S(C_2 r - C_1 W_d \xi) \tag{6.26}$$

The output y and error e can be written as:

$$y = Wu + W_d\xi = WS(C_2 r - C_1 W_d\xi) + W_d\xi \tag{6.27}$$

$$e = r - y = (1 - WSC_2)r - (1 - WSC_1)W_d\xi \tag{6.28}$$

The cost function to be minimised is:

$$J = \frac{1}{2\pi j}\oint_{|z|=1} \left(K^2 Q\Phi_{ee}(z^{-1}) + R\Phi_{uu}(z^{-1})\right)\frac{dz}{z} \tag{6.29}$$

From equations (6.27) and (6.28), the control and error power spectral densities can be written as:

$$\Phi_{uu} = SC_2\Phi_{rr}C_2^* S^* + SC_1 W_d \Phi_{\xi\xi} W_d^* C_1^* S^* \tag{6.30}$$

$$\begin{aligned}\Phi_{ee} &= (1 - WSC_2 - W^*S^*C_2^* + WSC_2W^*S^*C_2W^*S^*C_2^*)\Phi_{rr} \\ &= (1 - WSC_1 - W^*S^*C_1^* + WSC_1W^*S^*C_1^*)W_dW_d^*\Phi_{\xi\xi}\end{aligned} \tag{6.31}$$

Let I be the integrand of the cost function, given by equation (6.29):

$$I = K^2 Q\Phi_{ee} + R\Phi_{uu} \tag{6.32}$$

$$\begin{aligned} &= SS^*(R + K^2QWW^*)(C_2\Phi_{rr}C_2^* + C_1W_d\Phi_{\xi\xi}W_d^*C_1^*) \\ &\quad + K^2(Q\Phi_{rr} - K^2Q\Phi_{rr})(W^*S^*C_2^* + WSC_2) \\ &\quad + K^2(QW_dW_d^*\Phi_{\xi\xi} - K^2QW_dW_d^*\Phi_{\xi\xi})(W^*S^*C_1^* + WSC_1) \end{aligned} \tag{6.33}$$

Defining the following spectral factors:

$$Y_cY_c^* = \frac{D_cD_c^*}{A_cA_c^*} = K^2QWW^* + R \tag{6.34}$$

$$= \frac{K^2BB^*B_qB_q^*}{AA^*A_qA_q^*} + \frac{B_rB_r^*}{A_rA_r^*} \tag{6.35}$$

$$= \frac{K^2B^*B_qA_r^*BB_qA_r + A^*B_r^*A_q^*AB_rA_q}{(AA_qA_r)(AA_qA_r)^*} \tag{6.36}$$

$$\Rightarrow D_c D_c^* = K^2 B^* B_q^* A_r^* B B_q A_r + A^* B_r^* A_q^* A B_r A_q \tag{6.37}$$

$$\Rightarrow Y_c = \frac{D_c}{A A_q A_r} \tag{6.38}$$

$$\Rightarrow Y_{f1} Y_{f1}^* = \frac{D_{f1} D_{f1}^*}{A_{f1} A_{f1}^*} = \Phi_{\xi\xi} W_d W_d^* = \frac{\Phi_{\zeta\zeta} B_s B_s^*}{A_s A_s^*} \tag{6.39}$$

$$\Rightarrow D_{f1} D_{f1}^* = \Phi_{\xi\xi} C C^* \tag{6.40}$$

$$\Rightarrow Y_{f1} = \frac{D_{f1}}{A} \tag{6.41}$$

$$Y_{f2} Y_{f2}^* = \frac{D_{f2} D_{f2}^*}{A_{f2} A_{f2}^*} = \Phi_{rr} = \Phi_{\zeta\zeta} W_s W_s^* = \frac{\Phi_{\zeta\zeta} B_s B_s^*}{A_s A_s^*} \tag{6.42}$$

$$\Rightarrow D_{f2} D_{f2}^* = \Phi_{\zeta\zeta} B_s B_s^* \tag{6.43}$$

$$\Rightarrow Y_{f2} = \frac{D_{f2}}{A_s} \tag{6.44}$$

Since $\Phi_{\xi\xi} = \Phi_{\xi\xi}^*$, $\Phi_{rr} = \Phi_{rr}^*$, $Q = Q^*$ and K is a constant, the integrand I can be written as:

$$\begin{aligned} I = {} & Y_c Y_c^* S S^* (C_1 Y_{f1} Y_{f1}^* C_1^* + C_2 Y_{f2} Y_{f2}^* C_2^*) \\ & -(K^2 Q \Phi_{\xi\xi} W_d W_d^* W^* S^* C_1^* + K^{2*} Q^* \Phi_{\xi\xi}^* W_d^* W_d W S C_1^*) \\ & -(K^2 Q \Phi_{rr} W^* S^* C_2^* + K^{2*} Q^* \Phi_{rr}^* W S C_2) \\ & +K^2 Q(\Phi_{rr} + \Phi_{\xi\xi} W_d W_d^*) \end{aligned} \tag{6.45}$$

Let

$$\Phi_1 = K^2 Q(\Phi_{rr} + \Phi_{\xi\xi} W_d W_d^*) \tag{6.46}$$

$$\Phi_2 \underline{\underline{\Delta}} K^2 Q \Phi_{\xi\xi} W_d W_d^* W^* \tag{6.47}$$

$$\Phi_3 \underline{\underline{\Delta}} K^2 Q \Phi_{rr} W^* = K^2 Q \Phi_{\zeta\zeta} W_s W_s^* W^* \tag{6.48}$$

Then the integrand I becomes:

$$\begin{aligned} I = {} & Y_c Y_c^* S S^* (C_1 Y_{f1} Y_{f1}^* C_1^* + C_2 Y_{f2} Y_{f2}^* C_2^*) \\ & -\Phi_2 C_1^* S^* - \Phi_2^* C_1^* S^* - \Phi_3 C_2^* S^* - \Phi_3 C S + \Phi_1 \end{aligned} \tag{6.49}$$

By completing the squares, I is split into parts which depend on the controllers and parts which do not,

$$\begin{aligned} I = {} & \left(Y_c S C_1 Y_{f1} - \frac{\Phi_2}{Y_c^* Y_{f1}^*} \right) \left(Y_c S C_1 Y_{f1} - \frac{\Phi_2}{Y_c^* Y_{f1}^*} \right)^* \\ & + \left(Y_c S C_2 Y_{f2} - \frac{\Phi_3}{Y_c^* Y_{f2}^*} \right) \left(Y_c S C_2 Y_{f2} - \frac{\Phi_3}{Y_c^* Y_{f3}^*} \right)^* + \Phi_0 \end{aligned} \tag{6.50}$$

where

$$\Phi_0 = \Phi_1 - \frac{1}{Y_c Y_c^*} \left(\frac{\Phi_2 \Phi_2^*}{Y_{f1} Y_{f1}^*} - \frac{\Phi_3 \Phi_3^*}{Y_{f2} Y_{f2}^*} \right)^*$$

which does not depend on the controllers.

Using equations (6.38), (6.41) and (6.47) we split $\frac{\Phi_2}{Y_c^* Y_{f1}^*}$ in equation (6.50) into stable and unstable parts:

$$\frac{\Phi_2}{Y_c^* Y_{f1}^*} = \frac{K^2 Q \Phi_{\xi\xi} W_d W_d^* W^*}{Y_c^* Y_{f1}^*} \tag{6.51}$$

$$= \frac{K^2 \Phi_{\xi\xi} \frac{B_q B_q^*}{A_q A_q^*} \frac{CC^*}{AA^*} \frac{B^*}{A^*}}{\frac{D_c^*}{A^* A_q^* A_r^*} \frac{D_{f1}^*}{A^*}} \tag{6.52}$$

$$= \frac{K^2 \Phi_{\xi\xi} B_q^* C^* B^* A_r^* B_q C}{D_c^* D_{f1}^* A_q A} \tag{6.53}$$

$$= \frac{G}{A_1 A} + \frac{F z^g}{D_c^* D_{f1}^*} \tag{6.54}$$

Comparison of equations (6.53) and (6.54) yields the following diophantine equation:

$$G D_c^* D_{f1}^* z^{-g} + F A_1 A = K^2 \Phi_{\xi\xi} B_q^* C^* B^* A_r^* B_1 C_z^{-g} \tag{6.55}$$

Therefore from equations (6.25), (6.38), (6.41) and (6.54)

$$Y_c S C_1 Y_{f1} - \frac{\Phi_2}{Y_c^* Y_{f1}^*} = \frac{D_c C_{1n} D_{f1}}{A_q A_r A (C_{1d} A + C_{1n} B)} - \frac{G}{A_q A} - \frac{F z^g}{D_c^* D_{f1}^*} \tag{6.56}$$

$$= \frac{C_{1n}(D_c D_{f1} - G A_r B) - C_{1d} G A A_r}{A_q A_r A (C_{1d} A + C_{1n} B)} - \frac{F z^g}{D_c^* D_{f1}^*} \tag{6.57}$$

$$= T_1^+ + T_1^- \tag{6.58}$$

where T_1^+ and T_1^- represent the stable first term and unstable second term of equation (6.57) respectively.

Similarly using equations (6.38), (6.44) and (6.48), we split $\frac{\Phi_3}{Y_c^* Y_{f2}^*}$ in equation (6.50) into stable and unstable parts:

$$\frac{\Phi_3}{Y_c^* Y_{f2}^*} = \frac{K^2 \Phi_{\zeta\zeta} Q W_s W_s^* W^*}{Y_c^* Y_{f2}^*} \tag{6.59}$$

$$= \frac{K^2 \Phi_{\zeta\zeta} \frac{B_s B_s^*}{A_s A_s^*} \frac{B_q B_q^*}{A_q A_q^*} \frac{B^*}{A^*}}{\frac{D_c^*}{A^* A_q^* A_r^*} \frac{D_{f2}^*}{A_s^*}} \tag{6.60}$$

$$= \frac{K^2 D_{f2} B_q^* B^* A_r^* B_q}{D_c^* A_s A_q} \tag{6.61}$$

$$= \frac{M}{A_s A_q} + \frac{N z^g}{D_c^*} \tag{6.62}$$

Comparison of equations (6.61) and (6.62) yields the following diophantine equations

$$M D_c^* z^{-g_1} + N A_s A_q = K^2 B_q^* B^* A_r^* B_q D_{f2} z^{-g_1} \tag{6.63}$$

From equations (6.25), (6.38), (6.44) and (6.62)

$$Y_c S C_2 Y_{f2} - \frac{\Phi_3}{Y_c^* Y_{f3}^*}$$

$$= \frac{D_c C_{2n} D_{f1} C_{1d}}{A_q A_r A_s C_{2d}(C_{1d}A + C_{1n}B)} - \frac{M}{A_s A_q} - \frac{N z^{g_1}}{D_c^*} \tag{6.64}$$

$$= \frac{C_{2n} D_c D_{f2} C_{1d} - M A_r C_{2d}(C_{1d}A + C_{1n}B)}{A_q A_r A_s C_{2d}(C_{1d}A + C_{1n}B)} - \frac{N z^{g_1}}{D_c^*} \tag{6.65}$$

$$= T_2^+ + T_2^- \tag{6.66}$$

where T_2^+ and T_2^- represent the stable first term and unstable second term of equation (6.65) respectively.

On substitution from equations (6.58) and (6.66), the cost function given by equation (6.29) becomes:

$$J = \frac{1}{2\pi j} \oint_{|z|=1} \left((T_1^+ + T_1^-)(T_1^+ + T_1^-)^* + (T_2^+ + T_2^-)(T_2^+ + T_2^-)^* + \Phi_0 \right) \frac{dz}{z} \tag{6.67}$$

As T_1^+, T_2^+ are stable and T_1^-, T_2^- are unstable $T_1^- T_1^{+*}$ and $T_2^- T_2^{+*}$ are analytic in $|z| \leq 1$. Using the Residue theorem and the following identities:

$$\oint_{|z|=1} (T_1^- T_1^{+*}) \frac{dz}{z} = - \oint_{|z|=1} (T_1^+ T_1^{-*}) \frac{dz}{z} \tag{6.68}$$

$$\oint_{|z|=1} (T_2^- T_2^{+*}) \frac{dz}{z} = - \oint_{|z|=1} (T_2^+ T_2^{-*}) \frac{dz}{z} \tag{6.69}$$

It is clear that the integrals of the cross terms $T_1^- T_1^{+*}$, $T_1^+ T_1^{-*}$, $T_2^- T_2^{+*}$, $T_2^+ T_2^{-*}$ are zero.

Hence the cost function simplifies as:

$$J = \frac{1}{2\pi j} \oint_{|z|=1} \left((T_1^+ T_1^{+*} + T_1^- T_1^{-*} + T_2^+ T_2^{+*} + T_2^- T_2^{-*}) + \Phi_0 \right) \frac{dz}{z} \tag{6.70}$$

Since T_1^-, T_2^- and Φ_0 are independent of the controllers, the cost function is minimised when T_1^+ and T_2^+ are set to zero.

Setting T_1^+ and T_2^+ to zero, from equations (6.57) and (6.65), obtain the controllers as:

$$T_1^+ = 0 \Rightarrow C_1 = \frac{C_{1n}}{C_{1d}} = \frac{GAA}{D_c D_{f1} - G A_r B} \tag{6.71}$$

$$T_2^+ = 0 \Rightarrow C_2 = \frac{C_{2n}}{C_{2d}} = \frac{M A_r (C_{1d}A + C_{1n}B)}{D_c D_{f2} C_{1d}} \tag{6.72}$$

To satisfy the implied equation:

$$D_c D_{f1} = H A A_q + G A_r B \tag{6.73}$$

which guarantees closed-loop stability, we need to multiply both sides by:

$$D_c^* D_{f1}^* D_c D_{f1} = D_c^* D_{f1}^* HAA_q + D_c^* D_{f1}^* GA_r B \tag{6.74}$$

Using equations (6.37), (6.40), and (6.55)

$$\begin{aligned} D_c^* D_{f1}^* HAA_q &= (K^2 B^* B_q^* A_r^* BB_q A_r + A^* B_r^* A_q^* AB_r A_q)(\Phi_{\xi\xi} CC^*) \quad (6.75) \\ &\quad -(K^2 \Phi_{\xi\xi} B_q^* C^* B^* A_r^* B_q CA_r B - FA_q AA_r Bz^g) \\ &= \Phi_{\xi\xi} A^* B_r^* A_q^* C^* AB_r A_q C + FA_q AA_r Bz^g \quad (6.76) \\ D_c^* D_{f1}^* H &= \Phi_{\xi\xi} A^* B_r^* A_q^* C^* B_r C + FA_r z^g \quad (6.77) \end{aligned}$$

Hence H must be solved from the following diophantine equation together with the diophantine equation (6.55):

$$D_c^* D_{f1}^* Hz^{-g} - FA_r B = \Phi_{\xi\xi} A^* B_r^* C^* B_r Cz^g \tag{6.78}$$

From the implied equation the optimal controllers given by equations (6.71) and (6.72) can be written as:

Feedback controller:

$$C_1 = \frac{GA_r}{HA_q} \tag{6.79}$$

Reference controller:

$$C_2 = \frac{MA_r D_{f1}}{D_{f2} HA_q} \tag{6.80}$$